Challenger at Sea

Challenger at Sea

A SHIP THAT REVOLUTIONIZED EARTH SCIENCE

Kenneth J. Hsü

PRINCETON UNIVERSITY PRESS • PRINCETON, NEW JERSEY

Library of Congress Cataloging-in-Publication Data
Hsü, Kenneth J. (Kenneth Jinghwa), 1929–
[Schiff revolutioniert die Wissenschaft. English]
Challenger at sea : a ship that revolutionized earth science / Kenneth J. Hsü.
   p. cm.
Revised translation of: Ein Schiff revolutioniert die Wissenschaft
Includes bibliographical references and index.
ISBN 0-691-08735-0
1. Submarine geology. 2. Glomar Challenger (Ship).
3. Deep Sea Drilling Project. I. Title.
QE39.H7813   1992
551.46′08—dc20   92-1142 CIP

This book has been composed in Times Roman with Helvetica

Princeton University Press books are printed on acid-free paper,
and meet the guidelines for permanence and durability of the
Committee on Production Guidelines for Book Longevity of the
Council on Library Resources

Printed in the United States of America

10 9 8 7 6 5 4 3 2 1

Contents

Contents

List of Figures

List of Plates

I received a letter from Hoimar von Ditfurth in the summer of 1979 when I came back to Chengtu from a vacation in eastern Tibet; he asked me to write a book on ocean drilling for general readers. Von Ditfurth had been a medical doctor, and his contact with patients caused him to lament the communication gap between scientists and the public. He quit his medical practice and became a free-lance publicist. His book on evolution was a best seller, he was the editor of a popular science magazine, and he was well known for his science programs on German television. I had contributed articles to von Ditfurth's magazine, and helped with his television programs depicting the Mediterranean desiccation and the extinction of the dinosaurs.

I had wanted for more than a decade to write a book on geology for laypersons. Like von Ditfurth, I was alarmed by the false images of scientists presented to society. None of my four children chose a career in science. My daughter did study biology, but quit when she decided that modern bioscience was dominated by the financial interests of the pharmaceutical industry and gene manipulation. The image of scientists was not helped by popular books such as *The Double Helix:* in contrast to the portrait of a selfless Madame Curie, who devoted her life to seeking truth, the latter-day chemists pictured themselves as career-makers, pirating ideas while refereeing research proposals, stealing data by raiding and ransacking the offices of colleagues, all in the name of a race for the Nobel Prize. Physicists too had not helped their own image, when a few of the less scrupulous in their ranks told half-truths to promote atomic power. I wanted to write a book to show that we geologists, at least, are nice guys!

In 1970 I had written a manuscript called *The Mediterranean Was a Desert*, but I had not been able to get it published. A publisher friend, Bill Freeman, told me that the story was fascinating, but that the subject matter fell, as they say in German, "between chair and bench"—there was too much science for a trade book, but not enough for a scholarly treatise. I asked von Ditfurth if the *Mediterranean* manuscript was what he had in mind. No thanks, he answered; he wanted the whole story of the ocean drilling by

Glomar Challenger, and he sent me a contract with a generous advance.

I wrote the first draft of this book in the spring of 1980, while we were out drilling in the South Atlantic, just as Caesar wrote *The Gallic Wars* in the heat of battles. The advantage of this approach is its immediacy. To make a work on science more readable for laypersons, I decided to interrupt the flow of the narrative with biographical sketches, stories of amusing incidents, excerpts from my diary or from the daily operations reports, and even philosophical digressions. I was able to consult the shipboard reports of all previous cruises, which are considerably more spontaneous and truthful than the operations reports published later in the *Initial Reports of the Deep Sea Drilling Project* series. It would not have been possible to recapture the atmosphere of Jackson's Leg 55 or Schlanger's Leg 61, if I had not been able to read their daily operations resumes, written in moments of intense emotion. Only veterans could appreciate that we entered a different world when we sailed away on the *Challenger*. For this reason, I treasured the passages added from my diary, even though an editor might find them interruptive or irrelevant. The immediacy was also a disadvantage: words of anger can be polemical, and they do not make good writing.

My original intent to write about the people on the *Glomar Challenger* produced a 200-page manuscript. Hoimar von Ditfurth did not accept that; he wanted to have more science and more pages. He also pointed out to me that I had a contractual obligation to deliver a 400-page work.

The first revision changed the book from a source book on the history of geology to a readable textbook of marine geosciences. In addition to discussing our experiences while chasing magnetic lineations, for example, I had to add a long chapter 4 summarizing the history of the earth science revolution of the 1960s; I had to explain geology, in addition to recounting what we did and how. Von Ditfurth was happy with the new version, more so than I was.

I wrote the text in English; it was translated into German and published in 1982. A Chinese translation of the original English manuscript appeared in 1984. Then I looked in vain for an American publisher; I began to appreciate Bill Freeman's statement that the book, especially after the revision requested by von Ditfurth, probably did fall between the cracks.

It was, however, good reading for those who had sufficient knowledge of the subject matter, and they told me their impressions. Al Traverse found it a most valuable reference when teaching beginning geology to non-majors; he copied numerous illustrations from the book as handouts for his class. My daughter, Elisabeth, liked the way I tried to make geology interesting. Not knowing anything about the subject, however, she had to persevere to follow my suspenseful style. (At her suggestion, I now include a short epigraph at the beginning of each chapter so that readers know what is coming their way.) A chemist from Basel and an engineering professor from Zürich told me they had enjoyed reading the book and learning what their colleagues in geology are doing. Other friends told me that I should give up any hope that the book will be leisurely reading for everyone; I might as well accept the fact that my readers will only be those who want to learn something.

The reactions of Chinese readers were similar. Many of them had missed the earth science revolution, and they were happy to find a summary in Chinese in a readable volume. Others appreciated the historical perspectives and the philosophical insights. Thousands of copies were sold in a few months, but almost all to students or to professional geologists.

The manuscript of the English text lay in a drawer for years, and might never have seen daylight again but for two events. Bill Menard wrote *The Ocean of Truth*, a personal history of the earth science revolution. This was to be the first of two volumes, telling the stories until 1968, but Menard died before he could start the second volume on the history of deep-sea drilling. Although my manuscript had been written before Menard's, it happens to be a natural sequel, picking up the threads where Menard left off, with a minimum of repetition. About the same time, the paperback edition of *When the Mediterranean Was a Desert* was published, and Ed Tenner wrote me that the Princeton University Press would like to publish more science for general readers.

I asked Tenner if he would reconsider my manuscript on the *Glomar Challenger*, which had not been accepted because of a split decision. Writing in the throes of acrimonious competition for shiptime, I had been too harsh on some of my JOIDES friends—who happened to be referees of the manuscript. Their marginal notes had made me aware of my one-sidedness. Now I promised a revision that would delete all inappropriate "bursts of passion." (I can afford to make such a promise, because those who might be interested can find those passages in the German and Chinese editions.) Al Fisher was the final referee. He told me, and apparently also Tenner, that

"*Ein Schiff* . . . [the German title of this work] has to be published." Fisher could read more between the lines, because he was a veteran of the *Challenger*.

I was very pleased to have this chance to revise the text ten years later. First of all, the original manuscript was completed in 1981, two years before the termination of the *Glomar Challenger* drilling. The least I could do was present a complete history, 1968–1983. Second, I was no longer involved in the science politics of ocean drilling, having turned to field geology in China. What appeared important to me at one time, especially when I was aboard the *Challenger*, now seemed so trivial; I could take a distant view. Not only the "bursts of passion" (of which there were not too many in the original text anyway), but also the petty critiques are now deleted. Besides, I could now use the comments of previous readers to make the book better.

Instead of accepting the opinion that this book falls "between chair and bench," I would take the more positive view that there is something for everybody. The many veterans of the *Challenger* may treasure the memory of life on a drilling expedition. Friends and colleagues in geology might be interested to learn in more detail what we did out there. Beginning students of geology might find a readable reference, a narration with a historical perspective that helps them to understand the hows and whys. For historians of science, it is a source book imbued with the atmosphere of a bygone age.

With these objectives and those readers in mind, I carried out the revision. The book is now mainly for geologists, students of geology, and historians of geology. It is not easy reading for general readers, although it is written in a language understandable to anyone with a college education. It is not arranged in an orderly fashion, and it is interspersed with side tracks. It would not make a conventional textbook, but it could be an interesting reference after one has acquired systematic knowledge from classroom instruction. It is not a scholarly treatise on the history of geology, and the references are not footnoted. It is the story of a participant, and the partisanship is undisguised. The science historian will take it with a grain of salt, like one does when reading Caesar's *Gallic Wars*.

The German title of the book is translated as *The Ship That Revolutioned a Science*. I took the view that Vine and Matthews, like Karl Marx, wrote only the manifesto; the Leninistic action of revolution was carried out by the *Glomar Challenger*. On the eve of the creation of the Deep Sea Drilling Program, there was a schism between continental geology and marine geophysics. Tension was in the air, and the danger of "civil strife" was imminent. The drill vessel came onto the scene at the most timely moment. Geologists like myself would not be convinced until the revolutionary geophysical theories were verified by classical methods of geology, stratigraphical and sedimentological. Without the _Challenger_ at sea, the theory of seafloor-spreading, like the theory of continental drift, would have divided, not united, the earth science community.

Organized around this central theme of revolution, the revised manuscript now includes four parts. The first four chapters, constituting Part One, describe the events leading up to the eve of the revolutionary action by the _Challenger_. The next eight chapters, making up Part Two, narrate the verification of the predictions by geophysical theories that triggered the revolution, as the _Challenger_ sailed from the Atlantic to the Pacific and from there to the Indian Ocean during the first and second phases of the Deep Sea Drilling Project, 1968–1973. The next four chapters, Part Three, tell the stories of discoveries on new ground after the breakthrough, mainly those made during the third phase of the Deep Sea Drilling Project, 1973–1975, when the _Challenger_ returned via the Southern Oceans to the Atlantic. The last four chapters, Part Four, are concerned with the mopping-up actions of the International Phase of Ocean Drilling, 1975–1983.

I would like to express my gratitude to the many people who assisted with the production of this book. Eva Pour typed the manuscript, Albert Uhr did the drafting, and Urs Gerber made the glossy reprints ready for reproduction. Ueli Brigel helped in many ways. I would also like to acknowledge my indebtedness to the many persons who gave me permission to reproduce illustrations. I appreciate the advice of many readers of the previous editions; their suggestions helped greatly in this revision.

Zürich, November 1990

Preface to the Chinese Edition

There have been two milestones in the history of the earth sciences: geology was founded in the late eighteenth century, thanks to James Hutton's uniformitarianism and William Smith's stratigraphy, and then underwent a second revolution during the 1960s and 1970s. Prior to the foundation of the science, the deluge described by the Bible was considered the first cause of all causes in geology. This was the basic philosophy of Abraham Gottlob Werner and his Neptunist students. The Hutton-Smith revolution was Baconian: observation was to be the basis of all conclusions. After the authoritarian influences of people like Roderick Murchison, geology became increasingly more doctrinaire. Charles Lyell's substantive uniformitarianism made the straitjacket even tighter: the past could not have been much different from the present. Carried to extremes, the doctrine said that past continents must have been situated where they are now. The theory of continental drift was thus considered fundamentally unacceptable by leaders of North American geology.

The revolution of the 1960s was spearheaded by marine geoscientists. Investigations of terrestrial magnetism, seismology, and geothermics had discovered a wealth of data that could not be explained by the doctrines formulated on the basis of observations made on land. The first cause of earth processes was now seen to be thermal convection in earth's mantle. The corollaries were seafloor-spreading, generation of seafloor magnetic lineations, transform faults, mid-plate volcanism, mantle hot-spots, and deformations on passive and active continental margins.

There was tension in the air, and the threat of civil (or uncivil) strife between the old guards of land geology and the young Turks of marine geophysics. The controversy was reminiscent of the battle between the Neptunists and Vulcanists of the late eighteenth century, or that between the "drifters" and "fixists" in the middle decades of this century. The *Glomar Challenger* was timely; using classic methods of stratigraphy, new postulates based upon fancy new theories were verified. The controversy was resolved, and the revolution of the earth sciences could be accomplished.

The Laplacian absolutism has become deeply rooted in many of us. It was thought that scientific truth would be within our grasp, if we just had enough observations, experiments, and data processing. One of the reasons I wrote this book was to delineate the subjectivity in science, and to point out the blind alleys. Analyzing the mistakes of my own career, I found the ultimate cause of many a poor judgment of mine in the Confucian philosophy of my upbringing.

I was once called a "traitor to the traditional cause," because I was a "fixist" for twenty years, and hoped to find in vertical movements the driving forces of all tectonic processes, before I was "converted" to the thinking of the seafloor-spreaders. Thinking it over, I realize that habits and emotions play a critical role in our scientific judgment. My Confucian respect for elders was reinforced by my emotional attachment to my teachers in America who helped me during dark hours of need; I defended their faith religiously. In chapter 5 of this book, I describe the agony of the realization that I had been wrong, that my beloved teacher had been wrong. I understood only all too well Werner's pupil, Citizen Jean François d'Aubuisson de Voisin, when he said:

The facts which I saw spoke too plainly to be mistaken; the truth revealed itself too clearly before my eyes, so that I must either have absolutely refused the testimony of my senses in not seeing the truth, or that of my conscience in not straightaway making it known. There can be no question that basalts of volcanic origin occur in Auvergne.

I too was staggered by what I saw, and there could be no question that the seafloor has spread.

Loyalty is a Confucian virtue, and vanity is a Confucian vice, but both are fatal to good judgment in science. Again in chapter 5 of this book, I tell of my hurt vanity when the selection committee chose Vine's theory of seafloor-spreading, not my observations on Franciscan mélanges, as a frontier lecture of the 1966 San Francisco Geological Society meeting. If I had swallowed my pride and attended Vine's talk, I might have realized then and there that ocean-floor subduction and the genesis of mélange were corollaries of his theory; I might not have stayed lost in a dead end for another three years.

The human element figures not only in our failures, but also in our successful endeavors. Courageous persistence in the face of adversity was exemplified by Sy Schlanger's explorations of mid-plate volcanism and by Dale Jackson's verification of the hot-spot theory. They were not the only virtuous ones; there were many acts of chivalry, generosity, modesty, and compassion by my friends who set sail on the *Challenger*.

This preface was written at the request of Professor He Qixiang. I have, therefore, emphasized aspects of our Confucian failings. I hope that the publication of this work in Chinese helps remedy an injustice—my Chinese colleagues missed the earth science revolution of the 1960s. In the midst of the so-called Great Proletarian Cultural Revolution, they had to read the *Selected Writings of Chairman Mao*, while their colleagues in the West were reading Vine and Matthews's manifesto on seafloor-spreading. Chinese scientists lost their chance to join this historic effort because of the evils committed by the "Gang of Four." Now that reform has been effected, my Chinese compatriots can march with their international colleagues in the pursuit of truth. I have high hopes for their eventual accomplishments, and I hope the lessons of history told in this volume will be helpful in a small way.

Peking, August 1983

Preface to the German Edition

This is a book about a ship, about a revolution, and most of all about the people who made the revolution with the ship. The ship is the drilling vessel *Glomar Challenger*. The revolution is the revolution in the geological sciences. The people are my teachers, my friends, my colleagues, my students, and, of course, myself. The book is, therefore, largely autobiographical.

The *Glomar Challenger* has been cruising since the summer of 1968, and hundreds, if not thousands, have participated on her cruises. The initial reports of the drilling cruises will include about one hundred volumes when they are all published, and each has more or less one thousand pages. In my book there will be omissions, obviously, and some aspects will be unduly emphasized. Having declared this book to be autobiographical, I hope I will be excused if I have concentrated on subjects I know more about, or on the people I know best, including myself. I can be excused, perhaps, because I was lucky enough to have been involved in many of the activities, and acquainted with most of the people, responsible for this revolution in geology. I hope my omissions are not essential to my story, and I hope my omissions will not be considered a slight of the many achievements by numerous persons not mentioned in this slim volume (slim compared with the hundred monographs that claim only to be the *initial* reports of the Deep Sea Drilling Project). If I have made serious omissions, I hope to be told, and to make corrections if there is a revised edition. Meanwhile, it should be remembered that I have written a personal memoir, not an official history of the research cruises of the *Glomar Challenger*.

One of the reasons that prompted me to write this book was that I wanted to present geologists as we think we are. Geologists in Hollywood movies or in detective stories are often odd people. Since geology is not usually taught in secondary schools, most people know little about the subject, and even less about the persons in the profession. Distinguished geologists, like distinguished persons in other professions, are devoted to their science, but most of them are not fanatics, and they are distinguished most of all by their modesty, even if they rarely suffer fools.

Another reason to write such a book for laypersons is to illustrate with one particular example how science is made. Science is a human endeavor, and scientific ventures can bring about frustration, disappointment, or even anguish or pain, but they can also lead to contentment, fulfillment, joy, and happiness. There is jealousy, selfishness, arrogance, littleness, but also generosity, modesty, humility, and fairness. Mistakes in science are made by all, and I have tried in this work to analyze the emotional makeups that led to mistakes, and certainly included a self-analysis of my own errors.

The _Glomar Challenger_ has sailed to all parts of the world's oceans, she has been active longer than a decade, and all matters of geology have been the subjects of research. It was not an easy task to decide if the organization of the book should be geographic, chronological, or thematic. The compromise— using all three formulas—has its imperfections. I started with _Glomar Challenger_'s inaugural journey from a port in the Gulf of Mexico, and continued with a narration of her cruises to the Atlantic to test the seafloor-spreading theory, to the Pacific and the Indian oceans to check out predictions of the theory of plate-tectonics, to Antarctica to explore the past history of climatic changes, and back to the Atlantic to investigate the newly discovered dramatic happenings in ancient oceans. In fact, except for a short incursion to the Pacific early in the project, this was more or less the ship-track of the _Glomar Challenger_ during the first three phases of the Deep Sea Drilling Project. Naturally, the narrative was not able to follow a strictly chronological order; the ship has been back and forth between the Atlantic and the Pacific repeatedly, and the same themes have been investigated over and over again. The last four chapters of this book summarize the last phases of the Deep Sea Drilling Project after the program was internationalized in 1975. New explorations were being made, and new problems were coming up, providing ammunition for a future revolution.

The manuscript was completed a year ago. Meanwhile, the _Glomar Challenger_ has been continuing her activities. I should not let this book go to press without mentioning at least a few of last year's successes. Off the West African margin, a sample of a 130-million-year-old salt was brought up, a feat that had been attempted several times before. East of Barbados, the drill string penetrated ocean sediments that had been thrust under the edge of continents—another major achievement. Finally, the 600-meter barrier, which had hindered all previous drilling into the ocean crust, was broken; the Leg 83 team managed to deepen a previously drilled hole in the Pacific and drill more than one thousand meters into the ocean crust! It has now become obvious that a drill ship is not just an *ad hoc* tool for testing some new ideas, to be abandoned after the fulfillment of a mission. Like Galileo's telescope, which revolutionized astronomy, and Lawrence's cyclotron, which revolutionized nuclear physics, a drill ship is an indispensable tool for the advancement of the earth sciences. I have been invited by the U.S. National Science Foundation to a meeting next week to discuss the future of ocean drilling. There are ambitions to use a big vessel and continue ocean exploration until the end of this century.

I would like to thank all who helped make the completion of this book possible, especially Albert Uhr and Urs Gerber for their preparation of illustrations, Carolina Hartendorf and Barbara Das Gupta for their secretarial assistance, and Dr. Ueli Briegel and my wife, Christine, for their help with minor revisions of the German text. I would also like to gratefully acknowledge the kindness of the persons and organizations that allowed me to reproduce their figures.

The book is written for my late wife, Ruth, and for the *Schlüsselkind* she had hoped to care for. It is written for my children, Elisabeth, Martin, Andreas, and Peter, so that they would be able to understand their father better. Most of all, it is written for my wife, Christine, for all that she has done.

Zürich, May 1982

Acknowledgments

In this book, numerous illustrations have been reproduced from, or redrafted on the basis of, illustrations in earlier publications. I would like to thank the following persons and organizations for permission to reproduce figures from their works.

Wolf Berger and E. L. Winterer, Plate stratigraphy and the fluctuating line, International Association of Sedimentology, Spec. Publ. no. 1 (1974): 11–48, for Figs. 19.4 and 19.5.

Joe Curray, The IPOD program on passive continental margin, *Philosophical Transactions of the Royal Society of London*, series A, no. 294 (1980): 17–34, for Fig. 18.1.

William Glen, *The Road to Jaramillo* (Stanford: Stanford University Press, 1982), for Figs. 4.6 and 4.10.

James Gilluly, Aaron Waters, and A. O. Woodford, *Principles of Geology* (San Francisco: Freeman, 1951), for Figs. 1.4, 7.2, 7.3, and 7.4.

John Imbrie and K. P. Imbrie, *Ice Ages: Solving the Mystery* (Short Hills, N.J.: Enslow, 1979), for Fig. 2.1.

H. P. Luterbach and I. Premoli-Silva, Note préliminaire sur une révision du profil de Gubbio, Italie, *Revista Italiana di Paleontologia* 68 (1962): 253–288, for Fig. 20.1.

H. W. Menard, *The Ocean of Truth* (Princeton: Princeton University Press, 1986), for Fig. 12.2.

Alan Smith and J. C. Briden, *Mesozoic and Cenozoic Paleocontinental Maps* (Cambridge, England: Cambridge University Press, 1977), for Fig. 6.1.

Walter Sullivan, *Continents in Motion* (New York: McGraw-Hill, 1974), for Figs. 1.1, 4.1, 4.3, 4.4, and 7.1.

AMERICAN GEOPHYSICAL UNION

W. E. Bonini and R. R. Bonini, Andrijy Mohorovicic: Seventy years ago an earthquake shook Zagreb, EOS (Trans. Am. Geoph. Union) 60, no. 41 (1979): 700, for Fig. 1.3.

H. R. Heirtzler, G. O. Dickson, E. M. Herron, W. C. Pitman, and X. LePichon, Marine magnetic anomalies, geomagnetic field reversals, and motions of the ocean floor, *Journal of Geophysical Research* 73 (1968): 2119–2136, for Fig. 4.11.

xxx

Acknowledg-ments

B. Isacks, J. Oliver, and L. R. Sykes, Seismology and the new global tectonics, *Journal of Geophysical Research* 73 (1968): 5855–5899, for Fig. 7.6.

Jason Morgan, Rises, trenches, great faults, and crustal blocks, *Journal of Geophysical Research* 73 (1968): 1959–1982, for Figs. 7.7 and 12.2.

CAMBRIDGE UNIVERSITY PRESS

D. A. Johnson, Abyssal teleconnections II, in K. J. Hsü and H. Weissert, eds., *South Atlantic Paleoceanography* (Cambridge, England: Cambridge University Press, 1985), for Fig. 19.2.

CONSERVATOIRE ET JARDEN BOTANIQUE DE GENÈVE

Gilbert Bocquet, B. Widler, and H. Kiefer, The Messnian model: A new outlook for the floristics and systematics of the Mediterranean area, *Candollea* 33 (1978): 287, for Fig. 16.3.

GEOLOGICAL SOCIETY OF AMERICA

H. H. Hess, History of ocean basins, in A.E.J. Engels *et al.*, eds., *Petrologic Studies: A Volume to Honor A. F. Buddington* (New York: G.S.A., 1962), pp. 599–620, for Fig. 4.5.

R. G. Mason and A. D. Raff, Magnetic survey off the west coast of North America, *G.S.A. Bull.* 72 (1961): 1259–1266, for Fig. 4.7.

Walter Pitman and M. Talwani, Seafloor spreading in the North Atlantic, *G.S.A. Bull.* 83 (1972): 619–643, for Fig. 6.3.

N. Shackleton and N. Opdyke, Oxygen-isotope and paleomagnetic stratigraphy of Pacific core V-28–239, *G.S.A. Mem.* 45 (1976): 449-464 for Fig. 16.2.

NATURE

F. J. Vine and D. H. Matthews, Magnetic anomalies over oceanic ridges, *Nature* 199 (1965): 947–949, for Fig. 4.9.

J. T. Wilson, A new class of fault, *Nature* 207 (1965): 343–347, for Fig. 12.4.

SCIENCE

F. J. Vine and J. T. Wilson, Magnetic anomalies over a young oceanic ridge, *Science* 150 (1965): 485–489, for Fig. 4.10.

VERLAG F. PFEIL, MÜNCHEN

P. G. Bianco, Potential role of the paleohistory of the Mediterranean and Paratethys basins on the early dispersal of peri-Mediterranean freshwater fishes, *Ichthyol., Expl., Freshwaters* 1 (1990): 147–184, for Fig. 16.4.

Acknowledg-

ments

I would also like to acknowledge the reproduction of illustrations from the following volumes of the *Initial Reports of the Deep Sea Drilling Project* (Washington, D.C.: U.S. Government Printing Office): vol. 42A, 1978, for Fig. 7.10; vol. 18, 1973, for Figs. 8.1, 8.2; vol. 60, 1981, for Fig. 9.3; vol. 7, 1971, for Fig. 10.3; vol. 33, 1976, for Fig. 10.2; vol. 55, 1980, for Figs. 11.1, 11.2, 11.3, 11.4; vol. 22, 1974, for Figs. 12.5, 12.6; vol. 29, 1975, for Figs. 13.1, 19.3; vol. 40, 1978, for Fig. 14.1; vol. 92, 1986, for Fig. 17.2; vol. 50, 1980, for Fig. 18.2; vol. 87, 1985, for Fig. 18.3; vol. 78A, 1984, for Figs. 18.4, 18.6; vol. 56/57, 1980, for Fig. 8.5.

One The Eve of a Revolution

1963–1968

1   Moho and Mohole

Harry Hess wanted to drill down to the Moho, the base of the earth's crust. The venture failed, but technological innovations developed for the Mohole Project made the Deep Sea Drilling Project possible.

Impressions of a Scientist

As they say in America, if you dig a deep enough hole, you will reach China. Nobody ever seriously prepared to do so, but Harry Hess once did persuade the Congress of the United States to invest millions of dollars in a hole to the mysterious realm of Moho, ten kilometers below the ocean.

I first came across Harry Hess in the Prudential Insurance Building in Houston, Texas. That was February 1954. I was fresh out of graduate school, and had just pulled down my first job, with the Exploration and Production Research Laboratory of the Shell Oil Company. Hess was already a well-known geologist, then chairing the Geology Department of Princeton University. He came to Houston on a fund-raising tour. For his talk to the Princeton Alumni Association, he chose to speak on *guyots*—the newly discovered sunken flat-top mountains of the Pacific.

Harry Hess had been the young commander of the troop-transport M/S *Cape Johnson* of the U.S. Pacific Fleet during the Second World War. While ferrying troops for landings in the Marianas and the Philippines and on Iwo Jima across the Mid-Pacific, the echo-sounder of his vessel registered a seafloor topography that was far from monotonous. Undersea mountains appeared one after another on the echogram; they rose thousands of meters above the deep ocean floor (Figure 1.1). They all had steep sides and a flat top, looking somewhat like mesas on the Colorado Plateau of the western United States. Crisscrossing over the mountains revealed a circular outline. This morphology indicated to geologist-commander Hess that those underwater features were sunken volcanic islands with their tops chopped off. He called them guyots, in honor of

Arnold Guyot, Princeton's first geology professor, who also lent his name to the building that houses the Princeton geology department. When the war ended, Hess could report the discovery of 160 of the drowned islands in the Pacific basin.

Oceanographic vessels returned to Hess's old hunting ground and were able to find more guyots. They also used a tool for dredging, a scissor-like hook at the end of a long steel cable, to grab and twist off pieces of rock from the steep sides of the sunken islands. They turned out to be mostly volcanic rocks, as Hess had expected.

Submarine volcanoes are not uncommon on the present-day seafloor; they are called *seamounts*. What made the guyots unusual were their flat tops. Hess assumed these were caused by erosion—the guyots were once volcanic islands, but their conical tops had been planed off by the pounding of ocean waves. The islands seemed to have stood firm while waves were doing their job of cutting down the parts above sea level. Then, for some reason, subsidence started, and the flat-topped islands sank three or four thousand meters to become guyots.

I was always quick to jump to conclusions. I had learned in school that volcanoes in the oceans should sink under their own weight, but the rate of subsidence is finite. When seamounts (underwater volcanoes) are active, molten liquids, or lavas,

1.1. Guyot. Using echo-sounding, Harry Hess discovered numerous flat-topped submarine mountains, which he named guyots. The one shown in this figure rises 2000 m above the surrounding ocean floor, and the top is about 1000 m below sea level. The diameter of this guyot is more than 20 km. Deep-sea drilling has verified the postulate that guyots are sunken volcanoes.

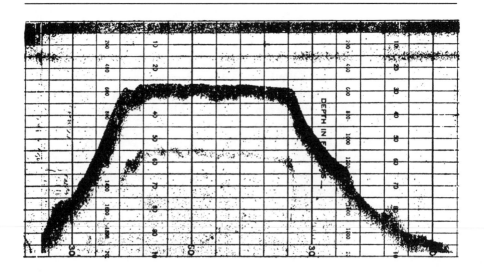

pouring out of the interior of the earth pile up so fast that subsidence cannot keep up with the build-up. The seamounts eventually rise above sea level, forming volcanic islands. However, volcanoes become inactive sooner or later. Then ever-present gravity does its work and pulls the volcanic islands back down to the abyss.

That a volcanic island should sink under its own weight in an ocean had been, in fact, a common assumption in geology. Charles Darwin invoked the idea to explain the origin of coral atolls. Therefore, I was a little surprised that Hess should be disturbed by the presence of guyots in the deep sea. Too shy to bring up the point during the discussions after the talk, I nevertheless wrote Hess a letter.

I received a prompt reply from Professor Hess, typed single-spaced on three sheets of stationery, apparently by himself. He apologized that he could not find time to give me a more detailed explanation; the usual activities of a newly begun semester had taken a heavy toll on his working hours. He was also apologetic about his less than professional skill in handling a typewriter. Just the same, he patiently explained to the slightly arrogant young man why the simple-minded scheme of subsidence under load cannot explain all the problems in connection with the origin of guyots.

The facet of the problem that bothered Hess most was the flat top. Waves did their work slowly. Why did the islands seem to have paused for such a long time, waiting until the waves had removed mountains and cut the volcanic islands down to sea level? When and why did the flat-topped mountains start their irrevocable journeys down to the abyss? After discussing the pros and cons of the various explanations at some length, Hess reiterated his puzzlement and questioned the possibility of a simple answer.

Years later both of us learned more about the oceans. Guyots did indeed sink under gravity. Hess's difficulty could be traced to his assumption that the flat top was due to erosion. When we look at the islands of the Pacific today, some with volcanoes rising hundreds of meters above sea level, it seems a formidable task indeed to cut those volcanic mountains down. However, flat-topped islands—such as Saipan, Tinian, and others that were used as air bases for U.S. B-24 bombers during the war against the Japanese—are not uncommon. They are flat not because waves pounded the mountains down, but because the topography has been evened out by the deposition of flat-lying sediments. Elsewhere, flat-topped islands may owe their lack of relief to horizontally accumulated lava flows or ash beds. If we accept Darwin's theory that atolls build on a sinking volcanic foundation, the sediments trapped in the lagoons of an atoll should gradually cover the tip of the volcano to make a flat top. Those atolls on which coral grows too slowly eventually sink to become guyots. From this point of view, guyots are simply coral atolls that died young. In the tropical Pacific, where coral growth could always keep pace with the sinking of the foundations, sunken volcanoes are now crowned by living reef corals, as Charles Darwin once theorized.

I shall come back to the story of guyots in a later chapter. What impressed me back in 1954 was not what Hess wrote, but that he did write. I was not convinced by his arguments then, but I did treasure the autograph from the famous man.

My second encounter with Harry Hess took place in Washington, D.C., a few years later. I was appearing before a distinguished national audience during the annual meeting of the American Geophysical Union to present my first professional lecture. It was scheduled as the first talk of the first session—a rather ticklish spot. Furthermore, I had sent in the theme in a moment of lightheartedness, and I came to realize only belatedly that I had chosen a poor subject for a start. Normally a young man beginning his professional career would choose to report on a fact-finding research project, in which he could make a good impression on his peers by demonstrating his intelligence and diligence. I had instead selected a theoretical subject—the origin of geosynclines.

The concept geosyncline was invented by James Hall, a nineteenth-century geologist, to designate ancient sites of sediment accumulation that seemed destined to become mountain chains. Hall was the State Paleontologist of New York, and he studied the Paleozoic rocks in the Appalachian Mountains and in the plateau country to the west of the mountains. The mountain rocks, which are deformed, are quite similar to those on the plateau, which are flat. However, Hall found that a rock formation of any given age is several times thicker in the mountains than on the plateau (Figure 1.2); strata seemed to have been crumpled where they are much thicker. Before the sedimentary formations were crumpled during the process of building mountains, they had been laid flat. The bottom of a sedimentary pile should thus be most depressed where the pile is thickest. Such a warped surface had been called a _syncline_ by geologists, and James Hall added geo- to the expression to emphasize its imposing dimensions. European scientists noted, however, that crumpled strata in the mountains were not always thicker than flat strata in plateau country, but the former seemed to have been deposited in deeper water. They too used the expression geosyncline to designate a site that was eventually to become a mountain chain.

The term geosyncline was to become a catchword, and appeared in everyone's theories even though no one seemed to know what a geosyncline was. There were no modern analogues. Instead, people speculated, somewhat idly, on why geosynclines sank, permitting the accumulation of thick piles of sediments. In the short discourse I planned to present orally, I would try to demythify the concept. I had some simple arguments to show that geosynclines are simply depressions on the surface of the earth, such as ocean basins, continental margins, and so on. After sediments were accumulated, the seafloor would subside under the added load, just as guyots sink under the weight of added volcanic piles.

The idea was not new, although I did give it a new twist with some recent data. The idea was also not all that bad; what I said

1.2. Geosyncline. James Hall in 1859 postulated that sediments of mountains form unusually thick sequences (a); his European colleagues found that such sediments were deposited in deep-sea troughs (b). Both types of depositional sites were called geosynclines, and a geosynclinal theory of mountain-building was introduced. This dogma led to a blind alley and hindered the progress of geology for a century, and is now replaced by the theory of plate-tectonics.

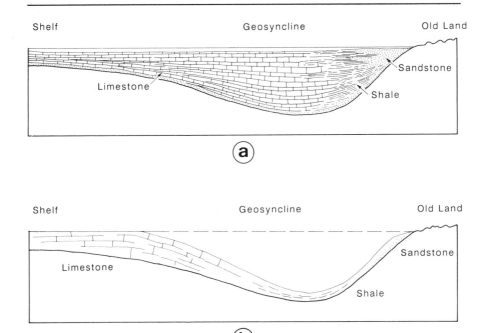

then is still valid today, although I did miss a point or two (see chapter 6). However, it was presumptuous for a young man to tell his seniors the obvious; I grew increasingly nervous as the time of the meeting approached.

At the appointed hour, I went to the General Services Administration Building in Washington, D.C., where the meeting was being held. I was early, half an hour early. Gradually people filed in, and many venerated scientists of the older generation took their seats in the front rows. Hess was to open the meeting and chair the first session. Eight o'clock struck, the hour I was to give my talk, but he was nowhere in sight. As the time passed, 8:05 ... 8:10 ... 8:15 ... 8:20, I sat there and became more and more edgy, gradually losing my nerve. Finally, at about half past eight, Hess ran in and climbed up to the rostrum. Wiping sweat off his brow, he murmured an apology—he had looked for the wrong GSA (Geological Society of America) Building, and nobody seemed to know where it was. (It was in New York City!) Catching his breath, he tried—without remarkable success—to pronounce my name, as he introduced the first speaker. By then I was completely overcome with stage fright. When I stepped before the microphone, I realized that I could hardly find my voice. I stuttered, and had to read my written script in broken sentences. I did finally stagger through, but was met by dead silence at the end of a miserable presentation. It was a "bomb," as they say in show business, and the lack of response was a manifestation of the hostility toward the brash young man. Hess, however, was genuinely sorry; he thought that people remained silent because they did not want to start discussions, as the session had been delayed by his late arrival. He apologized profusely to me after the meeting, when I was blaming myself for my less-than-brilliant "début."

As the years went by, I learned some self-confidence, while Hess remained modest. He did much for the earth sciences. His idea of seafloor-spreading started a scientific revolution, and his quest for an imaginative research project opened a new era of ocean exploration.

The last time I saw Hess was in the spring of 1969. We had just completed a drilling cruise in the South Atlantic (chapter 5), and I had come to Princeton to discuss the cruise results with my shipboard colleagues. This cruise has achieved fame as the expedition that verified the predictions of the theory first elaborated by Harry Hess. However, Hess was so modest that he actually seemed embarrassed to be proven right. Toward noontime, Hess and I walked from Guyot Hall to the Princeton Faculty Club for lunch. He did not mention the subject of seafloor-spreading, nor did we even discuss our findings. Instead, we chatted about the difficulties of privately endowed universities in the days of high inflation. He had just stepped down as chairman of the department, but he still had many national and international "obligations." He was overworked and seemed tired.

Hess died of a heart attack shortly after my visit. He will always be remembered by us as the one who made the first move that led to one of the most successful undertakings in the earth sciences—the Deep Sea Drilling Project (DSDP) of the Joint Oceanographic Institutions for Deep Earth Sampling (JOIDES).

Moho

One of Hess's many "obligations" had been to serve on a committee to help select research projects worthy of financial support by the U.S. National Science Foundation. Back in the fifties, when physicists were asking for and getting more and more funds for bigger and better accelerators, geologists seemed to content themselves with pocket money for making geological maps. It was said that Hess and his committee were going over one research application after another on a spring day in 1957. They grew increasingly restless and irritated after finding nothing exciting in the proposals, and began to ask themselves if anything worthwhile in geology was still left to be done.

The heroic age of geology, spanning the last decades of the eighteenth and the first half of the nineteenth century, had long since become a nostalgic memory of a far distant past, an age in which James Hutton founded geology as an observational science, William Smith discovered the value of fossils in ordering strata in temporal sequences, Charles Lyell demythified the book of Genesis and preached uniformitarianism, and Charles Darwin came up with his theory of evolution. It seemed that few discoveries of comparable significance had turned up during the twentieth century. Progress was, however, being made in geophysics, such as the discovery of the Moho.

On 8 October 1909, Yugoslavia was struck by an earthquake whose epicenter was some 25 kilometers south of the village of Papuspsko, near Zagreb. A local geophysicist, Andres Mohorovicic, made a routine study of the time it took the first earthquake waves to reach various registering stations—the so-called *first-arrival time*.

The shock of an earthquake sends out several different kinds of waves. Some are transmitted by alternately compressing and extending an elastic medium, similar to the way sound waves travel in air. These are called *compressional waves*, or *P-waves*. They are the fastest and should be the first registered by a seismograph at any station. In a homogeneous medium, the speed of propagation of the compressional wave, V_p, should be constant. Consequently the first-arrival time t should be directly proportional to the distance between the epicenter and the registering station, S, or

$$t = S/V_p .$$

This is, of course, a simple linear relation in Newtonian kinematics that we all learned in middle school. Traveling to a station 200 kilometers from the epicenter should take the wave twice as long as traveling to a station 100 kilometers from the epicenter. Plotting the arrival time against the distance traveled for the stations within 300 kilometers of Zagreb, Mohorovicic obtained a straight line through the origin, illustrating that the simple predicted relation was indeed confirmed by the seismic records (Figure 1.3). As our middle-school physics teacher tells us, the slope of the straight line in this figure, or the travel time divided by distance traveled, is the inverse of the speed of the wave propagation. Mohorovicic's calculations showed that the compressional waves were traveling at a speed of 5 to 6 kilometers per second when they reached those stations.

Records of the earthquake were available from other seismic stations, with the most distant one at Tiflis in the Caucasus, some 2,400 kilometers from Zagreb. To his surprise, Mohorovicic discovered that the first-arrival times of stations more distant than 300 kilometers did not fall on the straight line going through the origin. Instead, the arrival times at those stations, when plotted against distance traveled, gave another straight line, which did not go through the origin; the line also had a more gentle slope (Figure 1.3). The slope indicated that the fastest earthquake waves traveled to more distant stations at 7 or 8 kilometers per second. How was he to explain this faster speed? Was there another kind of wave that traveled faster, or did the same compressional waves take a faster route?

The example of a person driving an automobile can easily provide an answer. If someone wants to travel from Zürich to Kilchberg, some 10 kilometers away, he takes the shorter lake-shore highway, despite the 50 km/hour speed limit. If, however, he wants to drive to Weesen, some 60 kilometers away, in the shortest possible time, he takes a detour to get on the freeway, where he can drive at 120 km/hour. Mohorovicic used the same logic to explain his data on travel times: the fastest-traveling waves to all stations are compressional waves, but the ones that made a detour to the freeway became the first to arrive at stations more distant than 300 kilometers. In scientific terms, the "detour" is called "wave refraction," and the "freeway" is a deeper and denser medium that permits a faster propagation of waves than the shallower underground.

We have known since the time of Isaac Newton that the interior of the earth is denser than the near-surface ground. In the middle of the last century George Airy, the Royal Astronomer of Great Britain, speculated that the earth had a thin, lighter, solid shell enveloping a denser, liquid interior, from which hot lavas from volcanoes were derived. He called the outer, solid shell the "earth's crust." Lighter crust floating on heavier liquid should maintain a flotation equilibrium. Such a model of gravitational balance has been called Airy's model of *isostatic* (equal pressure) equilibrium (see chapter 4). Both the sinking

1.3. Moho. Mohorovicic noted that seismic waves traveling to more distant stations move at greater speed, as shown by this reproduction of his record of the 8 October 1909 earthquake in Yugoslavia. He concluded that those faster waves moved through a denser rock (the earth's mantle), which is separated from the upper rock-layer (the earth's crust) by a boundary, which is now called the Moho.

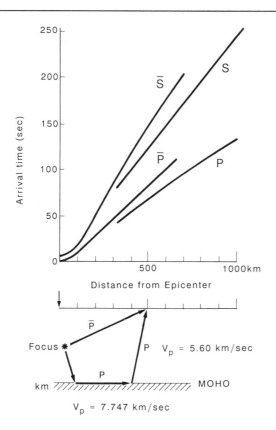

of guyots in the oceans and the subsidence of geosynclines under sedimentary load have been interpreted in terms of Airy's model. The now famous theory of continental drift advocated by Alfred Wegener during the early part of this century also invoked isostatic equilibrium to justify the "drift." Wegener's theory was rejected by geologists because they listened to their geophysicist colleagues, who maintained that the crust could not have been floating on a heavier liquid substratum, because the heavier rock material beneath the crust is a solid, not a liquid. What is the basis of this geophysical conclusion?

We have mentioned that an earthquake shock sends out several kinds of waves. The P-waves (compressional waves) are the fastest. Trailing behind are the slower-moving S-waves (shear waves). P-waves can go through solid and fluid alike; we hear sounds because acoustic waves are compressional waves and can be transmitted by air. S-waves, however, cannot go through a liquid. Through the analysis of seismic records, geophysicists have found that S-waves are not registered at some very distant stations, proving the presence of liquid in the earth's interior (Figure 1.4). However, the liquid interior is much deeper than Airy assumed; S-waves travel in a solid medium down to a depth of 2,900 kilometers below the surface. Farther down is the liquid core of the earth (Figure 1.4). This set of data permitted the recognition of an earth's core enveloped in a 2,900 km thick earth's mantle. Now Mohorovicic's data told him that the mantle has a lighter outer skin, which is the earth's crust; the crust under the continent is about 50 kilometers thick. P-waves going directly through the crust to nearby stations traveled at a speed of 5 or 6 kilometers per second, while P-waves making a detour through the mantle to more distant stations could manage a speed of 8 kilometers per second. Between the crust and the mantle a difference occurs in the physical properties of the earth's rock material—a finite difference, a discontinuous change. Because it was Mohorovicic who first recognized this discontinuous change in seismic velocity when a wave traverses the boundary between crust and mantle, geophysicists now use the term *Mohorovicic's Discontinuity* to designate the surface separating the crust from the mantle. With time, the term was shortened to Moho.

The rocks we see on or near the surface of the earth constitute the crust. One can make some deductions regarding what rock types are underground by studying the speeds at which various rocks transmit elastic waves and comparing them with the actual speeds of wave propagation through the crust and the mantle. The most common rock type hauled up by dredging in the oceans is *basalt*, which is solidified from lavas erupted by volcanoes. Basalt is a *mafic* rock—that is, its chemical composition is rich in magnesium and iron (ferrum). The minerals making up basalt are very-fine-grained, mostly small crystals (less than a millimeter long), because hot lavas erupted onto the seafloor are quickly quenched. If a molten liquid of the same composition as lava is placed in a crack or fissure one or two kilometers below the seafloor, the same minerals would crystallize out, but those crystals could grow slowly to larger sizes, several millimeters long. Such a slowly cooled rock is in all other aspects identical to basalt, but geologists choose the name *gabbro* to distinguish the coarse-grained variety. Since the speed of wave propagation through the ocean crust is about the same as that measured through basalts or gabbros, earth

1.4. Internal structure of the earth. Compressional waves can be transmitted through a solid medium; they can go through the liquid core, but they cannot reach the so-called shadow zone, as shown by this figure. Shear waves cannot be transmitted through the core, and thus cannot travel beyond the shadow zone. The structure of the earth has been determined from earthquake studies. Numbers on diagram denote minutes required for the travel of P-waves.

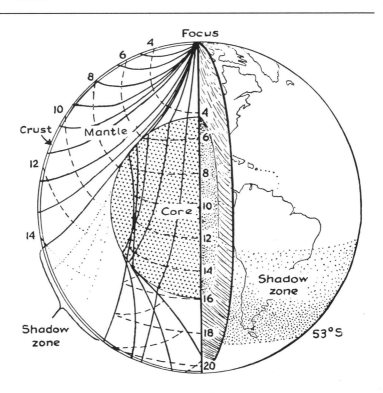

scientists agree that the crust under the ocean is made up largely of those rocks. The common rocks exposed under sedimentary strata on continents are granites and gneisses. However, data on speeds of wave propagation suggest that the lower part of the earth's crust under continents may also be a mafic rock.

Earth scientists could not agree during the 1960s on what kind or kinds of rock are in the mantle below the Moho. The rocks of the mantle, which transmit elastic waves at decidedly higher speeds, are certainly something other than basalts or gabbros. Two possibilities have been suggested: eclogite, which has the same composition as basalt (but these same chemical compounds have been crystallized into minerals considerably denser than those in basalt or in gabbro), and peridotite, whose chemical composition is different from that of basalt. (Peridotite has less silica, less aluminum, and less calcium oxide, but more magnesium oxide, than basalt: it is also denser than basalt.) Seismic waves can travel through either eclogite or peridotite at a speed similar to their speed through the mantle.

Earth scientists were interested in the mantle, because a knowledge of its composition would influence our interpretation of geological processes. Throughout the middle decades of this century, the overwhelming majority of North American geologists belonged to what is now called the "fixistic school"—they believed that the continents and oceans were permanent and had been fixed in their positions since the beginning of the earth. Of course, the evidence could not easily be ignored that ancient continents had sunk to become deep seas or that ancient seafloors had risen to become mountains. However, the "fixists" postulated that all such earth movements were predominantly up and down, and that there had been little horizontal displacement such as that advocated by "mobilists" such as Alfred Wegener with his theory of continental drift. For the "fixists," the assumption of a mantle made up of eclogite would solve all their problems. Basalt (or gabbro) and eclogite have the same composition, but basalt may become eclogite if it is subjected to a great pressure, or eclogite may be changed into basalt if it is heated up. When basalt is turned into eclogite, the volume decrease causes the ground to sink; continents would turn into oceans, and the subsidence could permit the accumulation of the so-called geosynclinal sediments. When eclogite is changed into basalt, the volume increase causes the ground to rise, and deep seas would be converted into plateaus or mountains, as shown by the rocks of the Alps.

The postulate of an eclogite mantle was favored mainly by theorists during the 1950s. Geologists working in the field found eclogite only very rarely. Instead, in places where a piece of an ancient seafloor has been raised into a mountain, one finds beneath the basalt and gabbro, which should have constituted the ancient ocean crust, not eclogite but peridotite. Therefore geologists, especially European geologists, were not particularly bound by the "fixistic" doctrine; they preferred the alternative, that the earth's mantle consists of peridotite. Hess was one of the few North American "mobilists" who could see beyond the prevailing dogmas.

Mohole

Returning to our account of that day in 1957 when Harry Hess, Walter Munk of the Scripps Institution of Oceanography, La Jolla, California, and others were swallowing their disappointment at the lack of imaginative research applications, Walter Munk, half in jest, came up with a somewhat outrageous idea that one should drill through the crust of the earth to discover the nature of the mantle rocks beneath the Moho. Yes, why not? Hess took Munk's idea seriously and proposed referring the project to the AMSOC (American Miscellaneous Society) for action.

The American Miscellaneous Society had been established in the summer of 1952 by Gordon Lill and Carl Alexis of the Geophysical Branch of the Office of Naval Research. The society had no statutes or bylaws, no official membership, no officers, no formal meetings, no proceedings. Any scientist having something to do with the ONR could become a member simply by saying so.

On 20 April 1957, during a breakfast meeting at the Munks' home, Harry Hess, Roger Revelle (Scripps), Walter Munk, Joshua Tracy (U.S. Geological Survey), and Gordon Lill appointed themselves members of an AMSOC subcommittee and wrote a proposal to the National Science Foundation to study the feasibility of drilling a hole through the Moho (the project was nicknamed the Mohole).

The Moho lies some 30 to 50 kilometers under the continents. Technical considerations precluded drilling such a deep hole on land. However, geophysical investigations had shown that the Moho was much shallower beneath the ocean. If a hole could be spudded on the deep-ocean floor, the mantle would be reached after penetrating 5 kilometers through the ocean crust.

The idea of using drilling as a scientific method was not entirely new. Charles Darwin's theory that coral atolls were built on a foundation of sunken volcanoes was tested in 1897 by a hole drilled at Funafuti in the Pacific, although the final proof was not achieved until the 1950s, when a hole drilled on Eniwetok reached a basalt basement (see chapter 10). T. A. Jagger of the U.S. Geological Survey campaigned in 1943 for a drilling into the ocean beds, but his vision was far ahead of the financial and technical capacities of his time.

Times had changed. In 1957, two years before the Americans decided to send a man to the moon, nothing seemed to cost too much, and nothing seemed impossible. The AMSOC announced its plans for a Mohole Project during a meeting of the International Union of Geodesy and Geophysics that year, held at Toronto, Canada, and received enthusiastic endorsement. After the Soviet delegation dropped a hint that they were planning to drill into the earth's mantle, the United States government quickly accepted the phantom challenge of a race to the Moho.

The AMSOC Moho Committee became formally affiliated with the U.S. National Academy of Sciences and acquired thus both respectability and an official status. Applying for financial support from Congress through the National Science Foundation, the AMSOC soon received funding for exploratory drilling. On 23 December 1960, a contract was concluded between the NSF and the Global Marine Exploration Company of Los Angeles, and the drilling barge *CUSS I* was to be modified to do the first drilling tests, scheduled for early 1961.

The community of earth scientists was divided by the projected Mohole like the French intellectuals were by the Dreyfus Affair. Young radicals vociferously supported the project; levelheaded conservatives emphasized the impracticability of such a venture. I worked for an industrial research laboratory then, and I remember many debates I had with Ken Deffeyes, a Princeton graduate and a student of Hess's. I was against the Mohole, using prevailing arguments concerning technical difficulties: We saw no possibility of anchoring a drilling barge to the deep-sea floor. Furthermore, a drill bit would be worn out long before we could drill to the Moho, and it seemed impossible to insert a string of drill pipe back into the hole in the deep-sea floor after the string had been pulled out in order to change the drill bit. Deffeyes, however, was full of optimism, and he was to be proven right.

With typical American initiative and technical ingenuity, the proponents of the Mohole solved the first of the two technical problems right away. They developed a dynamic-positioning system (Plate I): a drill vessel is equipped with four motors, or side thrusters. After the vessel arrives on site, a beacon is dropped overboard so that it falls to a designated spot on the ocean bottom. The beacon emits acoustic signals, which are picked up by the receivers on the vessel. The data are then fed to a shipboard computer. When the vessel drifts away from the fixed spot, the computer reads the acoustic signals and determines the direction and magnitude of the drift. A message is then sent to the vessel's side thrusters to bring about the desired motion that will instantly correct for the computed drift. Over time, the system has been perfected to such a degree that drift can be held to within a radius of 50 meters during deep-sea drilling.

Even without drift, the drill vessel still moves up and down with the waves. Therefore, the segments of drill pipes, connected together to make a so-called drill string, cannot all be rigid, or else the "string" would be broken by a slight pitch and roll of the vessel. The engineers solved that problem also. They invented pipes called "bumper-subs" that can absorb up-and-down motions. Constructed like an automobile shaft, a larger, outer pipe could move up and down a smaller, inner pipe to telescope vertical motion, while the two sets of pipes were geared up to transmit the rotary motion needed for drilling a hole.

The barge *CUSS I*, now equipped with a dynamic-positioning system, drilled the first hole in 948-meter-deep water off La Jolla during early March 1961 and penetrated to 315 m beneath the seafloor. A site near the island of Guadalupe off the west coast of Mexico was then selected. In April 1961, the vessel was able to successfully hold position for three weeks, to drill five more holes. A string of pipe segments connected together was lowered into 3,558 m of water, and the deepest hole penetrated 170 m into sediments and 13 m into basalt. As *CUSS I* was towed back to Los Angeles, President Kennedy of the United States sent congratulations and called the experiment a "historic landmark."

The second technical difficulty was also not insurmountable. Less than a decade after the Guadalupe test was made, engineers from the Deep Sea Drilling Project successfully tested the reentry technique on 14 June 1970 in water more than 6,000 meters deep, 300 kilometers southeast of New York City. The reentry system consists of a high-resolution scanning-sonar system, a reentry cone 5 meters in diameter, and a drill-pipe-positioning system (Plate I). The reentry cone is lowered and set on the ocean bottom where a deep hole is to be drilled. The drill string is threaded through the cone. When the drill bit is worn out after several hundred meters of penetration into the seabed, the drill string is pulled up to the deck of the vessel, but the reentry cone remains on the seafloor above the bore hole. After the old drill bit is replaced by a new one, the drill string is again lowered to the seafloor. Then an underwater transmitter-receiver, one part of the scanning-sonar system, is lowered down inside the drill pipe until it protrudes through the central opening in the drill bit. The underwater package is connected to a control-display unit on the vessel by thousands of meters of electric cable. Information, in the form of underwater sound echoes, is received by the underwater electronic package, amplified, and transmitted to the control-display unit for processing and display on a sonar screen. This display tells the captain of the vessel the position of the reentry cone relative to that of the lower end of the drill string. He can maneuver the vessel to "home in" on the cone, until it is close enough so that a water jet from the drill-pipe-positioning system can place the lower end of the drill string directly into the reentry cone. The string can then slide from there into the hole.

The reentry technique was used many times during the last ten years of the Deep Sea Drilling Project, until the changing of the drill bit became almost a routine operation. Thus I was wrong in both of my objections to the Mohole. However, one source of trouble that was unforeseen proved to be the most difficult to overcome—the problem of removing drill debris away from a hole. This is never a problem on land. A fluid-circulation system pumps water or mud into the hole for cooling. The circulating fluid, on its return journey, brings drill cuttings back up to the rig floor to be disposed of. This is possible because the drill string on land is always housed in a casing, and the circulating fluid pumped down the drill pipe can find its way back in the space between the pipe and the casing. In deep-sea drilling we are not able to set casing from the wellhead through thousands of meters of water, to the rig floor of a drilling vessel. Drilling debris from the bottom of a hole brought up by cooling fluid is dumped on the seafloor around the opening of the borehole, so that the debris piles up like a large anthill. The "anthill" can be built up only to a limited height; eventually the debris will fall back into the hole. The friction of loose debris on the drill pipe will eventually stop its rotary motion, with the result that further drilling is not possible. To overcome this difficulty, engineers are now designing a riser system that can pump circulating fluid back up to the drill vessel. Considerable progress has been made during the last few years. However, until a successful test is carried out, we cannot be certain that we have the knowledge to drill to the Mohole.

Death of the Mohole

During the years after the Guadalupe test, the *CUSS I* was demobilized, but the AMSOC committee began formulating plans to construct a new and larger drilling vessel capable of reaching the Moho. On 28 February 1962, the National Science Foundation announced the selection of Brown and Root, Inc., of Houston, Texas, as the prime contractor for the Mohole Project. The total cost of drilling off Guadalupe had been 1.8 million dollars, and the early estimates for the Mohole were around 15 to 20 million. The cost escalated, and the estimate was placed at 68 million in late 1964 and 112 million in October 1965. Meanwhile, the lack of visible success and the ever-increasing cost estimates gave rise to new opposition. The opponents now realized that nothing was impossible if enough time and money were made available for engineering development. The relevant question was, therefore, not so much the technical feasibility, but the scientific merit of drilling the Mohole, when there were other, more worthwhile projects that would use less money for more science.

It was pointed out by some geologists that the earth's mantle is probably exposed on the steep wall of deep-sea trenches, and a piece of rock from the mantle could be hauled up by dredging. Meanwhile, many other questions about marine geology were emerging that could very well use the funding. Hollis Hedberg had in 1961 assumed the chairmanship of the AMSOC and the responsibility of studying the feasibility and desirability of drilling to the Moho.

Hedberg, professor of petroleum geology at Princeton, had many years of experience in the petroleum industry, which had developed the drilling techniques in marine waters. The Hedberg committee considered the options—should one try to reach the Moho in one giant stride, or through two stages of development? The "one big binge" approach was favored by impatient big spenders. However, the majority, led by the chairman, could see the risk of "putting all the eggs in one basket." It seemed unwise to invest tens and hundreds of millions of dollars just to get a chunk of rock, when we were not even sure if we were ever to get that chunk! The Hedberg committee in 1963 recommended an intermediate step: during a three-year campaign before the final assault on the Moho would be made, a dynamically positioned drill vessel of modest size could drill a number of shallow boreholes on the seafloor to gather experience; the cost of such a drilling campaign would be only about 9 million dollars per year.

The so-called intermediate stage of the Hedberg committee was eventually carried out in the form of JOIDES-DSDP. However, as Hedberg told me some years later, his proposal was not exactly welcomed by some establishment members of the National Academy. Meanwhile, the Mohole Project entered the arena of national politics. Congressman Albert Thomas from Houston, Texas, was then chairman of the House Appropriations Committee. Used to doling out billions for space research, the funding request for the Mohole was only a second decimal figure in his bookkeeping. With a view toward supporting the development of the drilling industry, which was centered in his hometown, Congressman Thomas was a strong supporter of the Mohole.

In 1966, Congressman Thomas died suddenly of a heart attack. Without its "godfather" in the Congress, the Mohole Project suffered a setback. On 5 May the House Appropriations Committee eliminated funds for Project Mohole from the NSF budget for the following fiscal year. After the Senate committee restored the funds, the full House voted on 18 August 1966, 108 to 59, not to go along with the restoration of funds. Project Mohole was killed, after 25 million dollars had been spent.

As Cesare Emiliani commented then and in years later, the best that can be said about Project Mohole is that it was premature. Looking back, however, we are thankful to Harry Hess and his AMSOC friends for their vision. The Mohole _was_ premature. We are not certain, even today, after more than two decades of experience in ocean drilling, if we could ever drill a hole down to the Moho, even if we were blessed with a billion-dollar budget. On the other hand, the dynamic-positioning system and other technological innovations resulting from the Mohole Project made possible the adventures of _Glomar Challenger._

Ice Age and LOCO

Cesare Emiliani was told to think big. He tried, and his modest request to get long cores to investigate the history of the Ice Age was the seed for the multimillion-dollar Deep Sea Drilling Project of the Joint Oceanographic Institutions for Deep Earth Sampling.

A Thermometer for Ancient Climates

In the summer of 1958, I was in Miami, Florida, on a business trip. My friend Bob Ginsburg arranged a luncheon and introduced me to Cesare Emiliani. Ginsburg and Emiliani were good friends, and Ginsburg insisted that I had to meet Emiliani, who was always full of ideas. At the time the debate on the Mohole was a favorite topic of conversation, and we soon entered into discussions on the merit of ocean drilling. Emiliani was a person who openly spoke his mind. He was incensed that millions of dollars were being poured into an impracticable project while he pleaded in vain for drilling shallow boreholes into ocean sediments in order to gather samples for studies of the Ice Age.

Emiliani had trained as a paleontologist at the University of Bologna. He was very proud of his native land and of his alma mater; he never lost an opportunity to praise the beauty of Tuscany or to advance Bologna's claim to have the oldest university in the world. At Bologna, Emiliani studied planktonic foraminifera (see Plate II). These small, one-celled animals once lived suspended near the surface of the ocean, and their calcareous skeletons are now preserved as microfossils in sediments on the ocean bottom. After the Second World War, Emiliani received a fellowship to engage in postdoctoral research at the University of Chicago. His horizons were broadened through his association with Hans Geiss and Sam Epstein, young scientists who worked in the Laboratory of Nuclear Chemistry with Nobel Laureates Harold Urey, Willard Libby, and Enrico Fermi.

Urey received the Nobel Prize for his discovery of deuterium, or heavy hydrogen. A normal hydrogen atom has a proton in the nucleus and an electron in orbit, with an atomic weight of one. Deuterium has a neutron in addition to the proton and the electron, and has, therefore, an atomic weight of two. Urey further discovered that oxygen atoms may include heavier varieties also. A normal oxygen atom has eight neutrons and eight protons in the nucleus, with eight electrons in orbit; the normal atomic weight is therefore sixteen. However, rare oxygen atoms (less than 1% of the total) have one or two additional neutrons (and the normal number of protons and electrons); they have atomic weights of 17 or 18 and are thus referred to as oxygen-17 (^{17}O) or oxygen-18 (^{18}O), respectively. Two atoms that have the same number of protons and the same number of electrons, but a different number of neutrons, have the same chemical properties but different atomic weights, and are called isotopes. Hydrogen and deuterium are isotopes of hydrogen, and the oxygen isotopes are ^{16}O, ^{17}O, and ^{18}O.

When oxygen combines with other elements to form compounds, a small proportion of heavy atoms, oxygen-17 and oxygen-18, will be present. The percentages of the various isotopes of oxygen in a compound are referred to as the oxygen-isotope composition of that compound. Commonly the oxygen-17 atoms are too few in number to be measured accurately; it suffices to express the isotope composition in terms of the relative proportions of oxygen-18 and oxygen-16. Normal ocean water, an oxygen-bearing compound (H_2O), has been homogenized to such an extent that its ratio of ^{16}O to ^{18}O is the same almost everywhere. We can choose this as a standard (standard mean ocean water, or SMOW). Oxygen compounds with more oxygen-18 atoms are said to have a positive $\delta^{18}O$ anomaly with respect to SMOW, and those with fewer have a negative anomaly. River water usually has fewer oxygen-18 atoms than SMOW—that is, a negative $\delta^{18}O$ anomaly.

Urey reported his findings during a talk in 1946 at the Swiss Federal Institute of Technology. Paul Niggli, the director of the Institute of Crystallography and Petrography there, immediately saw an application to geology. At the time, geologists had some difficulty in distinguishing limestone ($CaCO_3$) deposited in a lake from that deposited in an ocean. If freshwater has a negative $\delta^{18}O$ anomaly compared with SMOW, then the calcium carbonate precipitated in a lake should have a negative $\delta^{18}O$ compared with the calcareous skeletons of marine organisms that have precipitated out of ocean water. Niggli's reasoning was indeed correct. Eventually, a fossil belemnite (a remote ancestor of the squid) was chosen as a standard. Isotope compositions of other calcium carbonates can be compared with this solid standard, and $\delta^{18}O$ anomalies can thus also be expressed in terms of parts per thousand, P.D.B. (P.D. being the initials of the Pee Dee Formation in North Carolina from which the standard belemnite was collected; B stands for belemnite). As Niggli predicted, freshwater limestones have negative $\delta^{18}O$ as compared with P.D.B.

The rules governing the distribution of heavy oxygen atoms in a compound are, however, complex. Through more detailed studies Urey and his co-workers at Chicago found out that calcareous skeletons precipitated out of standard mean ocean water do not always have the same isotopic composition. Under normal circumstances, a skeleton precipitated from seawater at a cooler temperature has slightly more oxygen-18 atoms than a skeleton from the same seawater at a warmer temperature. In other words, the oxygen-isotope composition of a marine fossil species is temperature dependent; skeletons precipitated from warmer seas have a more negative $\delta^{18}O$ value.

Urey gave a talk in Columbus, Ohio, in 1949 on the potential applications of oxygen-isotope analysis to geology. He and his students at Chicago had analyzed a fossil belemnite from a Jurassic formation. These cylindrical shells have a stubby end and a pointed end, and their original skeletons have been preserved. When a section is cut, one can see well-defined growth rings. Samples from 24 rings were analyzed, and the oxygen anomaly range corresponded to temperatures of 14 to 20 °C. The temperature variation suggested that the Jurassic swimmer had lived through 3 summers and 4 winters after its youth and had died in the spring, at the age of 4 years.

I was a graduate student then, and was properly impressed that for the first time, earth scientists had found a way to read the temperature record of the ancient ocean—by measuring the isotopic composition of fossil skeletons. We had now passed beyond mere speculation, and were able to get some numerical data on the climate of the past.

The Great Ice Age

What did we know then about the climatic history of the earth?

During the second half of the eighteenth century and the beginning of the nineteenth century, scholarly views of the earth and of the earth's history were greatly dominated by theologians and philosophers, who interpreted the biblical Scriptures literally. The book of Genesis was accepted as the absolute truth, and Noah's flood was interpreted as one of the catastrophes willed by God to change the physical world and its living beings. Then came the days of enlightenment, when natural philosophers tried to liberate science from theology. Two British geologists, James Hutton and Charles Lyell, are considered the founders of geology, because they did much to furnish a scientific basis for the study of the earth and of its history. They pointed out, on the basis of numerous observations in the field, that the processes shaping the features of the earth are physical processes. Since physical laws are immutable, and do not change over time, we should expect that geological processes, including those of the past, should operate in accordance with the same physical principles since time immemorial. In other words, the processes we observe now should also have been operating during the geological past. This philosophical premise is called *uniformitarianism*, and was the generally accepted basic principle of geology since the middle of the last century, in contrast to the older view of catastrophism through divine intervention. Lyell, in his enthusiasm for the new doctrine, went to extremes and claimed that the conditions have also not changed greatly on earth throughout the geologic ages; processes have been operating at a uniform rate with more or less the same degree of intensity.

Every schoolchild in Switzerland knows, however, that the past climate was not always the same; there was an Ice Age in the not too distant past, in which the whole country was buried under ice like the continent of Antarctica today. They are even taught that there were four glacial stages, namely Günz, Mindel, Riss, and Würm, which were separated by interglacial stages of warm climate. But how did we find out about the Great Ice Age?

The discovery is a Swiss story. In the valleys and on the plains of the Swiss Midland, farmers are used to digging up stones and rocks of various sizes, some of which weigh many tons. The boulders are made up of rocks that are quite different from those cropping out in the nearby countryside; they are exotic, or *Findlings* (Figure 2.1). Their nature is so alien, diluvialists of the eighteenth century thought they had found some relics of the Flood; the exotics were believed to have been carried down from mountains by huge floods at the time of the Deluge.

Charles Lyell had a more realistic estimate of the stream power; he postulated instead that there might have been ice-rafting during the time of Noah—stones and boulders frozen in drifting icebergs were dispersed by the Flood and deposited as exotics when the ice melted away.

Lyell had not seen glaciers in his youth, and he had no idea of the power of glacial transport. Furthermore, he did not believe that conditions on earth during the past had been so different that glacial ice could have advanced to the Swiss Midland. Swiss farmers were not bound by such philosophical dogmas, and they were used to seeing the advance and retreat of mountain glaciers. They took for granted that the erratic blocks were brought down at a time when Switzerland was much colder; they were brought down by glaciers coming from the mountains to the plains. Jean-Pierre Perraudin, a mountain guide from the Valais, communicated this opinion to a number of his academic acquaintances. Since some of the *Findlings* were found on the sides of valleys, hundreds of meters above the valley floor, Perraudin had to advance the bold idea that those Alpine valleys were once filled completely with glacial ice! Such an imaginative postulate from a layman did not sit well with the academics; they dismissed the idea lightly as the tall tale of an illiterate.

Ignace Venetz, a Swiss civil engineer, was the first to listen attentively to Perraudin, and was the first to speak before a scholarly audience on the origin of the exotic blocks in the Alps, the Jura, and northern Europe. On the basis of his observations in the Valais, Venetz ventured a daring hypothesis at the 1829 meeting of the Swiss Society of Natural Sciences (SNFG) at the Hospice of the Great St. Bernard—he proposed the existence of an Ice Age in the not too distant past. Immense glaciers coming down mountain valleys coalesced and formed a great ice sheet that covered practically all of Switzerland, as

2.1. *Findlings.* Ignace Venetz accepted the belief, common among Swiss mountain farmers, that large boulders (called *Findlings*) in Swiss meadows, such as the one shown in this figure, were brought down by former glaciers. His 1829 postulate formed the basis of Agassiz's Theory of the Ice Age, first published in 1840.

PIERRE DES MARMETTES.

well as other parts of central Europe. This seemingly outrageous hypothesis was rejected by all except Jean de Charpenthier, and Venetz's paper was published only several decades later, long after the idea of an Ice Age had become an established scientific theory, because of its historical value.

Charpenthier spoke at the 1834 Lucerne meeting of the Swiss Society in support of Venetz. Charpenthier admitted that the past climate was once warmer, as indicated by the presence of fossil palm trees in the Molasse formations of the Swiss Midland, but then the climate got much colder. The proponents of the new theory thus claimed that the climate on earth was not always the same as it is now.

Historical records showed that Swiss mountain glaciers grew at times when the climate was colder. During the "Little Ice Age" of the seventeenth century, for example, mountain glaciers expanded to such an extent that their snouts came down and deposited layers of moraine down in the valleys where grass and wild flowers now flourish. At some more remote time during the Ice Age, it was even colder, and all of Switzerland was covered by an ice sheet. Large and small boulders were pushed to the valleys and plains by advancing glaciers, and they were left standing in the meadows as *Findlings* (Figure 2.1) long after the ice had melted and the glaciers had retreated back to the mountains.

The theory was very sensible, but the geological profession was dominated during the nineteenth century by the English, especially by Lyell, honored by many as the father of modern geology. The postulate of an Ice Age stood in gross contradiction to the Lyellian dogma that conditions on earth have been the same since the very beginning of time. Talks by Charpenthier found little echo outside of the small circle of provincial scholars who attended local meetings of the SNFG. The theory of an ice age might have remained buried for decades but for the energetic intervention of Louis Agassiz.

Called to Neuchâtel as professor at the age of 25, Agassiz attained international recognition as a paleontologist and was elected two years later president of the Swiss Society of Natural Sciences. Through the good offices of William Buckland, professor of geology at Oxford, Agassiz was granted a stipend from the Royal Society to study fossil fish, and he acquired influential friends in England. At the 14 September 1838 SNFG meeting in Basel, Charpenthier again talked about his theory of glacially transported exotic blocks. This time, the president of the society threw his support behind the heretic. His actions were to arouse the wrath of the masters of his day:

Elie de Beaumont found it necessary to rise up and dismiss such rubbish.

Agassiz wavered. He decided to take an excursion to examine the moraine deposits left behind by the Aar Glacier in the early months of 1840. After observing the striated boulders and ice-polished pavement at places from which the glacier had only recently retreated, the brilliant young man was no longer in doubt, and he gave his famous address on the theory of the Ice Age at the 1840 Neuchâtel meeting of the SNFG.

Agassiz's British patron, Professor Buckland, was horrified by the pronouncement of his protégé. Buckland traveled to Switzerland, intending to talk the young rebel out of his folly, but the elder gentleman was himself persuaded after he was taken to see the Aar Glacier. Agassiz then brought the new theory to Great Britain, where he spoke at the August 1840 Edinburgh meeting of the British Association for the Advancement of Science.

Lyell was still dragging his feet, and was slow to change his mind. Buckland and Agassiz came to visit Lyell at his Kinnordy Manor in northern England. They found a beautiful cluster of glacial moraines within two miles of the house. Lyell finally had to change his mind and abandoned the ice-rafting theory of the origin of exotic blocks. The theory of the Ice Age was accepted. Agassiz later went to North America, where he discovered that a large part of that continent had also once been covered by glacial ice.

Germany was slow to accept the new theory. _Findlings_ are not uncommon in northern Germany, but professors in Berlin who had never seen glaciers clung to the Lyellian postulate of iceberg transport, long after Lyell himself had discarded the mistaken notion. The German resistance lasted almost four decades, until 1875, when a Swedish Arctic researcher, Otto Martin Troell, came to Berlin and convinced some of the young Germans that the striations on the _Findlings_ had been scratched by glacial action. Toward 1880, the "old soldiers" faded away, and Albrecht Penck, who had in his youth spent many seasons in the Bavarian Alps, where glaciers abound, swept away the last agnostics with his masterful analysis of the history of the Alpine glaciation.

**Only Four
Glacial
Stages?**

Penck was a geographer and a student of landforms. Working in southern Germany, he noticed that modern streams meander weakly in valleys underlain by gravel deposits. He reasoned that modern streams did not have enough power to transport gravel, and that this coarse debris was brought down by rivers coming out of glaciers during an ice age. He further found that stream terraces on both sides of a valley are underlain by the same type of gravel deposit as that in the modern valley. These gravel deposits must have been laid down during earlier ice ages. Since he was able to distinguish four levels of terraces, he postulated that there must have been four ice ages. Using the names of the streams where he had made his observations, Penck and his associate Brückner came up with the names of four *glacial stages:* Günz, Mindel, Riss, and Würm, names that are known to every European schoolchild. In between were the *interglacial stages*, when the climate was as warm as, or even warmer than, that of the present.

Geologists in North America used a different approach to arrive at the same answer. They pointed out that glaciers push rock and mud debris forward, aside, and under as they advance, forming end, side, and ground *moraines*. T. C. Chamberlin of the University of Chicago and his associates counted the number of ice ages on the basis of the number of ground moraines, or *tills*, that have been deposited on the plains of North America. Like Penck, they also found four ice ages, which were named after the states in which the moraines are best exposed, namely Nebraskan (oldest), Kansan, Illinoian, and Wisconsinan. It was generally assumed that the four American stages corresponded to the four European stages.

Neither the American nor the European classic approach promised a complete record of glaciation history. Moraine deposits or tills laid down during an earlier glaciation may have been eroded away by a later glacier. In Switzerland, for example, the Riss glaciers seemed to have been the most powerful, and they managed to remove much of the earlier record. Also, the recognition of river terraces and their correlation from one stream valley to another presented problems and controversies. In fact, Penck at first thought that there were only three terraces, corresponding to three glacial stages, before he adopted his fourfold division. Later on, other scientists found more than four river terraces. Perhaps we had more than four stages of continental glaciation. Nevertheless, theories tend to become "incontrovertible truth" once they are written in schoolchildren's textbooks. We have to preserve our Günz, Mindel, Riss, and Würm, even if we have to invent "substages" Riss I, Riss II, Würm I, Würm II, Würm III, and so on to retain our magic number.

The ocean is a receptacle of sediments and should be able to provide a complete record of the history of the Ice Age. The problem there is sampling. When the HMS *Challenger* cruised around the world from 1870 to 1874, grab samples of ocean sediments were obtained, which told what is now down on the sea bottom, but gave little information about the past history of the ocean. During the 1925–1927 German _Meteor_ expedition to the Atlantic, gravity cores were taken, the earliest type of cores. A steep, hollow tube is hooked at the end of a cable, which is lowered into the sea; the momentum of the descent drives the tube into the seabed, and a sediment core is extracted from the tube after it is raised on board the vessel. Such gravity cores are about a meter long.

Wolfgang Schott studied the foraminiferal fossils in the _Meteor_ cores from the Equatorial Atlantic and made an interesting discovery. In the topmost part of the cores, among the many foraminiferal species present is one called _Globorotalia menardii_, which typically lives in tropical waters of the world. This was to be expected, because the cores came from equatorial waters. However, this warm-water species is not present in any of the sediments from the middle part of the cored section. Schott concluded that the sediments in this middle unit must have been deposited during the last glacial stage (Würm), when the Equatorial Atlantic was much cooler than it is now. Farther down, in the sediments near the bottom of the core, he found _Globorotalia menardii_ again; this seemed to correspond to the interglacial stage before the last glaciation (Riss-Würm interglacial).

Schott's results verified a prediction of the theory of the Ice Age. But a meter-long core represents a history of less than 100,000 years. Ocean sediments in the tropics accumulate at a rate of more than a centimeter per thousand years; only longer cores could give a complete record. However, the gravitational force of descent cannot force a corer much deeper than a meter or so into the seabed without considerably disturbing the core sediments. Charles Piggot tried using dynamite to drive a corer down with explosive force. He was not always successful, and the quick intrusion of the coring tube almost invariably disturbed the sediments in the cores. A major breakthrough in the study of ocean sediments was made by Börge Kullenberg, who designed a piston corer and used it successfully during the Swedish Deep Sea Expedition, 1947–1948. A piston is pulled out of a coring tube as the tube pushes down into the seabed; the displacement of the piston makes room for and sucks up the incoming sediment. With this type of corer, now named the Kullenberg corer, much longer undisturbed cores could be obtained, extending our knowledge of past climates much farther back in time.

The method, as pioneered by Schott, of counting skeletons of *Globorotalia menardii* can be applied to long Kullenberg cores. Using samples obtained from the Atlantic Ocean and the Caribbean Sea by the Lamont Geological Observatory, Palisades, New York, David Ericson and Goesta Wollin found that four episodes of a cooler climate during the last two million years were indicated by the absence of the *G. menardii* species; they thought that the four episodes corresponded to the four glacial stages on the continents of Europe and North America (Figure 2.2).

About the same time, Gustaf Arrhenius at Scripps reasoned that climate can exert a critical influence on ocean circulation, and this influence should be recorded by the chemistry and mineralogy of ocean sediments. He found, for example, that the calcium-carbonate content of the tropical deep-sea sediments fluctuated widely with time, and he considered such cyclic fluctuation a record of the past changes in climate. His colleagues at Scripps, Fred Phleger and Frances Parker, used this technique to study the cores from the Atlantic Ocean collected by the Swedish Deep Sea Expedition. They did not find the classic pattern of four glacial stages; instead, there seemed to have been at least nine repetitions of intensely cold climate.

A controversy arose: were there only four glacial stages during the last Great Ice Age, or were there more frequent cyclic changes? The theory of the Ice Age was heading toward a crisis during the late 1950s. Cesare Emiliani got into the act, and his proposal to obtain a long core to decipher the climatic history during the last Ice Age—known to us geologists as the Pleistocene epoch—was the seed of an idea that eventually evolved into the JOIDES Deep Sea Drilling Project.

Thinking Big   Cesare Emiliani told the story years later, in his history of a new global geology in the seventh volume of *The Sea*, edited by him:

> On January 1, 1957, I joined the Marine Laboratory of the University of Miami. One day in early February 1962, I was chatting with Bob Ginsburg on the lawn in front of our main building when Bob said to me, "You should be thinking big." I asked what he meant, and he said, "Just think big." And then he walked away. So I went back to my office and started thinking big.

Emiliani, using cores from Caribbean Sea, was a pioneer in applying isotope geochemistry to paleoclimatology. In 1955 he found, like Phleger and Parker, that there might be many more glacial stages during the Pleistocene than the four suggested by Penck. When he went to Miami, he intended to continue the work he had started at Chicago. To measure the isotopic composition of oxygen in compounds, one needs an instrument called a mass spectrometer. Emiliani himself was not an experienced instrument-maker, so he persuaded his friend Hans Geiss, who is now director of the Physics Institute

2.2. Climatic changes during the Ice Age. Two different methods were used to estimate ocean temperature during the late 1950s and early 1960s. Ericson and Wollin found that the percentage abundance of the foraminifera species *Globorotalia menardii* varies. Emiliani determined the paleo-temperature of ocean water by analyzing the oxygen-isotope composition of foraminifera skeletons. The latter method is now routine.

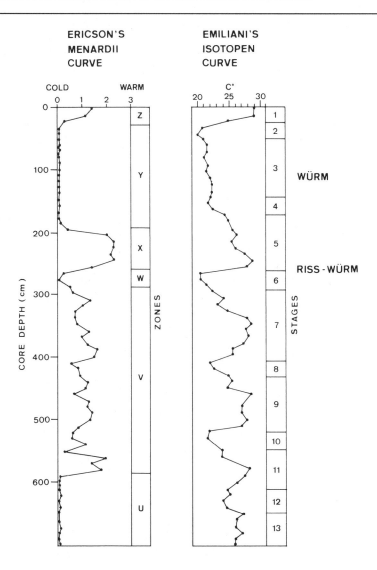

at the University of Bern, to come and help. They borrowed and tinkered until they scraped together a mass spectrometer, and they proudly showed me their laboratory when I visited Miami in 1958. Emiliani talked excitedly about his new findings. He had by now found seven cyclic changes in the Pleistocene climate (Figure 2.2), and correlation with Penck's chronology was difficult. Ericson and Wollin were wrong, he declared. Evidently we were on the threshold of a new era, on the eve of a scientific revolution. To carry out the revolution, we needed more sample materials, particularly long cores.

Emiliani did not want to reach the Moho; he only wanted to drill through the sediments of the Ice Age so that the rhythms of the climatic changes could be compared with astronomical cycles. I can still vividly recall the scene when he said, "Give me just a long core of 100 meters of ocean sediments. We can make more science with it than with a billion-dollar hole to the Moho!"

If my memory is correct, Emiliani erred when he wrote that he started thinking big in 1962; he had been thinking big since 1958. Ginsburg's passing remark merely propelled him into action. Emiliani telephoned R. F. Bauer, the president of Global Marine, and inquired about the feasibility and cost of drilling two holes, each 1,000 m deep, in 4,000 m of water south of Puerto Rico. It was feasible, he was told, and it would cost 2 million dollars. Next he called John Lyman, who was then program director for oceanography at the National Science Foundation, and asked for financial support; he was encouraged. On 26 March 1962 John Lyman received from Emiliani a proposal for a grant to support plans to obtain long cores from the deep-sea floor. By now Emiliani was really thinking big. Not only did he want long cores to resolve the Ice Age controversy; he proposed further that

geophysical, geochemical, micropaleontological, and mineralogical analysis of the cores will yield information of great importance on the conditions prevailing on the ocean floor, in the water column above, at the ocean surface, in the atmosphere, in neighboring continents, and even in outer space and in the sun, during the time of sediment deposition, i.e., during the past 100×10^6 years or so.

This is, of course, what the scientists of the JOIDES Deep Sea Drilling Project eventually achieved while working with cores obtained by the *Challenger* at sea.

From LOCO to JOIDES

The possibility of using a dynamic-positioning vessel to core ocean sediment was in fact not overlooked by proponents of the Mohole. The chairman of AMSOC, Gordon Lill, had recommended to the U.S. National Academy of Sciences as early as 1961 that a program of exploring the sediment and upper crustal layers of ocean basins should be carried out prior to the drilling of the Mohole, but he felt that it was up to leading oceanographic institutions to establish such a program, scientifically and financially separate from the Mohole Project.

Emiliani's initiative thus came at a most opportune time. However, it was obvious that such a major undertaking demanded resources far beyond the means of a single individual or a single institution. Maurice Ewing, then director of Lamont, joined in and campaigned actively in 1962 for an ocean-sediment coring program. In May 1962 the National Academy of Sciences committee on oceanography appointed at Emiliani's request, a group of scientists to discuss the drilling and coring project. This group had its first meeting on 10 June 1962 in Miami and endorsed Emiliani's March proposal; additional drill sites in the North Atlantic were suggested. Emiliani was elected chairman of the LOCO (Long Cores) committee to guide the implementation of the proposal.

The second LOCO meeting took place on 13 September 1962, again in Miami. During that meeting, as noted in the minutes,

following a suggestion by John Lyman, it was decided to ask the four leading institutions [Woods Hole, Columbia University, the University of Miami, and the University of California] to establish a nonprofit interuniversity corporation to build, own, and operate a vessel especially designed to drill and core oceanic sediments.

In fact, such a corporation was not established then, and the Deep Sea Drilling Project would eventually rent, not own, a drill vessel. When the Joint Oceanographic Institutions, or JOI, Inc., was finally constituted many years later, more than the original four institutions were included.

In 1964 I moved from New York to California. My old friend from student days at Los Angeles, Jerry Winterer, came to visit me in Riverside. He had since branched out into oceanography and was a Scripps representative on the committees exploring the feasibility of deep-sea drilling. From him I learned of the maneuvering for position in the game of science politics. The partnership in the new venture became a game of musical chairs. Miami and Princeton were dropped at first, when LOCO became CORE (Consortium for Ocean Research and Exploration). Later, Miami rejoined Lamont, Scripps, and Woods Hole; these four became the founding members of the program of Joint Oceanographic Institutions for Deep Earth Sampling

(JOIDES), named by Roger Revell, which was formally established at a meeting in Washington, D.C., in May 1964. Still later, the University of Washington became the fifth JOIDES member when congressional politics convinced the National Science Foundation of the need to increase regional representation and include an institution from the Pacific Northwest.

While the Mohole Project announced a planned drill site in the Pacific, a dynamically positioned vessel, *Caldrill I*, on loan to JOIDES drilled fourteen holes in the Atlantic in April and May 1965. Their impressive results convinced more and more people that an oceanwide program of sediment sampling was a preferable alternative to the "one-shot deal" of uncertain success. When the Mohole Project was finally being laid aside in 1966, a contract for a Deep Sea Drilling Project (DSDP) was awarded to the Scripps Institution of Oceanography. The DSDP organization was to carry out scientific programs formulated by JOIDES advisory panels. Taking advantage of the experience gained during preliminary Mohole and JOIDES drillings, a new drill vessel was to be built specifically for this project by Global Marine, Inc., of Los Angeles. This vessel, D/V *Glomar Challenger*, was launched in March 1968 and named after an illustrious predecessor in ocean exploration, the HMS *Challenger*.

3

The *Challenger* Goes to Sea

The Inauguration of *Glomar Challenger*

"Doc" Ewing had the ability to make people feel that what he wanted was not only what he wanted but what the Good Lord would have done. So the Challenger *was constructed and went to sea, and "Doc" had the honor of leading the inaugural cruise of the Deep Sea Drilling Project.*

A New Drill Ship

Glomar Challenger had a derrick amidships, 45 meters tall (Plate III). It was thus a landmark in any harbor, and we never had any difficulty finding it when we joined the vessel for a cruise (called a "leg" by the JOIDES community). The derrick had been designed to lift one million pounds, or a drill string seven thousand meters long. The height of the derrick had been limited so that *Glomar Challenger* could pass under bridges when entering all major seaports of the world. I myself had two fascinating experiences watching the tower slide under a bridge. The first time was in 1970, during Leg 13, when we sailed out of Lisbon down the Tagus River at night. The second was five years later, as the *Challenger* entered the Black Sea via the Bosporus; on this occasion the margin was so narrow, the city of Istanbul had to cooperate by stopping all traffic across the suspension bridge that unites Europe and Asia.

The top of the drill rig could be reached through a series of steel ladders, and the view from up there was extraordinary. I never ventured up, having developed a healthy respect for heights after I fell down the Gran Sasso once and broke my shoulder. However, the top of the derrick was an ideal retreat. During Leg 6, Al Fischer, the chief scientist, decided to hide there to escape for a moment or two the duties incessantly required of him. It so happened that at that moment he was being sought for an emergency meeting and was nowhere to be found. Alarm was sent out; it was feared he had fallen into the ocean.

There was great commotion until he came down. Ever since then, nobody has been allowed to climb the derrick except by special permission of the captain.

Directly under the derrick amidships was the moon pool, so named because it is a cylindrical hole in the derrick through which the drill string could be lowered down. Of course, the wall of the moon pool was watertight. Looking down from the upper deck, one could see crystal-clear ocean water and might be tempted to jump in for a swim on a hot summer day, but this was strictly forbidden. A scientist swam in the moon pool once when the vessel was on station in the Indian Ocean; he was considered mentally unbalanced and was sent home immediately to La Jolla after the ship docked at Mauritius.

Glomar Challenger was a completely self-sustaining unit and carried sufficient fuel, water, and stores to be able to remain at sea for a maximum of 90 days without replenishment. This logistical consideration is probably the reason why most *Glomar Challenger* cruises lasted less than two months. The longest, Leg 39, from Amsterdam to Cape Town, was 71 days at sea, but the cruise was interrupted for a "pit stop" at Recife, Brazil.

The vessel had 70 berths on board, with an additional bed in the dispensary that had been used more than once when an extra specialist was indispensable. Two crews, each including about 50 officers, seamen, and members of a drilling team, served on alternate cruises. The crew was joined by 10 marine technicians who were employed by DSDP, and by 9–16 scientists. The scientists were noted specialists from various academic, government, or industrial organizations; the co–chief scientists were selected for each cruise by the planning committee of JOIDES. The scientific team of a cruise was assembled by the chief scientist of DSDP, in consultation with the co–chiefs designate.

The two co–chief scientists were responsible for the scientific success of the expedition. They might also have been the proponents of the cruise, and may have participated in the planning. Two chiefs were necessary, because the drilling and coring was a 24-hour-a-day operation; chief scientists always had to be on the alert and make decisions when called upon. Usually each worked a 12-hour shift, but both were present when important steps were to be taken. Like generals in a campaign, they were supposed to follow the strategy and guidelines given by JOIDES and DSDP, but they were permitted, especially in the early phases of DSDP, a great deal of latitude in solving tactical problems. They chose the exact spot for drilling at each site. They determined how much of a drill section should be sampled with a drill core. They ordered the termination of a hole, in consultation with an operations manager who represented DSDP on board. The hectic life of a chief scientist on board *Glomar Challenger* has been described in my book *The Mediterranean Was a Desert*, also published by the Princeton University Press. Indeed, the record of more than twenty years of drilling has proven that the choice of the co–chief scientists greatly affected the success of many legs. However, even the best of them could do nothing in the case of technical difficulties or unforeseen bad weather.

Glomar Challenger was not anchored on station. The vessel could maintain a position within a circle of small radius during drilling operations through the help of a dynamic-positioning system, first developed for the proposed drilling of the Mohole (Plate I). The technique of deep-sea drilling was otherwise similar to rotary drilling on land. Two drilling teams, each on two six-hour shifts, worked around the clock. Each team had three roughnecks and a tool-pusher; a drilling superintendent was on board to supervise both teams.

After the *Challenger* reached a drill location, the drilling team would start to assemble the drill string as soon as the vessel's positioning had stabilized. Seven thousand meters of drill pipes were stored in pipe racks (Plate IV). A single-joint pipe was 9.5 meters long, and the double- and triple-joint pipes were two and three times as long. The very bottom of a drill string was a pipe called the drill collar. A drill bit was screwed onto the lower end of the drill collar, which was then picked up by a weight-lifting Bowen unit attached to the derrick, and lowered into the water through the moon pool. After the drill collar was tightly secured by a clamp on the derrick floor, its upper end was freed from the Bowen unit, so that the giant claw of the Bowen unit could be used to pick up the next pipe segment from the pipe rack. This segment was then lifted up and its lower end linked with the upper end of the drill collar. Then the drill string, with its newly added drill pipe, was again lowered through the moon pool to be clamped on tightly again, so that the Bowen unit would be free to pick up the next segment and link it to the last drill pipe (Plate V). Through this procedure, one segment after another was added and the drill string grew in length. By repeating this procedure hundreds of times, the drill string could reach an ocean bottom up to 6,000 meters deep and start drilling. The roughnecks were strong men who worked with the precision of ballet dancers on the derrick floor, while the tool-pusher operated the various mechanical gadgets. Naturally, it was important to clamp the top of a drill string very tightly while it was being detached from the Bowen unit. Once during an early cruise (Leg 4) the clamps failed to hold, and a string of more than four thousand meters of drill pipes was lost. It was a million-dollar mistake!

As we mentioned previously, some special drill pipes, called bumper-subs, made up the bottom part of a drill string. Bumper-subs could lengthen or shorten with the motion of the vessel, and they could transmit the rotary motion needed to turn the drill bit at the end of the drill string.

The drilling teams were experienced—some of the team members had been with the project since its inception—and they were quick and efficient. A pipe segment could be added on in about six minutes, and a drill string 4,000 m long could be assembled in less than 10 hours. The work on the derrick floor was dangerous, but the skill of the roughnecks kept disabling accidents to a minimum. Only one person ever died accidentally while on duty (see chapter 17).

To take core samples, a core barrel 9.3 meters long was placed inside the drill collar. Sediment was intruded into the inside of the barrel as the drill bit made its way into the seabed. Since it took about a day to assemble and take apart a long drill string, it would be impractical to haul the drill collar up every time a core was taken; instead, a fishing technique called wireline coring was used (Plate VI). The core barrel had a hook at the top. After a core was taken, a sandline with a catch (overshot) was sent down and hooked to the core barrel inside the drill collar, and the barrel was brought up. Then the drill string was unscrewed near its upper end. The barrel with the sediment core was taken out, and an empty barrel was then pumped down the drill string and seated inside the drill collar. Usually the whole operation took about an hour, coring soft sediments in 3,000-m-deep water. Thus cores might be coming up once every hour, adding to the stress of shipboard scientists who had to handle and study the samples. Worse, of course, were the occasions when nothing came up, when frustrated scientists again and again saw daylight at the other end of the core barrel.

"Doc"

On 24 June 1966, the Scripps Institution of Oceanography took over the management of the Deep Sea Drilling Project, and received a 2.6 million dollar contract from the U.S. National Science Foundation for 18 months of ocean drilling. Mel Peterson was the chief scientist and later the project manager. Terry Edgar was the staff-coordinating scientist, and later the chief scientist.

The first phase of DSDP started when _Glomar Challenger_ departed Orange, Texas, on 20 July 1968 for scientific drilling into the seabed of the Gulf of Mexico. Phase I was divided into nine legs, and the chief scientists for the first five cruises were senior staff from the five JOIDES institutions, namely Lamont, Scripps, Woods Hole, Miami, and Washington. The honor of leading the inaugural cruise was given to Maurice Ewing, then director of the Lamont Geological Observatory, who had been one of the chief proponents of JOIDES.

Maurice Ewing was a physicist, a geologist, a seaman, and a teacher. He was affectionately called Doc by all who knew him. Doc did not make a deep impression on me when I first heard him talk, during a professional meeting in Los Angeles. That was back in 1951, when I was a graduate student at UCLA. Several of us drove across town to the campus of the University of Southern California, where the meeting took place. The classroom was full, and a tall, gray-haired Texan was standing behind the speaker's rostrum. He mumbled, paused, and stuttered. Then a lantern slide was flashed on the screen and Ewing mumbled again in his Texas drawl. Sitting in the back row, I could not understand a word he was saying, and decided to walk out in the middle of his lecture. That was a mistake.

Maurice Ewing had been talking about the wonderful things he did with his newly developed tools of marine seismic surveys. As a young man Ewing had worked for the oil industry, exploring oil on salt domes with seismic measurements. After receiving a physics degree in 1931 from the Rice Institute in Houston, he went east and found a job at Lehigh University during the worst Depression years. Geophysics was then interdisciplinary, and marine geophysics had yet to be started. Geologists knew very little about the oceans, and they were still talking about land bridges or sunken Atlantis.

Young Ewing was sent to sea by Dick Field of Princeton to embark on a study of the ocean floor. After a short tour with the survey ship *Oceanographer*, Dick Field persuaded Woods Hole to give Ewing a chance on their R/V *Atlantis*. Ewing made his seismic profiling in the shallow water of the continental shelf off Cape Henry. Interpreting arrival times of shock waves produced by artificial explosions and refracted by subbottom layers beneath the seafloor, Ewing and his associates discovered a sedimentary sequence 3.8 km thick above the basement of the Atlantic margin.

Ewing's first major report of work on the Atlantic Coast done in waters less than 100 fathoms deep was published in 1937. A year later, he was making seismic refraction measurements down to 2,600 fathoms. He moved to Woods Hole in 1940, where he worked for the U.S. Navy during the Second World War. In 1946 he was invited to join the staff of Columbia University, and in 1949 he founded the Lamont Geological Observatory, later known as Lamont-Doherty. Together with his students, he continued the seismic exploration of the North Atlantic, and those results came out in a series of publications after the Second World War. He was presenting findings from one such investigation at that 1951 Los Angeles meeting.

In those days two vessels were needed to carry out seismic studies. On one vessel, sticks of dynamite were lit and thrown at frequent intervals into the sea to produce shock waves. The signals were received by hydrophones on the other vessel, some distance away. The exercises were not without danger. Once an assistant was killed when a grenade blew up in his hand as he was investigating the delayed reaction of a fuse. Sometimes I wondered if Doc had been more exuberant before the accident. I knew Doc only in his later years; his subdued manners completely belied his boyish enthusiasm and dedication to science.

Maurice Ewing was a most conspicuously effective member of the scientific community. As his friend Sir Edward Bullard wrote in a memorial shortly after his death, Ewing passionately believed that what he was doing was worthwhile, and he had great skill in the techniques of coercing bureaucracies and seizing opportunities. "He had the ability to make people feel that what he wanted was not only what he wanted but what the Good Lord would have done," Bullard told us in his eulogy. Ewing wanted JOIDES, and his persuasive powers did much to effect the metamorphosis of LOCO into JOIDES/DSDP.

My second encounter with Doc took place in Pensacola early in 1954. I had just joined the Research Division of the Shell Oil Company in Houston. Ewing had persuaded Shell and other companies to support his work at Lamont, and a schooner, the *Vema*, had been converted into an oceanographic vessel that had sailed to a Gulf port to start another round of investigation of the Gulf of Mexico. Doc was speaking to industry representatives, and with this chance to look at him at close quarters I acquired a respect and admiration for this dedicated man. On that occasion, he told us that they had just discovered salt domes under the abyssal plains of the Gulf of Mexico!

Salt domes are an array of pillarlike structures of salt thrust upward into sediments from deep-lying salt beds (Figure 3.1). Salt domes are common on the Gulf Coast of the United States. That was not surprising, because orthodox theories told us that salt beds were deposits in coastal lagoons, and one could expect lagoonal salts to be associated with the coastal sediments of the Gulf Coast. However, the presence of a salt bed under a deep-sea floor was difficult to believe; one could hardly imagine that the Gulf of Mexico had dried up so as to produce an ocean of salt.

Ewing's first discovery was made with the seismic refraction survey; he had to explain some anomalous results on the speed of transmission of seismic waves. There were thus considerable uncertainties in his interpretations. Several years later, in 1961, Lamont's research vessel, the *Vema*, went back to the Gulf of Mexico with a newly installed continuous seismic profiler.

The seismic profiling technique was an outgrowth of the echo-sounding principle. During the memorable voyages of the HMS *Challenger* a hundred years ago, the depth of the ocean bottom was measured by the ancient method of lowering a 100-kilogram weight on a hemp line; each sounding took hours. After the turn of the century echo-sounders were invented. Acoustic signals sent out by a vessel are reflected back by the seafloor; the echos are recorded by a needle-point pen on a moving scrill of paper, producing a depth profile of the seabed as the vessel sails forward. After the Second World War, it was perceived that signals with wavelengths longer than those used for echo-sounding could penetrate a sediment-covered ocean bottom, and be reflected after impinging upon a hard subbottom layer. During the 1950s various attempts were made to produce an adequate energy source. One of the most successful instruments was the airgun, towed behind a moving vessel. Acoustic signals are sent out when compressed air is periodically released; these signals, after they have been bounced back from bottom and subbottom reflectors, are picked up by a string of hydrophones, towed by the same vessel (Figure 3.2). The continuous seismic profiler can thus be regarded as a super-echo-sounder; it can register not only the bottom profile, but also the structure of subbottom layers.

3.1. Salt domes. Salt domes detected by continuous seismic profiling of a ten-mile-wide section of the Balearic abyssal plain in the western Mediterranean. Some protrude as knolls above the seafloor; others are completely buried.

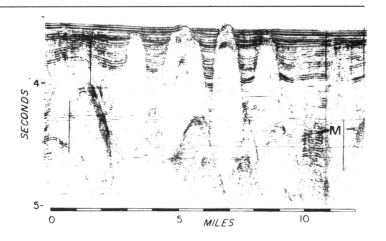

Salt is considerably more compact than soft sediments. The contact between salt and sediments is thus a reflecting surface, and can be depicted on a recorder. With the continuous seismic profiler, the existence of dome-shaped structures under the ocean floor was no longer in doubt. However, skeptics still argued that dome-shaped structures under the Gulf Coast are "mud domes," because some boreholes that penetrated into those structures found mud, but no salt, in the domes. Was it possible that Ewing's "salt domes" were in fact mud domes? Yes, indeed, Ewing agreed, they could be.

Dome under Challenger Knoll

On 19 August 1968, the day of reckoning finally came. The *Glomar Challenger* was positioned over one of the domelike structures; this one had been protruded as a small round hill above the abyssal plain of the Sigsbee Deep. The feature was named Challenger Knoll, in honor of the vessel that was to drill it.

Since hydrocarbons are trapped under the caprock of numerous salt domes, drilling into them is risky business. In 1964, a drilling platform sank into 190 meters of water because the buoyancy of the sea was reduced by hydrocarbon gas released

3.2. Continuous seismic profiling. A source emitting acoustic signals, called an airgun, is towed by a ship. The signals can penetrate the soft seabottom and are reflected by a hard layer (reflector) beneath the seafloor. The reflected signals are registered by the so-called "eel," also towed by the ship, which consists of a series of electronic gadgets called geophones.

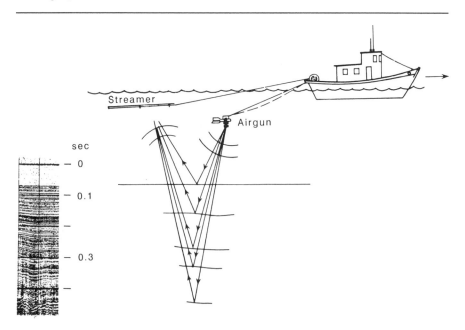

Streamer

Airgun

sec

— 0

— 0.1

—

— 0.3

—

from a high-pressure reservoir in a salt dome that had been penetrated by drilling. Such an accident is less likely to happen in deep water, where gas bubbles would drift away before they could collect into one giant gas bubble beneath the drill vessel. Of more concern to all was the danger of spilling oil into the sea: there would be no way to plug the hole after an accident to avoid polluting the gulf. However, the sediment above the Challenger Knoll was very thin, and dangerously excessive pressure was not expected. Nevertheless, Ewing was instructed to stop drilling if the drill string hit a hard layer! Under no circumstances was he to drill through the caprock of the salt dome.

After the position of the _Challenger_ was stabilized, the drill string was lowered to a depth of 3,600 m, where it hit bottom. After penetrating to a depth of 136 m, the drill bit hit something hard. The drill string was disconnected, and a core barrel was sent down before the string was reconnected. Rotary motion cut a core one and a half meters long. When the core was raised to the deck, it was found to consist mainly of porous limestone, somewhat oil stained; the rock was not sufficiently dense to seal a high-pressured hydrocarbon accumulation. Ewing sent the barrel down again, but caution prevailed. Only a 30-cm-long core was cut. This core was found to consist entirely of crystalline gypsum, a typical caprock; it was also oil stained.

Doc and his associates were satisfied. The drilling had proven beyond reasonable doubt that the elevation on the deep-sea floor, now named Challenger Knoll, is a salt dome and that a salt deposit is present under the deep-sea floor of the Gulf of Mexico. Not intending to push his luck too far and take unnecessary risks, Doc ordered the crew to raise the drill string. He had waited 15 years; now he had his proof. Ewing had to wait another year or two (after the deep-sea drilling of the Mediterranean) before an adequate explanation of the salts under the deep-ocean floor could be given. Nonetheless, with the help of _Glomar Challenger_, he was the first to make the revolutionary discovery that salt beds could be laid down upon ocean crust!

Horizon A

One of the other justifications of deep-sea drilling was to identify prominent acoustic reflectors under the seafloor. In fact, few reflectors have a dome-shaped outline on the profiling diagram; most are nearly horizontal. Lamont scientists discovered several such reflecting horizons in the western Atlantic. They have a very wide distribution: they were identified from beneath the seabed at the continental rise (i.e., the foot of the continental slope) offshore eastern North America, and found to extend eastward to the Bermuda Rise. Those acoustic reflectors were, however, not found on the Mid-Atlantic Ridge.

Reflecting horizons must be the top of hard layers, off which acoustic signals can be bounced back. But ocean sediments are commonly oozes. Doc dearly wanted to know what his reflectors were. Not knowing their age, he used code designations A, A*, and b for the three most prominent reflectors of the western Atlantic.

Piston cores taken from where Horizon A is exposed had yielded Cretaceous fossils, but the fossils were not from a hard rock and thus could not verify the age of the horizon. Now that a drilling vessel was available, Doc could obtain a sample of the hard rock from a borehole.

After his triumph at Challenger Knoll, Doc sailed the vessel east of the Bahamas and thence to the Bermuda Rise area to investigate the mystery of Horizon A. The _Challenger_'s crew was not yet ready to deal with hard rocks; only a few cores were obtained, and the holes had to be abandoned at a shallow depth because of worn-out drill bits. However, enough chips and fragments were found in DSDP Holes 6 and 7 on the Bermuda Rise to tell Doc that Horizon A is a layer of radiolarian chert of middle Eocene age.

Radiolarians are one-celled animals that swim in surface waters. They are abundant in fertile equatorial waters where nutrients are abundant. Their skeleton consists of silica, and radiolarian sediments are commonly found on the bottom of equatorial oceans as soft oozes (Plate VII). In places, the skeletons are recrystallized and the oozes have been turned into a hard rock—a rock called radiolarian chert, or radiolarite. The discovery of this old radiolarian chert in a temperate latitude was a surprise. Either the equatorial belt of high fertility was very broad during the Eocene, or the Atlantic seafloor had moved northward since that time.

The dating of reflectors has added the time dimension to marine geology. With seismic profiling we could identify reflecting horizons, beds or strata and map their distribution in space. When drill samples become available, we could describe the age and nature of the beds and strata. The science of seismic stratigraphy was born.

On 23 September 1968 the _Challenger_ returned to Hoboken, New Jersey, after drilling 11 holes at 7 sites. Some long-standing puzzles had been resolved, and new questions had come up. The adventures of _Glomar Challenger_ had begun with a resounding success: Doc Ewing and his shipmates were given a hero's welcome, and his "oil discovery" in the oceanic seabed was reported, with much fanfare, by the _New York Times_. Those were the days before Greenpeace!

Doc and his scientists remained modest, and the general comments section of their cruise report stated:

It is evident from the results of this pioneering cruise in deep-ocean drilling that the *Glomar Challenger* and its system work well, and that it can provide a vast wealth of data which will significantly improve our understanding of the earth on which we live and from which we extract our natural resources.

They knew we geologists had a new tool, and that the ship was going to revolutionize the earth sciences, just as Galileo's telescope changed astronomy.

End of a Career

Doc left Lamont shortly before his retirement, to start a new oceanographic institution in Galveston, Texas. I ran into him on numerous occasions in connection with JOIDES business. Also, I heard a good many stories about him from my Lamont friends. Doc was almost fanatic in his devotion to science. He was up in his office on the hill at seven each morning, and often worked late into the night. This habit of hard work started at Woods Hole and was to become a tradition of Lamont scientists.

The last time I saw Doc was in San Antonio, in the spring of 1974. The discovery of oil-stained rocks during the inaugural cruise had raised the question of possible pollution during deep-sea drilling. An Advisory Panel on Hydrocarbon Prevention and Safety was constituted by JOIDES in 1970; they had to examine all drill proposals and ascertain that no deep-sea drilling holes would be drilled into an oil or gas reservoir, causing pollution of the sea. (None ever have, thanks perhaps to this caution!) Doc was a member of the panel. I went to San Antonio with proposals for the second drilling cruise to the Mediterranean Sea. I was full of confidence, but ill prepared. A youthful member of the safety panel was antagonistic to the point of being aggressive, and I was equally short tempered. Doc uttered hardly a word during the hot debate, which lasted the whole evening. Toward the end, near midnight, he finally spoke up and effected a solution; his sympathy helped me out of a very unpleasant situation. I was thankful to Doc, and saddened to learn that he died a week after our San Antonio meeting.

Like Harry Hess, Ewing died suddenly of a heart attack; he had been overworked. Walter Sullivan was to dedicate his book *Continents in Motion* to Ewing and Hess; he properly recognized that they had performed a more essential role than any others in their contributions to the modern revolutionary discoveries of the earth sciences.

4 The Earth Science Revolution

Have the poles flipped? Have the continents drifted? Is there thermal convection in the earth's mantle? What is the origin of rift valleys on the ocean bottom? What is the origin of the earth's magnetic field? When did the earth's magnetic field reverse itself? What are those stripes of magnetic anomalies on the sea-floor? What is the origin of ocean fractures? Why are the ocean sediments so young? We had to have a new Leitbild, or guiding principle, namely the theory of seafloor-spreading, to solve our puzzles.

Have the Poles Flipped?

I did my dissertation at UCLA on hard, crystalline rocks. A standard procedure was to cut a slice of rock, glue it to a small glass slide, and grind it until the rock slice was 0.03 mm thick and could be examined with a light-transmission microscope. As a graduate student I spent a lot of time in the thin-section workshop. A young postdoctoral researcher from the Institute of Geophysics, Keith Runcorn, often came to use our rock-cutting saw because there was none in his institute. We tolerated the intrusion with friendly smiles until one evening when his huge rock samples broke our saw-blades.

In the spring of 1953, before Runcorn was to return to England, he gave us a lecture on his research. He was measuring the direction of magnetization, the so-called natural remnant magnetism (NRM), in rocks. He thought he could tell if the earth's magnetic poles had wandered or if the continents had drifted. We geology students did not believe in continental drift then, and we thought he was wasting his time on a discredited old idea. He did surprise us, though, with one of his claims: he told us in no uncertain terms that the earth's magnetic poles might have once been reversed. The earth's magnetic field has flipped—or has Runcorn flipped, we asked among ourselves.

Our sarcasm revealed only our ignorance. The polarity reversal was common knowledge to students of rock magnetism, but we geologists during the fifties were too specialized to know what had been done by our interdisciplinary colleagues.

The principle of studying rock magnetism is very simple. When we hang a magnetic needle on a thread in the air, the north-seeking end of the needle will point toward the north magnetic pole. Also, this end of the needle will dip down and make an angle with the horizontal; this angle is called the magnetic inclination. Meanwhile, the other end of the needle will point to the south and be tilted up; its inclination angle, of course, is the same as that of the north-seeking end, but it is given a negative sign because it is pointing upward. The angle of magnetic inclination varies latitudinally: in polar regions, the inclination is very steep, whereas the angle is almost nill at the equator.

When lava erupted from a volcano begins to cool, some iron-bearing minerals will crystallize out, and they acquire a magnetic property when the temperature is dropped below a certain point, called the Curie temperature of that mineral. For magnetite, a mineral of iron oxide, the Curie temperature is about 600 °C. When magnetic minerals acquire magnetism, the north-seeking pole of the mineral will point to the north and dip down with an inclination angle, exactly like the magnetic needle hanging on a thread. Furthermore, the acquired magnetic orientation, i.e., the natural remnant magnetism, is frozen in the rock when the lava is solidified into basalt. Therefore, the measured magnetic orientation of a rock _in situ_ should be exactly the same as the orientation given by a magnetic compass at the place where the rock is collected.

After many measurements were made, it became clear that the NRM of a rock at a place _A_ does not always show the polarity of the present magnetic field. Either the position of the earth's magnetic poles has changed, or place _A_ has moved away from where the rock acquired its remnant magnetism. The postulate that pole positions moved about is the basic tenet of the hypothesis of polar wandering. The interpretation that the geographic positions of various localities have shifted has been advocated by proponents of continental drift.

When the techniques of studying rock magnetism were primitive, only the relatively intense remnant magnetization of basalts could be measured. When Runcorn started his work under Lord Patrick Blackett, instruments had been constructed to measure the magnetism of sedimentary rocks. Sedimentary rocks, like solidified igneous rocks, could acquire a remnant magnetization, because detrital magnetic minerals in sediments, like little magnets, could also be oriented by the magnetic field at the time of sedimentation. Runcorn worked on red beds in the fifties; the rocks are red because of the presence of ferric oxide, which has a remnant magnetism.

Runcorn's results could be interpreted either in terms of polar wandering or in terms of continental drift. At the time he tended to believe in the former, but we were skeptical of both theories. We thought that the small discrepancy between the measured and predicted magnetism might be experimental errors. Nevertheless, the one kind of result that could not be easily dismissed was the apparent polarity-reversal: In this case, the little magnets in the rocks are oriented north-south, as expected in the present magnetic field. However, the north-seeking end is pointing to the south and is tilted upward in a rock collected from the Northern Hemisphere. The orientation is what one would expect if the North Magnetic Pole at the time of magnetization was located near the South Geographic Pole, while the South Magnetic Pole was then located near the North Geographic Pole. In other words, the NRM of those rocks indicated that the polarity of the magnetic poles should have been exactly opposite to that of the present—there had been a polarity reversal.

A rock magnetized during the time of apparently reversed polarity is said to have a negative polarity, or reversed magnetization. Runcorn found many reversely magnetized rocks among his collections.

Runcorn did not discover reverse magnetization. The first discovery was made long ago, in 1909, by a French geophysicist, Bernard Brunhes. Brunhes measured the natural remnant magnetism of volcanic rocks from the Massif Central in France and found reversed magnetization in basalts that were erupted in the not too distant past. The same observation of polarity reversal was made twenty years later by a Japanese scientist, Motonorie Matuyama, while studying Japanese volcanic rocks.

In 1954, when I joined Shell Research, we were interested in some oil-exploration problems that could be solved if we knew the *in situ* orientation of a rock sample from a drill core. This knowledge is difficult to obtain, because the drill string is constantly rotating when a core is taken. After the core is retrieved, one can no longer determine which part of the drill core originally pointed north. Consequently, we became interested in rock magnetism, because one could determine the *in situ* north-orientation of a drill core by measuring its natural remnant magnetism, assuming, of course, that there was neither polar wandering nor continental drift. I was a member of a team that made measurements on many thousands of specimens. We also found many instances of reversed magnetization. We could not understand this phenomenon, and we tried to seek more mundane explanations for this seemingly extraordinary fact. It was simply too difficult for us to envision that there were polarity reversals. Too arrogant to believe that the earth's magnetic field could actually have flipped, we dismissed the significance of our findings.

Runcorn was, meanwhile, giving talks at national and international meetings on his new findings. He, too, avoided the difficult subject of polarity reversal, concentrating on his interpretations of new data that proved continental drift: his measurements of natural remnant magnetism in rocks of various ages from Europe and America indicated that the two continents may have moved thousands of kilometers away from each other during the last 200 million years.

Have the Continents Drifted?

The idea of continental drift was inspired by the surprisingly good fit of the coastlines on opposite sides of the Atlantic (Figure 4.1). Several geologists had speculated on this during the nineteenth century before a German meteorologist, Alfred Wegener, wrote a book that unleashed a great controversy on the subject early this century.

The idea of continental drift was disarmingly simple and was thus particularly appealing to laymen. Wegener envisioned a light continental crust floating on a heavier, but weak, basaltic substratum; continents could plow their way across the oceanic

4.1. Continental drift. The idea that continents have moved apart was inspired by the nearly perfect fit of the coastlines of the Americas and Africa. Antonio Snider made this map in 1858 as a demonstration. The idea became an attractive scientific hypothesis after Alfred Wegener presented, in the early 1920s, considerable geological evidence that can be explained only by assuming lateral displacement of the continents.

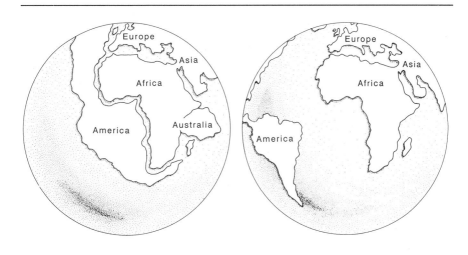

basalt, like icebergs drifting in seawater. According to Wegener, some time about 200 or 300 million years ago, all the present continents were linked together to constitute a single supercontinent, which he called *Pangea*. Eventually the Americas drifted westward, away from Eurasia and from Africa, giving rise to the Atlantic Ocean.

Wegener gathered a multitude of evidence, geological, paleontological, paleoclimatical, and geodetical. Some measurements turned out to be mistaken, such as the alleged geodetic proof that Greenland was moving westward at a speed of 35 meters per year. Other arguments could not be easily dismissed; it was especially difficult to explain the dispersal of ancient faunas and floras and the distribution of ancient glacial deposits without assuming drifting continents.

I first learned of the hypothesis back in 1944, as a college freshman in China. Chinese geologists liked the idea. Wegener's maps of drifting continents were made into large posters and hung on the wall in the laboratory where we did our weekly exercises. Nobody talked much about it, but we all thought that it was a foregone conclusion.

Thus I was surprised when, entering graduate school in the United States in 1948, I was told that the hypothesis had been proven wrong; it was not even talked about by serious students of geology. This was not what I had learned from reading the works of J. J. Joly, Emil Argand, Rudolf Staub, Arthur Holmes, and other European geologists. They all talked about drifting continents, and referred to continental collision when they explained the origin of mountains.

I went to talk to Ed Spieker, my professor at Ohio State University. Yes, Wegener's theory was dead, he told me. The continents are not floating on a heavier substratum like icebergs in seawater, because the substratum is not a fluid but a solid of finite strength (see chapter 1). Furthermore, the granitic crust of the continents could never have floated on liquid basalt, because granite has a lower melting temperature than basalt. It is physically impossible for a crust of granite to remain solid and to drift, if the temperature is high enough to render the basalt substratum sufficiently molten to permit the drift. Physical laws are immutable, and Wegener must have erred. Geophysicists told us that the theory of continental drift had been disproven.

Another problem was providing a force big enough to move drifting continents. Pushed to provide answers, Wegener speculated on a *Polfluchtkraft*, and he appealed to tidal drag. Having been educated as a meteorologist, Wegener was unduly impressed by the influence of the earth's rotation on the movement of air masses. He naively assumed that the same rotational forces might make the continents in high latitudes move toward the equator (*Polflucht*) and could induce them to drift westward (tidal drag). Wegener's reckless pronouncement was exactly what the noted geophysicist Harold Jeffreys needed to demolish the hypothesis. Jeffreys's computations showed that the postulated forces were several orders of magnitude too small to push continents about. Again a geophysicist told us that continents could not possibly be moved around.

Many geologists, however, liked Wegener's theory. Many a dispersal pattern of fossil organisms could be explained if there had been continental drift. Paleobotanists, for example, found that the fossil plants of the world belonged to two major groups during the Permo-Carboniferous period some 250 to 300 million years ago. A *Gigantopteris* flora left fossils in many parts of Eurasia, whereas the *Glossopteris* are found in different parts of the world—in India, in Africa, and in other southern continents. The distribution of the latter was particularly puzzling, because a land flora cannot easily, if at all, be transplanted from one continent to another across the ocean. Eduard Suess, the great Austrian geologist of the last century, thought that India, Africa, Australia, South America, and Antarctica were once all joined together as one giant continent, named Gondwanaland. Wegener's theory could now explain the present position of the broken-up supercontinent.

The distribution of land animals has also been taken as evidence for continental drift. The occurrences of *Lystrosaurus* in the 200-million-year-old Triassic formations of China, Africa, and Antarctica tell a revealing story. Those creatures lived like hippopotamuses in lakes and swamps of warm regions; they could not be expected to swim across oceans. But it should not have been difficult for them to move from China to Africa to Antarctica when they were all part of Pangea.

Geologists in the "fixistic" school, however, could not bring themselves to accept the postulate of continental drift; they preferred some less imaginative explanations. One favorite idea around the turn of the century was to postulate isthmuses or "land bridges" connecting continental masses. A "land bridge" between Asia and North America almost certainly existed during the last Ice Age, when the sea level was low enough that the shallow bottom of the Bering Strait was exposed, enabling North Asiatic tribes to walk into the New World; they became the forefathers of the American Indians. It taxes one's credulity, though, to imagine land bridges several thousands of kilometers long across the world's oceans. During the early part of the century, geophysical techniques were refined enough that one could differentiate continental from oceanic crust. Yet, all efforts to search for former land bridges as sunken continents ended in failure. The final blow was struck by marine geophysicists. Seismic surveys by Maurice Ewing and his associates showed that the Atlantic Ocean is underlain everywhere by ocean crust. By 1950, only a few aging biologists and some lay believers of "Atlantis" continued to talk about "land bridges" or sunken continents.

In the late forties and early fifties, George Gaylord Simpson, a Harvard professor of paleontology, came up with a different answer for the dispersal of land animals and plants. He talked about sweepstakes, or chance migrations. The immensity of geological time should permit some of the lucky ones to migrate across water and find a new home in another continent. There should have been enough driftwood to ferry land animals, and frequent intercontinental migrations of birds to ship seeds of plants over the ocean, not to mention the possibility of an occasional land bridge. Simpson was a persuasive speaker. I attended his talks in Columbus, Ohio, in 1950; I was convinced that it was not necessary to move continents when driftwood and birds could do the job.

A most critical piece of evidence presented by Wegener to support his theory is the distribution of sediments laid down by glaciers of the Permo-Carboniferous age. These glacial deposits have been found not only in Antarctica and the relatively higher latitudes of Australia, but also in the middle latitudes of Africa and South America, and in low northern latitudes of India (Figure 4.2). In places, rock-pavement polished by glacial motion has been found. The pavement is commonly decorated with parallel grooves and with striations, sculptured by angular debris dragged along at the bottom of moving ice. Those markings help indicate the direction of ice motion at the time. To cap surprise on surprise, the Permo-Carboniferous glaciers of India did not come from the north, where the Himalayas now stand, but from a flat region to the south, which is now tropical lowland. Wegener's theory provided a perfect explanation. India, Africa, South America, Australia, and Antarctica formed the southern continent Gondwanaland then, and this continent lay somewhere near the South Pole of that time. This continent was then covered by a giant ice cap. After the Gondwana continent was fragmented, all pieces except Antarctica drifted north. India has gone farthest, and met Eurasia in a collision that gave rise to the Himalayas. The glacial deposit of India was thus carried by continental drift from a position near the South Pole to a place north of the equator during the last few hundred million years!

Wegener's fascinating theory was too good to be true, I thought. Conditioned by my prejudice against this logical explanation, I found refuge in a convenient trap: the facts that I did not understand must not be important; I could pretend that those facts did not exist.

Convection Cells in the Earth's Mantle?

Time marched on, and new problems in geology kept popping up. One that troubled Harry Hess was the origin of the sunken flat-topped mountains of the Pacific (see chapter 1). Not content with the simple-minded explanation that the guyots sank under their own weight, he was to initiate the development of a new theory that would rescue Wegener from oblivion.

Hess was a young associate of Vening Meinesz, and he fol-

4.2. Permo-Carboniferous glaciation and continental drift. Glacial and glacier marine deposits, shown by stippled and wave patterns, respectively, have been found in South America, Africa, India, Antarctica, and Australia; the directions of glacier movement are indicated here by arrows. Such phenomena cannot be explained if the continents have stayed fixed in their present positions.

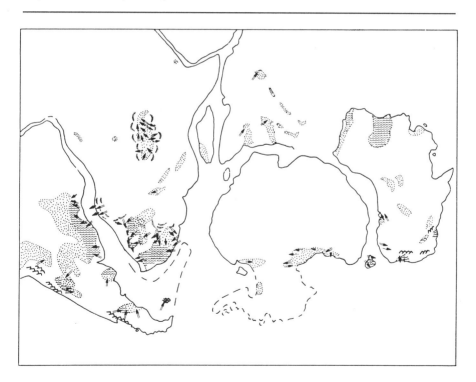

lowed the Dutch master during the 1930s to measure the earth's gravitational acceleration at sea in a submarine. Hess was also much influenced by his mentor's belief that heat transfer away from the earth's molten core must take the form of convection currents in the mantle.

Convection of fluids is a well-known phenomenon, caused by thermal stress induced by temperature difference. Anyone who has a fireplace in his living room knows that he can let in cold air through the windows to drive hot smoke up the chimney. Geologists trusting their own intuition could not imagine that solids in the earth's mantle could also move by convection, like a liquid. Experiments have shown, however, that a solid has little strength if it is subjected to a high enough temperature and pressure, or if it is deformed extremely slowly by stress of long duration. A solid could creep, or move like a fluid, under very small stress differences. The motion of glacial ice is a typical manifestation of the creep of solids.

Vening Meinesz reasoned that the earth's mantle is heated from below by the hot, molten core. The heat conduction by the solids in the mantle cannot transfer the heat away from the core-mantle boundary. Calculations showed that the mantle solids

4.3. Convection in the earth's mantle. Solid mantle material could creep under thermal stress and its movement form convection cells, as depicted in this sketch. Convection currents are now considered the basic driving force of tectonic activities on earth.

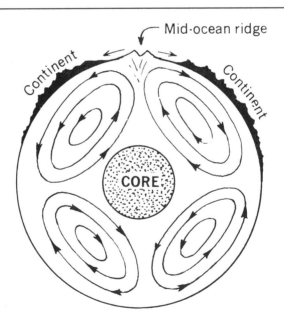

Mid-ocean ridge

Continent

Continent

CORE

are induced to move as convection currents long before sufficient heat is stored to cause mantle-melting (Figure 4.3).

One could test the idea by measuring the amount of heat coming out of the earth's interior; this heat flux transferred by conduction through the earth's crust to the surface is called surface heat flow, or simply heat flow. Where a convection current rises, it must give off much heat.

Shortly after World War II, Art Maxwell, a young naval officer retired from active service in the Pacific, set out from Texas to the West Coast in search of an opportunity to do postgraduate research. Maxwell had studied physics, and he wanted to go to Berkeley, the Mecca of nuclear physics. On his way, he stopped by San Diego to see his buddies in the Navy and looked into the Scripps Institution of Oceanography. He met Teddy Bullard of Cambridge, who had come to Scripps to work on a project to measure oceanic heat flow, i.e., the flux of heat through the seafloor. Prior to the war, Francis Birch of Harvard had made heat-flow measurements in Colorado, as had Bullard in Africa and in England; now Bullard had come to Scripps to look into the feasibility of making measurements on the ocean bottom.

The physical principle is simple. The heat flow is governed by the temperature gradient and thermal conductivity through the relation

Heat flow = temperature gradient × thermal conductivity.

Conductivity can be measured on samples in a laboratory, and the temperature gradient is determined by taking temperature measurements at different depths in a borehole.

The source of the heat flow should be the spontaneous disintegration of radioactive elements in granitic rocks, if there had been no convection currents in the earth's mantle. The continent has a thick crust and abundant granite, whereas the ocean crust is thin and is composed of basalt, which produces very little radioactive heat. It was thus expected that the heat flow reaching the ocean floor should be much smaller than that on land.

Bullard explained the project to Maxwell, who decided that an oceanographer's life on a vessel was much more to his liking than an indoor career smashing atoms. Bullard and Maxwell made a heat-flow probe, a steel rod or a piston corer with temperature-measuring devices called thermisters attached to it. The probe could be lowered to the ocean bottom and penetrate the seabed to determine the temperature gradient. Samples needed for conductivity measurements were obtained with the piston corer. In 1950 Maxwell went on the Mid-Pac expedition with Roger Revelle to make heat-flow measurements. They were surprised that the heat flow of the Pacific is about the same as that of a continent, or about 10 times more than what had been expected. The oceanic heat flow is too large to have come from radioactive minerals in the basalt crust; it must have come about by convective transfer from the hot liquid core of the earth.

Later on, two young scientists at Scripps, Richard von Herzen and Seida Uyeda from Japan, did many more measurements of heat flow in the Pacific and the Atlantic. They found that oceanic heat-flow values are even higher in certain more elevated belts of the ocean floor, such as the East Pacific Rise and the Mid-Atlantic Ridge. Also, the heat-flow values are lower than normal in regions of oceanic trenches. The pattern of heat-flow distribution indicated that convection currents move up under oceanic rises or mid-ocean ridges and sink below the trenches (Figure 4.3).

Vening Meinesz, Hess, and David Griggs all published articles during the 1930s and 1940s proposing convection currents as the driving force of mountain-building. Arthur Holmes from Edinburgh went a step further and suggested that the convection currents might be the motor for continental drift: convection currents in the mantle rising under a continent could tear the continental crust apart, first making a crack like the Red Sea or the Gulf of California, and eventually the crack could be widened into an ocean. Armed with such a vision, Holmes and his students at Edinburgh were far more receptive to the Wegenerian theory than their American contemporaries. I remember the many discussions I had with my friend Max Carman at UCLA, who had been a Fulbright Scholar at Edinburgh. Carman had seen the other side of the fence; he could not understand the almost hysterical persecution of Wegenerian heretics by the North American establishment.

Yes, the Americans kept the lid on tightly and laughed out of town anyone who dared to challenge conventional wisdom. When I was a student of David Griggs's in 1950, he used to tell us, after an evening of seminar discussions at his home, stories of his youth. As a brash young man from the Harvard Society of Fellows, he came to the 1939 meeting of the Geological Society of America with a working model and a movie film, to demonstrate in front of his elders the relation between mountain-making and thermal convection in the earth's mantle. Andy Lawson, a venerated geologist of another generation, was sitting in the front row, as was his wont; the septuagenarian could hardly wait until the talk was finished before squeaking in a loud voice: "I may be gullible. I may be gullible! But I am not gullible enough to swallow this poppycock."

Bailey Willis, another former president of the society, was sitting in the second row. He too got up, turned to face the audience, and announced solemnly: "All you here today bear witness. For the first time in twenty years, I find myself in complete agreement with Andy Lawson."

The Scripps discovery was a vindication. Mantle convection was now, for Griggs, the first cause, the driving mechanism of earth processes. From then on, he would preface his lectures with a remark that those who did not believe in thermal convection in the earth's mantle had no business sitting in the audience.

Rift Valleys on the Ocean Floor

With the blossoming of science after the Second World War, oceanographers on cruises did more than just geothermal measurements. Routinely, all research teams carried out echo-sounding surveys. Precision depth-recorders became an indispensable tool in every oceanographic vessel, and a tremendous amount of data accumulated. Bruce Heezen at Lamont, assisted by Marie Tharp, began a project to map the bathymetry of the seafloor.

Soundings carried out by the HMS *Challenger* during the last century had already indicated that a broad mid-ocean ridge seemed to exist, bisecting the Atlantic Ocean. The *Meteor* cruise during the 1920s revealed that the Mid-Atlantic Ridge had a rugged topography. A little later, a Danish expedition discovered a mid-ocean ridge in the Indian Ocean, called the Carlsberg Ridge after the donor who had supported the enterprise. (It is a joke in Denmark that natural scientists drink Carlsberg beer, while the social scientists drink Tuborg, to support their respective researches.) In the mid-thirties, the British vessel *John Murray* discovered through echo-sounding that the Carlsberg Ridge is split by a deep gully a few hundred meters deep in the middle. A similar rift valley is present on the Murray Ridge, which is the prolongation of the Carlsberg Ridge in the Arabian Sea.

Heezen and Tharp started their mapping of the seafloor in the early fifties. They soon found that the Mid-Atlantic Ridge is a continuous feature: extending from Spitzbergen on the north, the ridge joins the mid-ocean ridge of the Indian Ocean southwest of Africa. This mid-ocean ridge rises two or three thousand meters above the adjacent abyssal plains.

Like the Carlsberg Ridge, the Mid-Atlantic Ridge is also split by a median valley. Such a steep-walled valley on the mid-ocean ridges reminded Heezen and Tharp of the Rhine Graben and the East African Rift, which are tensional features, believed by geologists to have been pulled apart by tensional forces acting upon the earth's crust. Obviously working in unison with the force that lifts up the mid-ocean ridges is the force that tears them asunder! The action of the forces has also rendered the mid-ocean ridges an almost continuous belt of seismicity characterized by frequent earthquakes with shallow epicenters (Figure 4.4).

Heezen at first thought that the rift valleys in mid-ocean ridges were a manifestation of an expanding earth, which had been postulated by S. W. Carey of Australia to account for the fragmentation and drifting of continents. Hess thought, how-

4.4. Earthquakes and mid-ocean ridges. Heezen and Tharp surveyed the mid-ocean ridges of the Atlantic and Indian Oceans during the early 1950s and found a concentration of earthquake epicenters on the ridges (dark dots), as shown on this map compiled by J. P. Rothe.

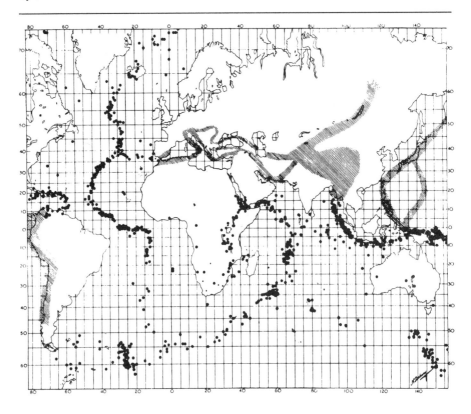

ever, that thermal convection in the earth's mantle was the explanation: mid-ocean ridges were where the convection currents rose. The heat brought along was sufficient to cause partial melting of the earth's upper mantle. The liquid fraction found its way to the surface, pouring out as underwater lavas and building seamounts and islands, while the crust of the mid-ocean ridge area was being split apart. Seamounts rose in time to become islands, with their tops chopped off, and they were swept by convection currents moving away from the mid-ocean ridge. They sank where the currents began to descend, and became guyots (Figure 4.5).

Hess's very speculative idea was presented as an invited paper in a Festschrift volume published in 1962 and dedicated to his retiring colleague A. F. Buddington. Hess modestly qualified his contribution as an essay in "geopoetry," quite fitting for a Festschrift article, which does not have to go through the normal refereeing procedures of a professional journal. Incidentally, the term "seafloor-spreading," which so nicely summed up Hess's concept, was coined in a 1961 paper by Robert Dietz of the Naval Electronics Laboratory of San Diego. As his friend Menard wrote, Dietz had come up with the same idea independently, although Dietz, who had consulted a preprint of Hess's article, gave the credit of priority to Hess.

A Dynamo in the Earth's Core

The idea by Hess and by Dietz might have gone the way of many other idle speculations, faded away and forgotten, if Fred Vine had not soon come up with an elegant synthesis relating seafloor-spreading to the newly gathered data on the magnetization of rocks under the seafloor.

Brunhes and Matuyama had produced data that suggested the

4.5. Seafloor-spreading. In 1962 Harry Hess postulated a theory of seafloor-spreading to explain the genesis of guyots. He showed in this sketch that submarine volcanoes were built up on a mid-ocean ridge (a), but they sank and became guyots (b, c) after their underlying ocean crust was moved by convection current away from the ridge.

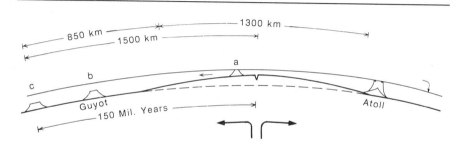

possibility of polarity reversal. The possibility is not all that farfetched if we understand the origin of the earth's magnetic field. The field is not produced by a giant solid magnet buried somewhere underground; the earth has a magnetic field because of the motions of molten iron within the core of the earth. Lord Patrick Blackett, a Nobel Laureate in physics, was the first to worry about the origin of the earth's magnetic field in connection with his studies of cosmic rays. He picked up an earlier idea, entertained by Albert Einstein, that the earth's magnetic field might be produced by the motion of a mass of material linked to the earth's rotation. Using gold he had borrowed from the Bank of England, Blackett constructed a cylinder weighing 15.2 kilograms. However, he was not able to detect a magnetic field in connection with the rotation of the gold cylinder. Later Bullard of Cambridge and Walter Elsasser of Princeton developed, independently of each other, the dynamo, or magnetohydrodynamic, theory of the earth's magnetic field.

The theory started out with a basic principle that one learns in middle-school physics: the motion of electrons, or electric current, through a wire can produce a magnetic field. The dynamo theory postulated the motion of electrically charged particles in the earth's core. As the field is produced by motion related to the rotation of the earth, Elsasser concluded that the magnetic poles should always be located near the spinning axis of the earth, and he believed, therefore, that the magnetic poles never did wander very far away from the location of the earth's geographic poles, which have remained at about the same places since the beginning.

The measurements of natural remnant magnetization of rocks in the fifties had suggested that either the magnetic poles had wandered or the continents had drifted. Since Elsasser's theory argues against polar wandering, then the continents must have drifted. With the establishment of the dynamo theory of the earth's magnetic field, Blackett and his student Runcorn became enthusiastic proponents of continental drift.

The dynamo or magnetohydrodynamic theory encouraged geophysicists to consider mathematical modeling, and different approaches were tried out. One experiment linked the reversals of the earth's magnetic field to convective movements of the turbulent and conductive fluid in the earth's core. The convection cells are likened to cyclones in the atmosphere. It could be shown mathematically that the configuration of such cyclone-like cells in the earth's core and the large fluctuations in their intensity determine when and if a polarity reversal of the earth's magnetic field is about to occur.

Although geophysicists could not quite agree on the exact cause, they had nevertheless done enough theorizing and modeling by 1960 to demonstrate that such reversal is not physically impossible. Agnostics like myself, incapable of understanding the magnetohydrodynamic theory, continued to reject the seemingly preposterous idea that the earth's magnetic poles could flip over. Experts on rock magnetism, however, began to ask questions as to the timing of polarity reversals: When did the poles flip over to their present positions? How often did the poles reverse themselves? How fast was the flip-over?

Dating of Polarity Reversals

In order to add a chronological perspective of a natural phenomenon, again we had to study rocks, and, in this case, the natural remnant magnetism of rocks. After the rocks are dated, the magnetic data give us a history of the polarity reversals of the earth.

There are two major methods for determining the age of a rock. Fossils, or paleontological dating, give a relative age; those are faunal dynasties and are expressed by such names as Jurassic period, Cretaceous period, and so on (see chapter 8). After the discovery of radioactive elements, it became possible to determine the absolute age of a rock that contains radioactive minerals, by measuring the relative proportion of the mother and daughter elements of radioactive decay.

Many volcanic rocks contain potassium-bearing minerals, and one of the potassium isotopes, ^{40}K, is radioactive and decays into argon ^{40}Ar and calcium ^{40}Ca. The determination of the ratio $^{40}K/^{40}Ar$ in a volcanic rock gives a measure of how much time has passed since the lava cooled, the ratio decreasing exponentially with time with the decay of ^{40}K.

Although potassium had been known to be radioactive since 1906, the dating technique was not developed until the 1950s. First there was the problem of the so-called branching ratio, namely, determining the relative proportions of calcium-40 and argon-40 yielded by the dual decay of potassium-40. This problem was resolved in 1955 when Gerald Wasserburg and Richard Hayden obtained a value close to the presently acceptable ratio of 0.117, i.e., 89.5% of the potassium-40 atoms decay to calcium-40 whereas only 10.5% decay to argon-40. The second difficulty was discriminating the radiogenic argon in the specimen from the argon-40 derived from air contamination, and this problem was particularly acute for younger rocks, which have yielded a relatively very small amount of the radiogenic product. The final technical breakthrough was achieved in 1954, when John Reynolds of the University of California at Berkeley built a new type of mass spectrometer with a very clean gas-extraction system.

West Coast geologists who wanted to know the age of Sierra Nevada granites were happy to get radiometric dates from Reynolds, Jack Evernden, and Garniss Curtiss. Geophysicists interested in dating magnetic reversals had to wait until the early sixties, when G. B. Dalrymple from the Berkeley lab joined Allan Cox and Richard Doell of the U.S. Geological Survey in Menlo Park. They finally published, in the June 1963 issue of *Nature*, the first polarity-reversal time scale. Everndern had meanwhile set up a Reynold-type mass spectrometer at the Australian National University, Canberra, so Ian McDougall and Don Tarling were able to present an improved time scale six months later. The data clearly indicated that the earth's magnetic field was reversed not only once, but repeatedly. By 1965, the magnetic time scale for the last 4 million years, as we now know it, was established (Figure 4.6).

4.6. Polarity-reversal chronology. A magnetostratigraphical time scale, reaching back to 4 million years before present, was established in the early 1960s by Cox, Doell, and Darymple, and modified and improved upon by McDougall and Tarling.

Age (million years)	Polarity epochs	Events	Pribilof Islands	Snake River	Sierra Nevada	Rome Basin	New Mexico	Hawaiian Islands
	Brunhes normal		M1 ×F1 •M2	M4	⊕+ G1 G2• M7	M10 •M11 •M12	M16 M17	M21 M22 M23 M24 M25 M30
1.0	Matuyama reversed	Oldu-vai	M3 ×F2	F3× F4× M5	⊙+G3 M8	°M13 °M14	M18	M26 M27
2.0								
3.0	Gauss normal	Mam-moth			M9	M15 •M19		M28 (•) M30
4.0	Gilbert reversed			×F5 M6 ×F6		F7×	M20	M29

When hundreds of thousands or millions of years go by without a reversal in the positions of the magnetic poles, the time interval is called a magnetic epoch. The present normal epoch is called Brunhes, and the last reversed epoch Matuyama, to honor the two scientists who first discovered reversed magnetizations in rocks. Short episodes (some tens of thousands of years) of quick changes in polarization were designated events in a magnetic epoch. Two such events occurring during the Matuyama epoch are the Jaramillo and Olduvai events, first discovered in the volcanic rocks in Jaramillo Creek, New Mexico, and in Olduvai Gorge, East Africa, respectively. The two magnetic epochs prior to Matuyama are the Gauss (normal) and Gilbert (reversed). Still earlier epochs discovered later have been given numerical designations, 5, 6, 7, and so on.

Stripes of Seafloor Magnetic Anomaly

Aside from measurements of polarity, the intensity of magnetization in rock samples can also be measured. A rock containing abundant magnetic minerals is intensely magnetized. This method has been used successfully by geophysicists to locate buried magnetic bodies, such as large iron ores. During the early fifties, when ocean exploration was being carried out on all fronts, Arthur D. Raff of Scripps developed a shipborne magnetometer to record magnetization of rocks under the seafloor. He then persuaded the authorities of the U.S. Coast Guard to let him use their vessel *Pioneer*, and in 1955 carried out, with R. G. Mason, a survey to map the intensity of seafloor magnetization with his new instrument.

Magnetic surveys are based upon the principle that the magnetic lines of force of a normally magnetized body in high northern latitudes point downward, parallel to the magnetic lines of force of the earth's magnetic field there. The two add up, and the sum, or the resultant field strength, at this spot is thus greater than normal, producing what is called a positive magnetic anomaly.

The *Pioneer*'s survey off the coast of the Pacific Northwest, in a region now called the Juan de Fuca Ridge, produced some interesting results. The magnetic anomalies are lined up in regular patterns: apparently very intense and very weak magnetizations of the seafloor are limited to alternate belts (Figure 4.7). Such a pattern has since been referred to as the magnetic lineations or magnetic stripes of the seafloor. That the anomalies should be linear is not necessarily puzzling, but the magnitude of the anomalies is surprisingly large, much greater than what one normally finds on land.

Some *ad hoc* explanations were put forward. Basalts have varying amounts of magnetic mineral; linearly arranged flat slabs of unusually magnetic rocks could theoretically produce the positive anomalies. Mason and Raff thought so in 1962. However, chemical variation of ocean basalts is limited, and even the most intensely magnetized basalt cannot be expected to produce the magnitude of the seafloor anomalies observed. An alternative was to postulate that magnetic stripes could be produced by parallel ridges of strongly magnetized rocks. However, there is no simple correlation of magnetic anomalies with seafloor topography.

While the origin of magnetic stripes under the ocean floor remained unknown, shipborne magnetometers became standard equipment on oceanographic vessels, and many thousands of kilometers of profiles were obtained. Magnetic lineations have been found in all the oceans.

When Mason and Raff's report was published in 1962, Fred Vine, then a graduate student at Cambridge University, was puzzling over the geophysical survey results brought back by Drum Matthews from the HMS _Owen_. Matthews had just made a detailed magnetic survey over a central part of the Carlsberg Ridge in the Indian Ocean. Vine used a digital computer to in-

4.7. Seafloor magnetic lineation. Mason and Raff discovered in 1955, off the coast of the Pacific Northwest, that apparently very intense and very weak magnetizations of the seafloor are lined up in alternate belts; those are now called seafloor magnetic lineations or stripes.

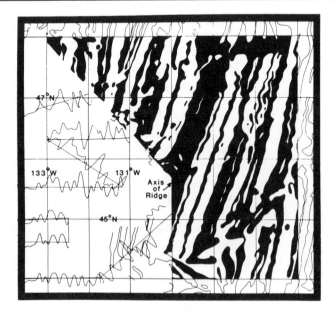

terpret Matthews's geophysical data—a new technique. He departed from tradition and asked for the first time whether a part of the seafloor could be reversely magnetized. Soon his results were showing clearly that stripes on the Indian Ocean floor must have been magnetized during epochs of negative polarity in the not too distant past.

Vine wrote a draft and hoped to win Ted Bullard as coauthor, but the professor politely declined. Vine then went to Drum Matthews, his collaborator and thesis advisor, who gave it more substance; their joint paper was sent to *Nature* in July 1963.

Many years later, Vine was to tell me his rationale. The key to this solution was almost as simple as "the egg of Columbus." To get his answer, he had only to ask the right question: What would be the field strength of the seamount area if the rocks are reversely magnetized?

In the Southern Hemisphere, the magnetic line of force of a normally magnetized seamount should strengthen the earth's magnetic field there, which is pointing upward, or negative; the result should thus be a strongly negative field (Figure 4.8). Instead Vine found a small positive anomaly, with the line of force pointing downward! Now Vine asked the right question, and the answer is as follows: If the volcanic rocks piling up on that seamount were polarized during a previous epoch of reversed magnetization, there should be a small positive anomaly on the seamount! This answer is the answer he needed to solve the mystery of the stripes of seafloor magnetic lineation—they are stripes of basalt rocks with normal and reversed magnetization.

Fred Vine deservedly received the Arthur Day Medal of the Geological Society of America for this flash of genius. The idea was the key that would resolve many old puzzles on land and new puzzles under the sea. The article by Vine and Matthews on magnetic anomalies over ocean ridges was the spark needed to light the fire of the great revolution that swept across the earth science community during the late sixties.

The Seafloor Has Spread!

With the key to decode magnetic lineations, everything began to make sense. Accepting the postulate that the median valley of a mid-ocean ridge is a crack, produced as the seafloor is rifted apart by tensional forces associated with convection currents in the earth's mantle, lavas welling into the crack should be imprinted by the magnetic polarization at the time of their extrusion. If the old median valley was again ripped open in the middle, and if a succeeding generation of lavas was poured into the new crack to form a new seafloor after the earth's magnetic field had reversed itself, the central stripe of magnetic lineation should be flanked by two stripes of opposite polarization that had been imprinted on the old seafloor of the split-apart old median valley. This process must have repeated itself several times to produce the alternate belts of positive and negative magnetic anomalies, like those mapped by Mason and Raff during the 1950s.

Vine and Matthews pointed out that the process should produce magnetic anomalies with a symmetry about the axis of the median valley if the seafloor of the old median valley has always been split open exactly in the middle (Figure 4.9).

By 1965, Cox and his associates had determined the duration of the four most recent magnetic epochs. They found that the Brunhes epoch lasted about 0.7 million years, and that the durations of the four epochs are in the ratio 0.7:1.7:0.9:1.7. The

4.8. Magnetic anomalies. Strong magnetic intensity indicates that the line of force of a local rock body reinforces the regional magnetization of the earth's magnetic field. Weak intensity, or negative magnetization, indicates that the regional field strength is compensated by the line of force of a local magnetic body. This figure shows the magnetic anomalies resulting from such combinations of local and regional fields in the area of the Carlsberg Ridge (Southern Hemisphere).

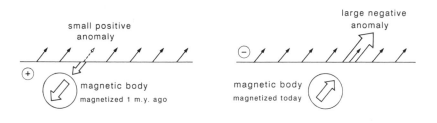

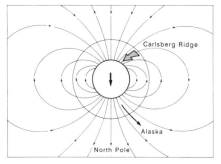
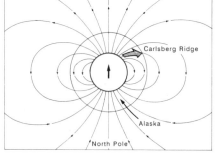

width of each belt of the magnetic anomalies, according to the new theory, should be proportional to the duration of the magnetic epoch during which the lavas bearing the magnetic imprint of that epoch were erupted, assuming, of course, that the seafloor has been spread apart at a constant rate.

Vine was soon to team up with J. Tuzo Wilson, who was visiting Cambridge in 1963, to verify the corollary of the seafloor-spreading theory. They gathered data from the Juan de Fuca Ridge, from the East Pacific Rise, and from the Reykjanes Ridge in the North Atlantic, and they found the linear correlation between the width of anomaly stripes and the duration of magnetic epochs predicted by the theory (Figure 4.10).

To take the matter one step further, Vine and Wilson used the width of the magnetic stripes to estimate the rate at which the seafloor has been spread apart. In the Reykjanes Ridge area, one half of the width of the magnetic stripe magnetized during the

4.9. Magnetic lineation produced by seafloor-spreading. In 1963 Vine and Matthews modeled the magnetization of normally and reversely magnetized stripes of seafloor as they moved away from the axis of the mid-ocean ridge where they were first emplaced, and produced this figure. The close approximation of the computed anomalies to the observed is a verification of the theory of seafloor-spreading.

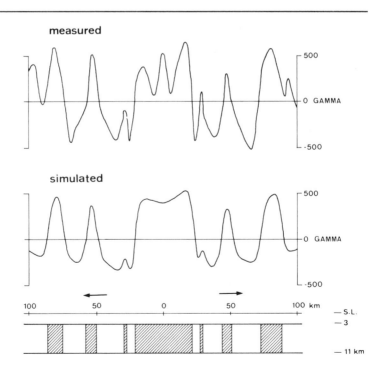

Brunhes epoch is 7 km, or 700,000 cm. This gives a rate of 1 cm per year to produce this strip of positively magnetized seafloor since the beginning of the Brunhes epoch 700,000 years ago. Each half of the magnetic stripe produced during the Matuyama epoch is 17 km wide, or 1,700,000 cm, again giving a half-spreading rate of 1 cm per year. The magnetic stripes of the East Pacific Rise are much broader, each approximately 4.4 times as wide as those of the Reykjanes Ridge area, because the East Pacific seafloor has been moving away from the central axis at a much faster rate of 4.4 cm per year.

The product of a brilliant graduate student's brainstorming was not universally accepted with enthusiasm. Hess, Tuzo Wilson, Bullard, and a few others were enchanted by the Vine and Matthews hypothesis. Doc Ewing and his associates at Lamont

4.10. Duration of seafloor-spreading and width of magnetic stripes. Assuming a linear rate of seafloor-spreading, Vine and Wilson calculated in 1966 the width of magnetic stripes on the basis of the polarity-reversal scale of Cox *et al*. Again, the computed width of the stripes closely approximates the observed width.

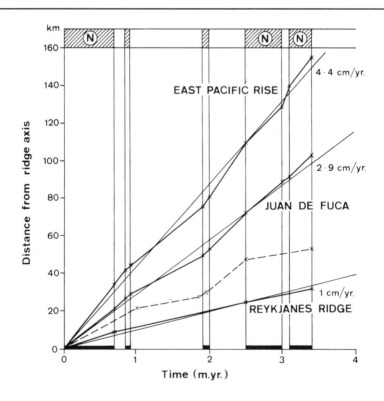

were, however, slow in accepting the somewhat preposterous idea. They continued to grind out many magnetic surveys, and the excellent data of the Reykjanes Ridge area used by Vine and Wilson to confirm the seafloor-spreading theory were in fact gathered by Lamont scientists.

A mass conversion to the new faith was triggered by chance. Two Lamont graduate students, Walter Pitman and Ellen Herron, went on the 1965 *Eltanin* cruises to the South Pacific. The magnetic surveys by the *Eltanin* 19, 20, and 21 were processed and plotted on profiles, and the similarity between them and the Juan de Fuca Ridge profile was apparent. The bilateral symmetry of the *Eltanin* 19 profile was the bomb that blasted the Lamont scientists out of their complacency. The story, as Bill Ryan told me later, went as follows. Two copies of the same profile were piled on a table by a technician while Walter Pitman was away. When Pitman came back and worked on the profiles, it was some time before he realized that the two copies had been placed face-to-face on top of each other. Since the bilateral symmetry is perfect, the profile looks almost exactly the same regardless of whether it is placed right side up or upside down. Pitman called in his colleagues, and the symmetry predicted by Vine and Matthews was there for all to see.

The *Eltanin* 19 profile persuaded not only the Lamont scientists; when it was shown during the 1966 meeting of the American Geophysical Union, the chairman of the session, Allan Cox, faced with the profile, exclaimed: "Good grief! Vine is right after all!"

The scientists at Lamont were quick to apply the Vine and Matthews theory to interpret their vast stores of data. In 1968, Jim Heirtzler, G. O. Dickson, E. M. Herron, Walter Pitman, and Xavier LePichon published an important paper on marine magnetic anomalies, geomagnetic field reversals, and motions of the ocean floor and the continents. They showed that all the magnetic profiles from the Pacific, Atlantic, and Indian Oceans have a bilateral symmetry about the axis of the mid-ocean ridges. On the basis of the width of the youngest magnetic stripes, the rate of seafloor-spreading in various regions can be calculated, as Vine did for the Reykjanes Ridge in 1966. The stripes of seafloor magnetic anomalies were given the numbers 1–34, and those anomalies could be dated on the basis of their distance from the axis of the mid-ocean ridges by assuming that the spreading rate has been constant during the last 30 million years. Furthermore, the seafloor data permitted Heirtzler and his co-workers to construct a time scale indicating the temporal sequence of polarity reversals of the earth's magnetic field. With the aid of this geomagnetic time scale, now known as the Heirtzler scale, the motions of the ocean floor and the continents could be determined. The computations easily proved the Wegenerian postulate that Africa has been moving away from South America for more than 100 million years, while the Atlantic Ocean was formed by seafloor-spreading.

Age of the Seafloor

With the establishment of the theory of seafloor-spreading, an earth science revolution was under way. The paper by Vine and Matthews has been likened by many to the manifesto of the revolution. The scientific revolution was necessary because many different methods of investigating the seafloor had turned out new data that could not be reconciled with the classic doctrines of geology—our new knowledge of bathymetry, our new knowledge of the history of the earth's magnetic field, our new knowledge of the geothermal state of the earth. However, most of the new methods were geophysical. Except for the enlightened few, there was a tendency for geologists working on land to take only a mild interest, if any, in articles published in the *Journal of Geophysical Research* by young people whom they did not know, summarizing results that they did not quite understand or appreciate. The revolution might not have succeeded at all, or there might have been a long period of schism in the earth sciences, if *Glomar Challenger* had not been built to provide a geological test of the seafloor-spreading theory.

The time scale of Heirtzler and others was the challenge. Geophysicists were supposed to tell us about the physical state of the earth. We geologists were the people who had a historical sense, who had a perspective in time. Now geophysicists were telling us the age of the seafloor (Figure 4.11). We could no longer ignore them. Either we had to use our time-honored geological methods to prove them wrong, or we had to swallow a bitter pill and serve the new revolutionary regime.

To date the seafloor geologically, we needed samples of the oldest sediments on the seafloor. To do this we had to drill holes into the seafloor, and *Glomar Challenger* was our tool for that. The scientific community was very much aware of its mission then. We might say that the emphasis of JOIDES had quickly changed from Long Core, or from deep-earth *sampling* to deep-earth *studies*, even before *Glomar Challenger* left for her inaugural voyage.

After Doc had the first crack for his pet problem in the Gulf of Mexico, the second cruise of the Deep Sea Drilling Project, headed by the project's chief scientist Mel Peterson, was scheduled to test the revolutionary theory. Unfortunately, the newly built vessel had mechanical trouble and had to spend much of the time assigned to Leg 2 in a shipyard. Peterson and his scientific crew waited impatiently in a New Jersey motel for several weeks before they were able to bring the vessel across the Atlantic to Dakar.

Five holes were drilled during Leg 2 in the North Atlantic, and three of those reached basement of predicted age. Their results were encouraging, but, as Peterson wrote, "mechanical failures aboard the vessel reduced the time available for drilling to such an extent that not all the program proposed for the testing of seafloor spreading in the North Atlantic could be completed." By chance, the third cruise of the Deep Sea Drilling Project was given the chance to perform the most significant experiment in the history of the earth sciences. By chance, I was to join that cruise and become a witness to the experiment.

4.11. Age of the seafloor. Heirtzler *et al.*, assuming a linear rate of seafloor-spreading, extended the magnetostratigraphic time scale to the late Cretaceous. The ages of the seafloor were dated by magnetic anomalies as shown on this map. The predicted ages (in millions of years) shown on this map have been largely verified by deep-sea drilling.

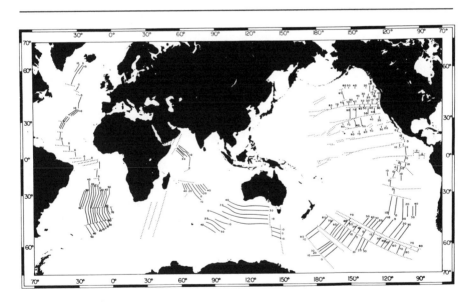

Two The Breakthrough

1968–1973

A Game of Numbers

Isaac Newton was not Chinese, because Confucian doctrines

cannot be disproven. Respect for authority and unwavering loyalty

to teacher and friends are strong enough emotions to resist any

revolution in science. Through a twist of circumstances I became

witness to a game of numbers, where the winner never once

missed. It was no longer a matter of chance, but a manifestation of

the predictive power of science: the deep-sea drilling cruises were

to verify again and again the predictions of the age of the ocean

floor by the theory of seafloor-spreading!

Schism in the Earth Sciences

The number of equations in problems of natural sciences is often less than the number of variables, and a unique solution is thus not always possible. Subjective reasoning can play a dominant role. Philosophers use the German expression *Leitbild*, or guiding principle. Naturalists of the eighteenth century commonly adopted the *Leitbild* that biblical statements are literally true; the sedimentary rocks are the record of the biblical Deluge. When a fossilized giant salamander was found in 1726, Jakob Scheuchzer, a Swiss naturalist, communicated excitedly to the Royal Society of London that he had found the bones of "one of the infamous men who brought about the calamity of the Flood."

James Hutton, a physician from Edinburgh, changed the *Leitbild*. He called attention to the different processes observable on the surface of the earth. There is volcanism; basalt lavas are quietly pouring out of volcanoes whereas ashes were thrown into the air by violent eruptions. There is weathering and erosion. There are rivers and ocean waves. Detritus was laid down on flood plains, on coastal shelves, and on the deep-sea floor. There are also earthquakes, manifestations of the working of the dynamic forces in the earth's interior. Mountains were made. Such processes, in Hutton's opinion, have been going on since the beginning of the earth. Physical laws, or laws governing the operations of the earth processes, have remained uniform. Hutton's *uniformitarianism* is the *Leitbild* of geology, as valid today as it was in 1788, when the theory was first published.

Hutton's *Leitbild* was modified half a century later by Charles Lyell, and his uniformitarianism is now called *substantive uniformitarianism*. The word substantive expresses existence, and substantive uniformitarianism postulates a uniformity of existence, in addition to the uniformity of physical laws. Lyell thought, and tried to make everyone believe, that the earth has always been what it is now. There are continents and oceans on the earth, and they have always been fixed in their present positions. A substantive uniformitarian cannot, therefore, accept continental drift. North American geologists, especially those educated during the first half of the twentieth century, have adhered to substantive uniformitarianism. I was a product of such an education, and I was a dogmatic substantive uniformitarian.

Geology as developed by Hutton and Lyell was a science based upon observations. Deductions simply followed. One saw molten lava cool down to form basalt; the deduction was simple: there was a molten lava where there is basalt. Simple deductions from singular observations could be correct. Going a step further, we saw deformed rocks on the side of mountains: originally horizontal sedimentary formations have been bent or broken into pieces. We made experiments and demonstrated to our own satisfaction that such deformation, such as bending or *folding*, as it is called by geologists, could result when a sedimentary formation is compressed; the folding of sedimentary formations is comparable to the folding of the tablecloth when one pushes on a covered table. When we observed such deformations, we deduced that they were induced on land by stresses within the earth's crust under the continent. What we saw on land could be explained by theories based upon observations on land. We thought that we did not need to infer interferences from the oceans, or to speculate on forces originating in the earth's interior under the oceans. That was the reason why great geologists like Andy Lawson and Bailey Willis demolished young Griggs when he tried to relate convection currents under the oceans to the making of mountains on land.

New tools of observation gave earth scientists a new territory—the oceans. Ocean sediments are thinner than expected, and they are young. Heat flux through the ocean floor was anomalously high. Stripes of magnetic anomalies were found, and they seemed to have been displaced by fracture zones. Large gravity anomalies could not be explained. Ocean trenches defied understanding. Then there was the concentration of seismicity in linear belts on the surface of the globe, and within inclined zones, now called the Benioff zones, dipping down and away from ocean trenches to hundreds of kilometers under continents. Those observations set up new equations to solve the old problems. The first reactions of geologists were predictable; many of us simply ignored the new findings and fought stubbornly against new ideas. The earth sciences were heading toward schism: land geology versus marine geosciences.

There is perhaps some inertia in all of us. If the new evidence is based on a methodology we are not familiar with, or principles we do not understand, we tend to be reluctant to accept the new triumphs. Sometimes we unwisely commit ourselves emotionally, when polemics replace logic and when one could never face the truth. In my personal case, there is one additional handicap: I was born in China.

Isaac Newton Was Not Chinese

Many people have wondered why Isaac Newton was not Chinese. Isn't the Chinese civilization one of the oldest and most glorious? Hasn't Needham compiled fifteen big volumes enumerating the great inventions and discoveries by the Chinese in science and technology? Haven't the Chinese proven much more effective than the Europeans in finding out about Nature and using natural knowledge to benefit mankind for fourteen centuries before the Copernican revolution? Nevertheless, this revolution occurred in "backward" Europe. China did not produce a Galileo, did not produce a Newton.

Why the Chinese failed to launch the scientific revolution is a favorite theme for dissertations in the history of science. Looking back at my own life, and experiencing the attitude of Chinese scientists during my lecture tours in China in the last decade, I began to perceive that there was no scientific revolution in China because the Chinese embraced Confucianism for two thousand years.

At the time of Confucius, the intellectual and social development in China paralleled that of ancient Greece. There were city states, and there were "a hundred flowers blossoming." Confucius was a teacher, and a politician in the State of Lu (in the present Shangtung Province). The eventual "monolithic" developments of Chinese intellectual tradition are traced back to the rise of the Han Dynasty, whose first emperors, in the two centuries before the birth of Christ, discovered Confucius's usefulness. Their predecessor, the First Emperor (of the Qin Dynasty), had unified China, but ruled only for two decades. His minister was a pragmatic, a so-called legalist. They burned books, especially those of Confucius; they ruled with an iron fist and suppressed rebellion with force. Nevertheless, the dynasty was overthrown a few years after the death of the tyrant.

The Han emperors adopted Confucius and Confucianism because of its ideological value to the stability of their rule. Confucius emphasized obedience and loyalty—the loyalty of son to father, pupil to teacher, soldiers to general, ministers to emperor. Deviation from this basic principle is unethical, immoral, even a crime punishable by death. The writings of Confucius became the bible. From then on, the Chinese turned dogmatic: Confucius was supposed to have found the Truth.

Chinese teachers are called _Shian-shen_, or earlier born. Those who are born after Confucius could have no more to contribute: they could only write footnotes, or second-guess what the Grand Master might have meant by his often obfuscating statements. Yes, one can invent paper, gunpowder, the compass, rockets, printing, the seismometer, rotary drilling, and so on. But those are technological innovations, not accomplished by _Shian-shen_, but by craftsmen for useful purposes, or done by people who were not concerned about philosophical truth. The philosophers could not be inventive; they could not rebel against Confucius. That the scientific revolution did not occur in China is not an idiosyncratic development; it was an inevitability. This tradition lingers on even today.

There were also tyrants in the West who demanded obedience and loyalty from their subjects. Rome had Caesar, but he was murdered. The Holy Roman Empire was founded by a Pepin, but the empire of fighting kings, dukes, and barons all but disintegrated after the demise of its founder. Even popes, with their centralistic Catholic ideology and power of excommunication, failed to extinguish the rebellious spirit that finally surfaced in the Renaissance and the Reformation. The great Copernican Revolution was a manifestation of the democratic faith that we are all fallible and prone to error. Galileo may have been silenced, but he never acquired the sincere respect for authority of his Chinese counterparts. The Bull on Papal Infallibility is no more than a bull, and was promulgated far too late to stem the advance of science.

What has led to the Chinese respect for authority? Is it an inevitability or an idiosyncrasy?

The secret of the traditional Chinese success with authority, I think, may be traced to the singular development of the Chinese language. For some unknown reason, Chinese invented and kept their ideograms as words, whereas the people of the West invented alphabets and phonetic writing. This may have been the most fundamental of all the reasons why China's history is so different from that of the West.

Linguists tell us that all the oldest writings started out as pictograms or ideograms, and each of those was assigned a syllabic phoneme. We need far more words than the number made possible by assigning a written word for each syllable. The ancient Chinese solved the problem by inventing various other ways of making words, and a form of phoneticization is only one of six: a phoneticized word consists of a phonetic part and a categorical part. *Tong* is the Chinese word and the phoneme for "togetherness." *Tong* accompanied by a metallic "part" means "copper." *Tong* accompanied by a woody "part" is a tree. In this mode of combining ideograms into words, the metallic or the woody "part" has no phonetic significance. With this measure, the Chinese marched halfway toward the dephoneticization of their writing.

Another common way of making Chinese words is to find meaning in two combined ideograms that become two parts of one new ideogram. For example, the Chinese word "promise" (*xin*) is made up of two parts, "man" (*ren*) and "spoken" (*ian*), because a person's spoken words should constitute a promise. In such a combination, the phoneticizations of the two parts *ren* and *ian* do not contribute in any way to the phoneticization of the written word *xin*, which had been spoken long before it was written. Syllabic symbols thus became divorced from their phoneticization, and the dephoneticization was total.

It occurs to one of my more observant friends that we Chinese tend to think of language as something written; our language is the writing. Language to a Westerner is spoken, or the speech. When we Chinese refer to languages, we use terms such as *Zhong-wen* (Chinese writing) or *Eng-wen* (English writing). I was horrified, for example, when my wife mentioned that she wanted to learn Chinese, but not Chinese writing. How could she learn Chinese if she did not learn Chinese writing? "Chinese" and "Chinese writing" are synonyms to me.

Until recently there was not a Chinese speech. There were thousands of dialects, of course, the Beijing dialect, the Nanjing dialect, the Canton dialect, and so on. But dialects are dialects, and they are not *Chinese*. The Swiss understand this paradox; they will tell you that there is no *Schwyzerdüütsch* (Swiss German), only *Baslerdüütsch, Bäärndüütsch, Züridüütsch*, and so on. But the Chinese writing is Chinese; Chinese cannot write down their different dialects like the Swiss do, because the Chinese writing is not phonetic. The one language, and the only language for the one people, is written but not spoken by all.

That the Chinese could have one Chinese "language" is probably one of the most important elements in the history of China. When foreign invaders came to China, they had no writing. In order to govern, they had to acquire a writing, namely, the Chinese writing. Eventually they forgot their own language, which was only spoken, and they talked in one of the Chinese dialects, which can be harmonized with the written language. They became absorbed, or assimilated, by the people they conquered. This happened to the tribal warriors who established the Wei, Liao, and Chin dynasties in ancient China. The Mongols were heading toward the same direction; they wrote a history with Chinese writing to phoneticize Mongolian speech, like the earlier development of the Japanese language. The Mongols escaped the fate of oblivion, because they did not stay in China long enough to learn Chinese, and because they eventually adopted the Korean practice of phoneticizing their language. The Manchus stayed too long in China. They did invent a phonetic writing, but few Manchus bothered to read or write in that imperfect language, when they had to learn Chinese to govern. After 300 years the Manchus lost their own speech as well, and the Manchurian language is spoken today only by a few specialists. The Manchus are no longer distinguishable from the Chinese; some would not even know they were Manchurian if they did not know their family history.

When the Germanic hordes descended onto the Roman Empire, some, like their Asian counterparts, adopted the spoken and written language of the conquered; they are the French, the Walloons in Belgium, the Lombards in Italy, the Burgunders ("Welsh") in Switzerland, and so on. Others were able to keep their own languages, because it was no big deal to invent phonetic writings of their own. Every young Swiss has no problem writing his Baslerdüütsch or Züridüütsch. With the phoneticization of writing, a person does not even have to share his language with his relatives in the next canton; he certainly feels no compulsion to bow to the authority of a foreigner who uses a strange language.

A Swiss relative of mine has a vacation house in the Jura. I used to be amused by the Jura Liberation movement. Its proponents took themselves seriously, even to the extent of committing terrorist acts such as setting off bombs in railroad trains. It seems crazy to a Chinese; why should the people of the Jura want to be independent of the canton of Bern? My European friends did not find it curious; they are used to that. There is the conflict between the Flemish and the Walloons. Then there are the nationalistic Scots, Welsh, and Irish who are not content to be British. The French, Spanish, Portuguese, and Italians are proud of their national heritages, even though their ancestors were the subjects of one empire and they used to speak and write in a common language.

History provides an undeniable record that phonetic languages are divisive. People drifted away from one another and lost the sense of common heritage when their spoken languages became formalized in different written phonetic scripts. It is perhaps no coincidence that the separation of the national states in Europe was cemented after the abolishment of Latin as the common written language.

The anti-authoritarian spirit of the Europeans seems to have come to them naturally. Thomas Aquinas may have been the greatest scholar of medieval Europe, but he was not an Englishman, and Francis Bacon did not suffer from a sense of betrayal when he proposed a different way of searching for truth. Nor did Copernicus worry about being sacrilegious when he proposed a theory of planetary motion completely different from that of Ptolemy, a Greek.

Through an idiosyncrasy in history, China evolved down a different path from that of Europe and adopted a different logic in constructing its writing. This quirk of fate has in turn led to a centralizing tendency and to the inevitable respect for authority.

I grew up in China. I was not taught Taoism, Buddhism, and certainly not Christianity. My *Leitbild* was the Confucian philosophy. Loyalty, constancy, gratitude—those are the virtues of a Confucian. I had a special reason to believe in those Confucian virtues, because of a personal attachment to a professor who was kind to me in my hour of need.

Cordell Durrell, my teacher at UCLA, had been a student at Cal (Berkeley), where Andy Lawson was his teacher. He learned not only geology and scientific philosophy from that great man, but also mannerisms; he learned to speak like Lawson. Wegenerian theory of continental drift was to him, like Griggs's idea of mantle convection was to Lawson, "poppycock." I heard him use the expression more than once.

Durrell was a great teacher. He taught us how to use a petrographic microscope. In 1950, when I first came to Los Angeles, I took a course on advanced petrography from him. We usually spent an afternoon per week in the laboratory peering through a microscope, learning how to figure out the genesis of a rock through identification of its mineral composition. My training in China had been deficient. Shortly after the exercise started, I asked for his help. He did something with the microscope and told me that the mineral was staurolite. After 10 minutes, I had problems again. It was kyanite, he said. Then I had problems again. This time, he lost patience and asked me if I knew how to use a microscope.

"I did," I murmured, "but I forgot."

"No, you didn't," he snapped back. "You did not forget. You never learned."

That was, of course, true. It was a painful process, but I did learn the logic of classification. Minerals are classified on the basis of mutually exclusive criteria. One asks a series of key questions to eliminate the alternatives and come up with a correct identification. Science is not a guessing game.

Durrell was a great petrographer, but he was only an average geologist. He was too much conditioned by respect for his teachers, and the theory of continental drift was dismissed as poppycock or charlatanism, simply because he never did find reason to be "disloyal" to his teacher. Durrell was a great human being, and he helped me in my hour of greatest need in 1953. I adopted him as my mentor when I was convinced that there is much littleness in striving to be great. My love, respect, and loyalty to him caused me to accept his judgments without subjecting them to impartial analysis. So continental drift was poppycock, convection current was poppycock, seafloor spreading was poppycock.

After I left UCLA I accepted employment in industry, and was a geologist working exclusively on land. Laboring under another misconception of Confucianism, I thought I should specialize; I almost took pride in the fact that my expertise had not been diluted through dilettante endeavors in marine geology. I read, of course, but the things other people did at sea, and the data they turned out, did not make an impression on me. My *Leitbild*, my outlook in geology, was derived from the sum total of my knowledge and experiences in land geology. I seldom looked up articles in *Nature*; I had not come across the article on seafloor-spreading by Vine and Matthews.

I worked on the geology of the California Coastal Ranges during the sixties. I thought I had made a very important observation when I discovered that the rocks that had been called Franciscan basement were not basement at all, but a mélange of large and small blocks broken and mixed in a giant shear (see chapter 8). In the autumn of 1966 the Geological Society of America announced a new policy of selecting a few persons to speak during the next annual meeting on their work in progress that might qualify as an "outstanding new contribution in geology." I was immodest enough to submit a summary of my most recent discoveries for consideration by the selection committee. It was painful to learn later that I was not a successful candidate. Fred Vine was a winner of the competition.

I went to the San Francisco meeting in November, and did look up Vine and Wilson's display of their work on the magnetics of the Juan de Fuca Ridge (see chapter 4). Not fully appreciating the multitude of interrelations involved, I dismissed their findings as mere of coincidence. I did not even bother to listen to Vine, and I formed a mental picture of him as a brash young man seeking sensation.

Was I too proud? Or did I wish to avoid the painful experience of admitting that I was in fact among the second best in a competition? Loyalty is a Confucian virtue, and vanity is a Confucian vice. I was probably afraid to listen to Vine, because it might have been too painful for me to change my opinion and to admit that my prejudice was wrong, or even dead wrong.

Several years later I got to know Fred Vine, and we became friends. I realized belatedly that I could not have been more wrong about Fred Vine and his theory. Vine, contrary to my image of him as a braggart or charlatan, is quiet and modest, and has a most genuine personality. Intelligent, thoughtful, and mature in judgment, he was not at all like some of the other "Teddy's boys"—the students of Sir Edward Bullard's at Cambridge whose brilliance included intellectual intolerance and one-upmanship. If I had attended his talk, I might have been spared two years of futility in my professional life.

A Quirk of Circumstance

In the early spring of 1967, I left California to join the faculty of the Swiss Federal Institute of Technology. One of my old professors gave a farewell party for me. Jerry Winterer, an old UCLA classmate who had gone into marine geology at Scripps, was holding court, praising the achievements of Vine and Matthews. I edged into the group, interrupted him, and vented my polemics. Winterer remained graceful and smiled that I would eat my words some day; he was going to see to it. "Not a chance," I screamed back. The matter might have ended there, had it not been for the Soviet invasion of Czechoslovakia.

My wife, Christine, and I were in Prague attending the International Geological Congress when the Soviets marched in on 23 August 1968. To see the brutal suppression of the "Prague Spring" was a deeply emotional experience, and we all tried to do our part to help those Czech scientists who fled. Mr. and Mrs. S., two geologists from Bratislava, who found temporary shelter in Switzerland, were looking for employment. I wrote a number of inquiries to my friends in the United States. An offer came from the Deep Sea Drilling Project for immediate engagement in the upcoming cruises. However, the offer came after our Czech friends had already accepted permanent positions in Australia. Writing to thank my friend Jerry Winterer, who had expended considerable effort to secure the placements, I apologized and indicated that I was willing to volunteer my services if the unexpected turn of events should have left DSDP in a bind.

I was just being polite, and did not expect that Winterer would call my bluff. In early November 1968, I received a telegram from Scripps inviting me to join *Glomar Challenger* at Dakar for the Leg 3 cruise across the South Atlantic. I was invited because one of the shipboard scientists had had to cancel at the last minute because of health reasons. It was this chain of coincidental events which turned my attention from the land to the sea, and which eventually changed my whole outlook in geology. Or perhaps it was not coincidental; perhaps my friend Winterer did not forget our debate at that farewell party, and wanted to teach me humility.

I went to Dakar on the first of December 1968, and easily located *Glomar Challenger* at the pier. The 45-meter derrick tower of the vessel had become a landmark overnight to local taxi drivers after her arrival. Most of my shipmates had already arrived. On the way to my cabin, I saw on the ship's deck a tall and slender man in a red baseball cap unpacking a wooden case. He looked like a lab technician. At the meeting of the scientific staff the next day in the core lab, I asked the Scripps representative, Gene Boyce, if he was the chief.

"No, I am the Scripps representative," Boyce said.

"My name is Maxwell," the man in the red baseball cap interrupted. "I am the chief scientist."

Art Maxwell had gone to work for the Office of Naval Research after he completed his heat-flow dissertation at Scripps. He then became director of the earth sciences division at the Woods Hole Institution of Oceanography, and also the chairman of the JOIDES planning committee. On those early cruises, the chief scientist was assisted by a deputy. The co-chief of Leg 3 was Dick von Herzen, who was also from Woods Hole and who had also worked on heat-flow studies at Scripps. Jim Andrews, who had just received his Ph.D. from Miami, and I were sedimentologists. The paleontologists were Tsuni Saito of Lamont, Steve Percival of Mobil Oil Company, and Dean Milow of Scripps. Gene Boyce, the Scripps representative, made physical-property measurements of cores.

The scientific staff on the *Glomar Challenger* stayed together for six to eight weeks, the normal duration of a "leg." After Leg 9 the chief scientist and his deputy were replaced by two co–chief scientists of equal rank. They assumed the responsibility of carrying out the scientific objectives of the drilling, as instructed by the JOIDES planning committee. On board, they worked closely with the captain to navigate the vessel to the targeted location. They gave directions to the operations manager and the drilling crew on the "coring schedule." In some places, a hole was cored continuously, so that a history was obtained without missing pages. At other localities, only "spot coring" was scheduled, to save time, if vital information only at certain predicted intervals was desired. The co-chiefs also decided, with the advice of the operations manager, when to terminate drilling at each hole. Like a conductor in an orchestra, they may not haved played any instrument themselves, but their coordinating effort was indispensable.

Sedimentologists, with the help of technicians, handled the cores after they were brought up on deck. The cores were enclosed in plastic liners inside a steel barrel; they were pulled out of a 9-meter barrel, and cut into six 1.5-meter sections. Each section was sliced in half; one half was described, photographed, and stored in a constant-temperature (4 °C) locker as archives, first on board ship, and eventually at Lamont (for Atlantic cores) or at Scripps (for Pacific and Indian Ocean cores). All samples needed for scientific investigations had to come from the other half, the "working half."

The work of the sedimentologists on Leg 3 was rather tedious, because the ocean sediments do not have much variety. Andrews and I split the cores, noted down the colors of the sediments, examined with a microscope the sediment samples smeared on glass slides, and determined their composition. Those ocean sediments consist mainly of tiny skeletons of one-celled plants, called nannoplankton (Plate VIII); these tiny rod-like, disk-like, or star-like fossils may make up 95% of a sediment (Plate II). Other constituents include skeletons of one-celled animals known as foraminifera (Plate II). Also present in oozes are traces of clay minerals, or other very fine detritus derived from continents.

The physical-property specialist had to do a number of routine measurements, such as sediment density, gamma-ray radiation emitted by the sediment, sonic velocity, thermal conductivity, shear strength of the sediment, and so on—useful data, but data that does not usually lead to spectacular results. The spectacular results of Leg 3 were to be the work of paleontologists, who could date ocean sediments. During the Leg 3 cruise, we depended on the paleontologists to check if the age of the ocean floor was indeed that predicted by the Vine and Matthews hypothesis.

The major subdivisions of geological time are eras: Cenozoic, Mesozoic, Paleozoic, and Proterozoic, and Archaean. Eras are divided into periods. Only sediment of Cenozoic and Mesozoic periods was penetrated by the deep-sea drilling. The periods of the Mesozoic are called Triassic, Jurassic, and Cretaceous; of the Cenozoic, Tertiary and Quaternary (Table 5.1).

The geologic periods have been further subdivided into epochs. The Jurassic and the Cretaceous both have more than ten epochs, with names that are familiar only to specialists. The

Table 5.1. Geological Time Scale

Period	*Epoch*	*Date Began (millions of years ago)*
Cenozoic Era		
Quaternary	Holocene	0.01
	Pleistocene	1.8
Tertiary	Pliocene	5.3
	Miocene	23.7
	Oligocene	36.6
	Eocene	57.8
	Paleocene	66.4
Mesozoic Era		
Cretaceous		144
Jurassic		206
Triassic		245
Paleozoic Era		
Permian		286
Carboniferous		360
Devonian		408
Silurian		438
Ordovician		505
Cambrian		570
Precambrian Eras		

Tertiary epochs are Paleocene, Eocene, Oligocene, Miocene, and Pliocene; Pleistocene and Holocene are the two Quaternary epochs. The interval of time defined by the Pleistocene was about the same as the duration of glaciation in the Northern Hemisphere, but strictly speaking, Pleistocene is not a synonym for the Ice Age.

Identification of geological periods and epochs gives only relative ages. The numerical age has to be determined on the basis of radiometric dating of radioactive minerals in geological formations (see chapter 4). We now know, for example, that the Tertiary period started 65 million years ago and that the Pleistocene epoch ("Ice Age") had its beginning 1.7 million years B.P. (before present).

Geologists used large fossils such as shells of mollusks or bones of dinosaurs to determine geological ages of sedimentary formations on land. Very small fossil skeletons, nannoplankton, foraminifera, diatoms, and radiolaria, were used to date ocean sediments. Foraminifera and nannoplankton commonly have a skeleton of calcium carbonate, or a calcified remain. Saito was a specialist on foraminifera and Percival on nannofossils. In some parts of the seafloor, especially in equatorial and polar regions, microorganisms may have a skeleton of silica, SiO_2. Those siliceous fossils are diatoms or radiolarians (see chapter 3). Dean Milow was a specialist on radiolaria, but as we encountered very few siliceous fossils during Leg 3, drilling in middle latitudes, Milow had little to do.

"A Game of Numbers"

Originally, two Atlantic cruises were scheduled to test the theory of seafloor-spreading. The work of Leg 2 in the North Atlantic was encouraging, but the results were not sufficiently convincing to all skeptics. The hopes and the limelight were now centered on the Leg 3 cruise. It was our turn to drill a series of sites near the 30° S latitude. We departed Dakar on 3 December. After drilling one hole west of Sierra Leone, we crossed the Equator on 12 December and arrived at Site 14 eight days later (Figure 5.1).

Magnetic lineations have been given numbers to facilitate reference. We were to have our first test of the Heirtzler time scale at Site 14, where Anomaly 13 was easily recognized. According to the Vine and Matthews theory, older seafloor marked by older anomalies was pushed farther and farther to the side, as new seafloor was spread out along the axis of mid-ocean ridges. If the rate of seafloor-spreading has remained constant, we have a linear relation:

Age of seafloor = half-spreading rate × distance to ridge axis.

The half-spreading rate for this part of the Atlantic, as deduced from the width of the youngest magnetic lineations of known age, is 2 cm per year. Since Site 14 was 760 km from the ridge axis, the formula told us that the age of the seafloor at Site 14 should be 38 million years (Figure 5.2). Now a number had been picked out, and it was up to Saito or Percival to tell us if the prediction was true.

I did not believe in seafloor-spreading. I did not believe in the simple formula. I did not believe that the sediment at the bottom of the hole would be 38 million years old; I was secretly hoping that it would be much, much older. Not only would we be able to look farther back in the history of the ocean; we also would disprove once and for all the ridiculous theory by the brash young graduate student from Cambridge.

It was a maddeningly slow process to string a drill bit, a drill collar, and some bumper-subs together to make a bottom-hole assembly, to join the assembly to segments of drill pipe, and to lower the drill string down through the "moon pool" in the middle of the ship. All night long we stayed up and watched with impatience. The drill bit touched the bottom at 4,346 m early in the afternoon of 21 December. After a core was cut, a steel rope (sandline) with a hook at the end was sent down inside the drill

5.1. Location of the Leg 3 drill sites in the South Atlantic.

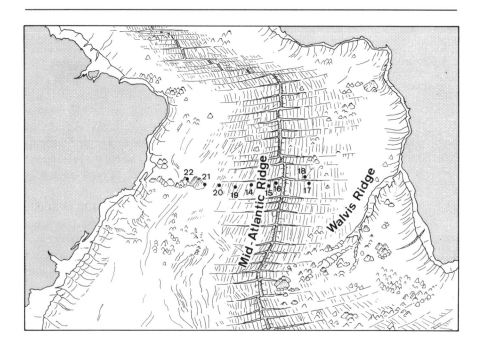

pipe to fish out the core barrel. Another hour of anticipation. Finally, the 9-meter barrel was up.

The plastic liner was cut, and the sediment was a pale brown ooze. Percival took a tiny bit of the sample, rushed down to the paleontology laboratory, and examined the sample under a microscope. In less than five minutes, he came up with an answer.

"The age is earliest Miocene."

"What's the age in terms of millions of years?" Art Maxwell asked.

"About 20 or 25 million years."

Everyone except me had a long face. This represents a sedimentation rate of less than 0.5 meters per million years. Seismic-profiling records indicated that we had more than 100 meters of sediments. If the sediment was already older than 20 million years at the 9-meter depth, the age of the ocean floor could be as old as 200 million years, not 38 or 40 million years.

Fifteen minutes later, Saito also came up to the core laboratory. Processing foraminifera samples took a little longer, but Saito confirmed the "ill tidings" of Percival: the sediment was indeed early Miocene!

Luckily, I kept my elation to myself, because the next series of cores told a different story. In fact, I began to get uneasy as soon as the second core came up. It was late Oligocene. Probably the sedimentation rate was not nearly that slow for the deeper sediments! Then one core after another was hauled up,

5.2. Mid-Atlantic Ridge, Southern Atlantic Ocean. The theory of seafloor-spreading predicts that the ocean crust is older at sites more distant from the ridge axis. Leg 3 was an "experiment" to test the theory.

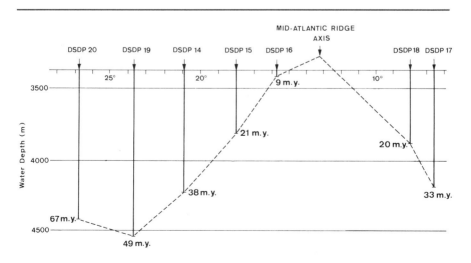

at regular intervals of every three hours. We seemed to make no progress. It was late Oligocene, late Oligocene, and late Oligocene!

Every time a core came up, I had to revise the average sedimentation rate. If the trend should continue, I told myself, the prediction would come true after all. Somewhat discouraged, I went to bed shortly after midnight on the twenty-first. We were already halfway down to the basalt basement of the seafloor at this site, but we seemed to have gotten stuck in late Oligocene sediments of about 30 million years of age.

I tried to get some rest, but I could not sleep. So I got up again and went back to the core lab to wait. Still cores were being brought up every three hours. I now felt like a gambler who was losing and betting against the odds. My heart sank with every new haul of sediments, when the revised estimate indicated that the prediction by the seafloor-spreading theory could not be very far wrong. Middle Oligocene, early Oligocene, and finally the last core came up. It was latest Eocene, or 38 million years, exactly as the theory had predicted!

Saito and Percival both agreed on this crucial age determination. The joy of the shipboard staff was great. I alone was subdued, murmuring to myself: "Wait till the next hole, you lucky fellows."

I thought the first test was an improbable coincidence, and I waited eagerly for our second trial. Site 15 was positioned over Anomaly 6, at a distance of 420 km from the ridge axis, and the age of the oldest sediments should be 21 million years, in the earliest Miocene.

The experiment started on Christmas Eve, 1968. The basement depth was estimated to be about 150 meters. The first core came up in the afternoon. It was Quaternary ooze, also pale brown in color. The fossil assemblage indicated that the sediment was deposited during the last Ice Age, less than a million years ago.

I began to hope again, because the sedimentation rate seemed to be too fast. At this rate, we would be getting into upper Miocene sediment 8 or 10 million years of age, not the 21-million-year-old lower Miocene. When the fifth core came up from a depth of about 100 meters below the sea bottom, I broke out in a broad smile. The sediments were still late Miocene, 6 or 7 million years of age. At the prevailing rate, I thought, the age of the oldest sediments could not be more than half of what had been predicted.

Cheered by my eternal optimism, I felt like a gambler in a

casino. The rolling of the drum coiling up the steel wire with a core barrel at its end reminded me of the spinning roulette wheel. When the "wheel" stopped, the barrel came out, and Percival and Saito would give a number. I could win with each play of the "roulette." The game would be won if the next barrel, or the one after next, brought up a number to kill the "preposterous" idea of seafloor-spreading. I remembered Karl Popper and his mistrust of inductive reasoning: the first 100 swans you see might be white, but should the 101st be black, the theory of white swans is proven wrong.

I had a lot of chances! I would have won if the constant rate of sedimentation had kept up. Well, I was due again for a surprise. The next barrel came up in the afternoon of Christmas Day, 1968. Below a red clay were ocean oozes. To my chagrin, I did not get my number. It seemed as if the casino owner had cheated. Instead of 7 or 8, as one might expect from the sedimentation rate, the number was 18! There was a gap in the record, and some 10 million years of missing pages. The sediment at the bottom of the barrel was 18 million years of age! Now I was running out of chances; we were just about 20 meters above the basement. My hopes had been devastated by this unexpected gap that "ate up" some 10 million years, just enough for my opponents to win.

It was no gamble for my shipmates; they were already convinced of the validity of seafloor-spreading and were waiting contentedly for their victory. The last core at this site came up at nine o'clock in the evening, and it was early Miocene, 21 million years again, exactly as the theory had predicted.

This "success" on Christmas Day called for double celebration. The cook baked a huge cake for us, and the chief scientist dug up a bottle of whiskey. We all drank a toast to the theory of seafloor-spreading, and to the many absent colleagues who made "our success" possible. I went along with a forced smile. Still I would not give up. Vine must be very lucky, I told myself. Wait till the next time!

Site 16 was positioned on Anomaly 5, 190 kilometers from the axis of the spreading ridge. The basement age there should be 9.5 million years, early in the late Miocene epoch. So we started my roulette game again. The sediment thickness here was almost 200 meters. The first four cores brought the hole 60 meters down, and the sediment was already early Pliocene, about 4 or 5 million years of age. I began to hope again. At the going rate, the bottom sediment should be about twice as old as the predicted age.

Although I had learned by then to refrain from optimism, still my hopes soared when the next core from the 90-meter depth (below the seafloor) yielded a fossil assemblage of late Miocene age. Already in late Miocene, and we still had to penetrate another 100 meters of sediment section. I was bound to win this one, I told myself.

What happened during the next 24 hours completely broke my spirit, once and for all. Late Miocene! Late Miocene! Late Miocene! Late Miocene! Late Miocene! Late Miocene! Late Miocene! Seven times Percival pronounced the age of the cored sediment late Miocene! Seven times Saito pronounced the age of the cored sediment late Miocene!

"Could you be wrong? Could it be middle Miocene or early Miocene?" I would ask Percival or Saito, timidly.

"No, not a chance! We cannot be wrong. You can look for yourself. You see, the fossil assemblages from the last seven cores are practically identical!"

To me it was like losing bets on a roulette game seven times in a row. I had always bet even, but the ball always went into an odd number. When the last core was up in the early morning of 28 December, Percival and Saito told the chief scientist that the seafloor at Site 16 is about 9 million years of age. Vine had won again!

As *Glomar Challenger* steamed toward the next drill site on New Year's Eve, 1968, I lay on my bunk and did some soul-searching.

"What is science?" I asked myself.

Science is a human endeavor to sort order out of chaos, to come up with a simple relation to explain a multitude of apparently unrelated facts, and to *predict*.

Yes, the ability to predict is the essence of science. Even though I did not yet fully appreciate all the implications of the seafloor-spreading theory, I was by then very much impressed by its ability to predict. What is the prediction of the alternative theory of permanent ocean basins? That classic theory predicts that the bottom of the sedimentary sequence at every hole is the beginning of time. This prediction had failed miserably at every site! Was there a chance that the basalt at every hole is not the bottom of the sedimentary pile? Nothing is impossible, of course, but there was no indication whatsoever of the presence of any sediment beneath the basalt; seismic profiling had told us that.

As New Year's Eve, 1968, was dawning, I was getting to be 40 years old. For the first time, I began to appreciate what a friend at UCLA had told me—that I might have been a better lawyer than I would ever be a scientist. Lawyers win cases, whatever the truth may be. Scientists seek truth, and will yield even if the truth hurts. Suddenly it dawned on me that I had to take a positive attitude toward the new theory if I wished to continue in a scientific career. I should perhaps try it out once, applying the theory to problems in land geology. I made the New Year's resolution to eat the imprudent words that I had uttered two years ago in Los Angeles, if the results at the next site should again confirm the prediction of the theory. Then, Vine had to be right, Jerry Winterer had to be right. And all those marine geophysicists had to be right!

Yes, the prediction was confirmed at the next site, at the site after the next, at the site after that, and so on. The prediction was confirmed at all our Leg 3 drill sites (Figure 5.2) and at almost all deep-sea drilling sites during the next two decades!

5.3. Verification of the seafloor-spreading theory. DSDP Leg 3 in 1968 verified the age of the ocean crust and confirmed the linear rate of seafloor-spreading as predicted by Heirtzler's magnetostratigraphic time scale.

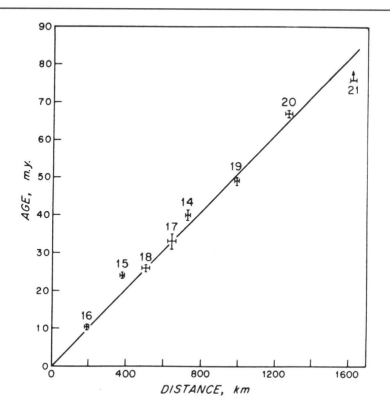

At first I still had difficulty keeping myself from playing the game of roulette, but I played the game with a weak heart and did not bet very heavily, to spare myself further emotional shock. However, the repetition of Vine's triumphs was convincing. My conversion was complete when the *Challenger* entered the harbor of Rio de Janeiro in early February 1969.

From Saul to Paul

The drilling campaign of Leg 3 was one of the greatest triumphs in geology. The theory of seafloor-spreading is right. The unbelievable assumption of linear rate of seafloor-spreading is correct, at least for the last 70 million years (Figure 5.3). I was lucky to be there, and to make a conversion from Saul to Paul.

Later, some of my colleagues accused me of being a traitor to the "cause." I was even pictured as an opportunist who jumped belatedly onto a fashionable "bandwagon." My detractors were not mandarins, but they did not yet perceive that loyalty has nothing to do with science. If we all had been loyal, we would still cite papal infallibility to prove that the sun goes around the earth. There would have been no scientific revolutions, no progress. My critics did not know the painful experiences I had had to go through on the *Challenger* during that Christmas in 1968. If I had refused to accept the overwhelming evidence, I would have avoided the label of weather vane, but the alternative was to become a conceited bigot who could not admit that he could ever be wrong in his scientific judgment.

Maxwell's numbers game verified the prediction of a theory that revolutionized the earth sciences. The drilling experiment, a tool of geologists, had yielded results understandable for a geologist. The schism in the earth sciences was no longer. Maxwell's numbers game also effected a metamorphosis, a scientific revolution, within me. Twenty years after I left China, I finally learned to think like an American, like a European.

6 Atlantic and Tethys

The ophiolites of the Swiss Alps are similar to the basement rocks under the Atlantic seafloor. Their sedimentary covers recorded a parallel development. The Atlantic Ocean and the Tethys Ocean (between Africa and Europe) first came into existence when Africa moved away from North America and Europe. The Gorringe Bank of the Atlantic is underlain by the same gabbro as that on Allalinhorn in the Swiss Alps.

Gabbro on Allalinhorn

My late wife, Ruth, was Swiss, from Basel. A city girl, she did not have much chance to try out mountain climbing. I myself was born on the plains of the Yangtze River. I never saw a snow-covered peak until I was twenty, when I drove across the Rocky Mountains on a tour of the American West. But we both liked mountains.

In 1961 the family went to Switzerland for our summer holidays. We rented a chalet in Saas-Fee, in the Alps. The September weather was good. Every day we could look out of the living room window and see the Allalinhorn. The temptation was great to get a closer look at this white pyramid. Finally, after a couple of weeks, Ruth's venturesome spirit triumphed over my inertia. She sought out a mountain guide, who was to take us up this four-thousander. We shied away from the beginner's route over the glaciers. Instead we overnighted at the Britannia Hütte and climbed the Allalinhorn from the Allalinpass. On our way, we had to scale a rocky cliff. Everybody noted that the terrain was not particularly suited for climbing. The rocks were massive, so there were few hand-holds. Also they weathered badly, and our guide warned us not to rely too much on the crumbly stuff.

The rocks of the Allalinhorn are gabbro. The Allalin gabbro has the same chemical and mineral composition as basalt, but the mineral grains of gabbros are coarser than those in basalt (see chapter 1). Basalts and gabbros are mafic rocks, and they occur together with ultramafic serpentinites in the Alps. Collectively they are called ophiolites, because these rocks are spotted in various shades of green like the color of snakes (Gr. *ophis*).

The sedimentary cover of the ophiolites in the Alps consists almost invariably of red or green cherts and white limestones. The cherts are rocks made up of fine-grained quartz; the flint used by Stone Age men for their arrowheads was a kind of chert. The Alpine cherts are different; they are called radiolarites because they consist of fossil skeletons of the small, one-celled organism radiolaria, although the fossils have been largely recrystallized and altered to quartz grains some 10 micrometers across (see chapter 3). The limestones are also very fine grained, but they were originally nannofossil oozes made up largely of skeletons of nannoplankton. Living radiolarias and nannoplankton swim in ocean water, and sediments made up of dead bodies of such and similar swimming organisms are called pelagic sediments (Lat. *pelagicus* = of the open sea).

After the cruise of the HMS *Challenger* in the last century brought back samples of ocean sediments, several Austrian and German geologists noted a remarkable similarity between the Alpine rocks and the modern sediments of the deep-sea floor. Gustav Steinmann, Eduard Suess, and others started to postulate that the Alpine radiolarites and limestones were the pelagic sediments of an ancient ocean, and the ophiolites constituted the earth's crust beneath the ocean floor.

That the ocean sediments and ancient ocean floor once submerged under more than 4,000 meters of water should be raised to form mountains more than 4,000 meters above sea level indicates that something remarkable has happened. That process of lifting up seafloor to form mountains has been called *orogeny* (genesis of mountains) or mountain-building. Geologists did not doubt that mountains are made, but there was little agreement as to their origin. Why should mountains be made? What are the forces acting upon the earth's crust to cause such great upheaval? Why are the mountains where they are?

Volcanoes make mountains, but the great mountain chains of the world are noted for their crumpled (folded) and broken (faulted) sedimentary formations. Astronomers and physicists of the last century thought that the earth had been cooling down from a molten fireball, and that the earth's outer shell had shrunk and been crumpled because of a volume contraction.

This simple "contraction hypothesis of mountain-building" was put to question after the discovery of radioactivity around the turn of the century. Not knowing the balance of heat generation by radioactivity and heat loss to space, scientists are no longer certain if the earth is cooling down or warming up.

Several geophysicists in the first half of this century, such as Vening Meinesz, Hess, and Griggs, envisioned a steady-state thermal regime. Heat from the earth's core, including a radiogenic component, is transferred to the base of the earth's crust by convection currents in the mantle, and this mantle convection is the driving force for mountain-making processes (see chapter 4).

Geologists were less concerned about the physics of mountain-building forces; they wanted to know why the mountain chains stand where they are. The orthodoxy that dominated our thinking for more than a century was the "geosynclinal theory of mountain-building," first formulated by a prominent American geologist, James Hall. Mountains are where geosynclines were; geosynclines were thus considered precursors of mountains. With this "geosynclinal theory of mountain-building" the emphasis of geological inquiries was shifted from the mountains to the mythical "geosynclines" during the first half of this century (see chapter 1).

The geosynclinal theory of orogeny was in fact a step backward, reducing geology from an observational science to speculative dogma. We all know what mountains are, and we can make all kinds of observations in the mountains, but nobody knew what geosynclines were. There was the American concept equating the geosyncline with a site for thick shallow marine accumulation, or the European idea that geosynclinal sediments had to be deep marine. The consequence was a tendency to attach a label of "geosynclinal" to sediments in all the mountains. Through this trick of semantics, the geosynclinal theory of orogeny became a self-evident, or tautological, truth.

Wegener had no use for geosynclines. Mountain-building was a consequence of the drifting of continents; Alpine mountains are where the continents came together, whereas circum-Pacific mountains are caused by the resistance of the ocean floor to drifting continents.

This idea that geosynclines were oceans had been anticipated by geologists of the last century. Edward Suess, a grand master of the last century, had a global view of things, and he had postulated the existence of a vast ocean between Eurasia and Africa. This equatorial ocean of the Mesozoic was called

6.1. The Bullard fit. This map shows a reconstruction by Bullard, Everett, and Smith in 1965, using computer mapping to restore the position of the continents (defined by the 500-fathom contour along the edge of the continental shelf) before the genesis of the Atlantic Ocean. Black areas represent overlap and shaded gray areas are gaps.

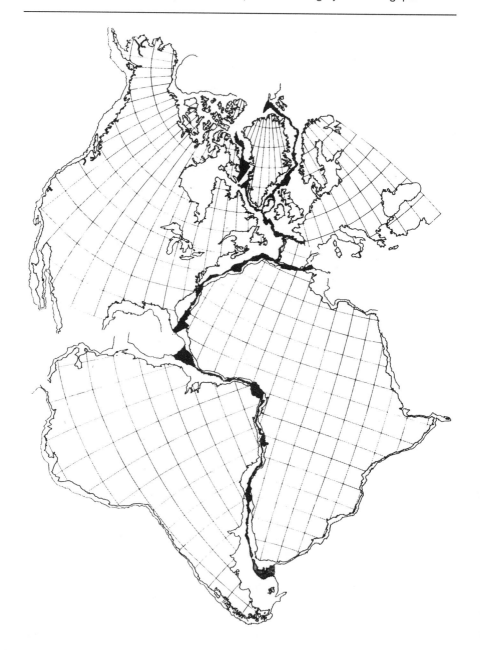

Tethys. The sediments of this ocean were the radiolarites and nannofossil limestones now cropping out in the Alps and in the Himalayas, and the ophiolites were the ocean crust. The shallow marine sediments of the Alps were merely deposits on continental margins.

The concept of Tethys was modified by Swiss geologists during the first half of this century. Emile Argand and Rudolf Staub, among others, accepted the Wegenerian theory of continental drift, and the concept of a supercontinent, Pangea (Figure 6.1). The Mesozoic Tethys came into existence because Europe and Africa drifted apart. Later on, Europe and Africa approached each other, and the Alps were formed when Africa (spearheaded by Italy) met Europe in a gigantic collision.

North American geologists, however, continued in the tradition of James Hall, and refused to recognize ancient oceanic sediments in the mountains. They remained true to their *Leitbild* of permanent continents and ocean basins. They could not accept a theory of mountain-building by continental collision, because they did not think that any ocean had ever been destroyed.

Although I myself was a product of American orthodoxy, I had seen enough of the Alps to recognize the deep-water nature of "geosynclinal" sediments in the Alps, and to accept the Swiss interpretation of Tethys. The ill-fated talk I gave in the 1958 session chaired by Hess was in fact an attempt to debunk the myth of the geosynclinal theory (see chapter 1). There are no geosynclines, I tried to say to my elders; there are only depressions on the surface of the earth, or subaqueous sites of sedimentation. "Geosynclinal" sediments have been deposited in inter-mountain basins, on deltas, on continental margins, or on deep-sea floors. The one common denominator among all geosynclines is the fact that they are underlain by an earth crust thinner than the usual continental crust, like the crust on continental margins or under ocean basins. There should thus be no need to have a name for a hypothetical entity that serves merely to hide our ignorance and to promote fuzzy thinking.

While I was not yet ready in 1958 to make the big jump that a wide ocean could be swallowed up by continental drift, I did not deny the existence of Tethys; only, I accepted the conventional view that the "Alpine geosyncline" was a deep-sea trough. The miogeosynclinal limestones in the High Limestone Alps of Switzerland were shallow marine sediments on the northern shore, and the dolomites of the Eastern Alps were

Gabbro

under the

Gorringe

Bank

deposits on the southern shore of this narrow seaway. The eugeosynclinal sediments were laid down in, and the ophiolites were the oceanic crust of, this Tethys. I did not imagine that I was to see the likeness of the Allalin gabbro during the drilling of the Atlantic.

When I was finally converted to the new school, on board *Glomar Challenger* in 1968, my mind wandered back to problems of geology on land. I began to ask myself if the Tethys Ocean, like the modern oceans, also owed its origin to a process of seafloor-spreading. The Tethys is gone, but its daughter, the modern Mediterranean, is still present between Europe and Africa. What is its origin? Was it a part of the Tethys, which was what was left over after Africa and Europe collided, or was the Mediterranean born after the demise of Tethys and the rise of the Alps? Losing little time during those last days of the Leg 3 cruise, I tried my persuasion on Art Maxwell, lobbying for an ocean-drilling expedition to the Mediterranean Sea.

Maxwell was at that time the secretary of the JOIDES executive committee and the chairman of the planning committee. JOIDES had in 1969 five member institutions, and each was represented by a member on the executive and planning committees. Scientific proposals for ocean drilling were submitted to JOIDES advisory panels for review and selection, before approval by the planning committee. Three site-selecting panels, responsible for the Atlantic, Pacific, and Indian Oceans, had been established. After the initial contract of 18 months for 9 legs of ocean drilling, the program was extended for another two and a half years.

With the addition of 15 more legs to the drilling schedule, JOIDES was ready to try out new experiments. My proposal thus came at an opportune moment. Through the good offices of Maxwell and the recommendation of the JOIDES planning committee, the executive committee decided in August 1969 to constitute a Mediterranean Advisory Panel to study drill proposals for this inland sea, and I was named a panel member.

The chairman of our panel was Brackett Hersey. He was a veteran leader of several Woods Hole cruises to the Mediterranean Sea, but had since become a busy administrator at the U.S. Naval Oceanographic Research Laboratory. Much of the planning was thus left to his student, Bill Ryan, and to me. We convened two informal meetings in Zürich, seeking the advice of senior scientists from European institutions, and we formulated a drilling program of some 15 holes for 7 weeks of drilling in the Mediterranean Sea. The program was accepted by JOIDES, and the first Mediterranean project was designated as Leg 13 of DSDP. Ryan and I were named co–chief scientists for the cruise.

I went to Lisbon on 9 August 1970 and was greeted the next day by old friends among the ship's crew after the *Challenger* docked. We had an international scientific staff. Sedimentologists were Vladimir Nesteroff (University of Paris), Guy Pautot (French Oceanographic Center at Brest), and Forese Wezel (University of Catania, Italy). Paleontologists were Maria Bianca Cita (University of Milan), Herb Stradner (Austrian Geological Survey, Vienna), and Wolf Maync (consultant, Bern, Switzerland). Jenny Lort, a graduate student from Cambridge, England, was to handle the physical-property chores.

Ryan arrived in Lisbon on the tenth. Terry Edgar, the new chief scientist of DSDP, also came to give us some last-minute instructions. Prior to the start of this cruise, the scientific community had become aware of the risks of marine pollution. There are many salt domes in the Mediterranean Sea, and they could be structures in which hydrocarbons have been trapped. The JOIDES Safety Panel had been constituted at almost the last moment to make sure that we should not find oil (see chapter 3). Ryan and Edgar had presented the proposals to the panel and gotten their approval.

On the twelfth of August, the day before our departure, Ryan suddenly pulled out of his briefcase a new drilling proposal for a hole west of Lisbon to investigate a fracture zone in the Atlantic. My European colleagues and I were anxious to sail into the Mediterranean; we were reluctant to give any shiptime away for an Atlantic drill site. However, Ryan and Edgar told me that the recent Leg 11 cruise drilling on the west side of the Atlantic had discovered a sedimentary sequence very much like the one well known to us in the Alps! In fact, my Swiss colleague Daniel Bernoulli from Basel was studying those sediments. It seemed that the history of the Alps and the Mediterranean was very much tied to the history of the seafloor-spreading of the Atlantic Ocean.

Not being familiar with the problem, I balked. On Friday the thirteenth, the eve of our departure, Edgar gave a farewell dinner for some of the shipboard scientists, and we all went to a "cave" afterwards to taste the local port. He and Ryan brought out the proposal again. It still seemed to me a last-minute afterthought, scribbled hurriedly by Xavier Le Pichon in his hard-to-decipher handwriting on two sheets of a yellow pad. Le Pichon had done much to help Ryan and me in the planning of the cruise, and his institute had chipped in with some indispensable surveying of proposed drill sites. Also, Ryan was very much convinced of the merit of the proposal. I was never very good at saying no. Not wishing to disappoint a friend, or to start a cooperation on a sour note, I gave in, but we set a limit of 36 hours of drilling time for this extraordinary hole.

The *Challenger* departed Lisbon at midnight of the thirteenth and steamed west-southwest to our first drill site. Our target was a submarine ridge called the Gorringe Bank, rising from a 5,000-meter-deep abyssal plain to a shallow bank 800 meters in depth (Figure 6.2). The relief of the bank is thus comparable to that of a four-thousander in the Alps. Geophysical studies suggested that the ridge was a fragment of an uplifted ocean floor. The pattern of magnetic lineations indicated that the seafloor there was created some 140 million years ago, shortly after Africa had broken away from North America. Although the nearby seafloor was underlain by young sediments several kilometers thick, the basement under the Gorringe Bank was covered by a thin sedimentary drape. The location thus offered an opportunity to drill through the sedimentary sequence and sample the seafloor.

It was my first tour of duty as a co-chief, and I had never navigated a vessel before. The site chosen was positioned on the steep northern flank of the ridge, and a precise positioning was necessary. Unfortunately, both the precision depth recorder (PDR) and the satellite navigation (sat-nav) were out of order on our first day at sea. Ryan went with a technician to do some repair work. Captain Clarke and his crew had to resort to the classic methods and navigate with a sextant. They were not too exact with their dead reckoning. I felt helpless, like a blind man searching for a needle in a haystack. Fortunately, the sat-nav was finally repaired, just in time to prevent us from going too far astray in our course. The PDR, however, still refused to function properly; we had to trust an old echo-sounder of pre–World War II vintage to guide our progress. With a touch of luck, we found the spot, dropped the beacon, and started operations.

The drill string penetrated young Cenozoic sediments, and then Cretaceous black shales. There was no record of sedimentation for the period from 20 to 100 million years ago. A gap like this is called a hiatus or an unconformity. Deep-sea drilling eventually established that this hiatus is very widespread on both margins of the Atlantic because of erosion by strong currents at the bottom of oceans near margins of continents (see chapter 19). Thanks to this hiatus, however, we were able to penetrate into the basement at the Gorringe Bank in 36 hours of drilling. In fact, we had been planning to abandon the hole and go to the Mediterranean, as we thought we could not possibly

6.2. Gorringe Bank. A last-minute addition to the Leg 13 program in 1970 was to drill the Gorringe Bank, which rises several thousand meters above an abyssal plain. Gabbro at the basement 4 km below sea level under a thin sedimentary cover of the Gorringe Bank looks just like that on the Allalinhorn 4 km above sea level.

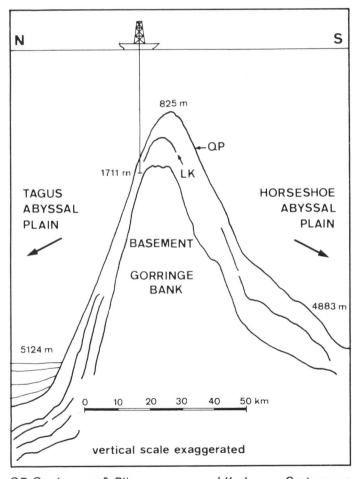

OP Quaternary & Pliocene   LK   Lower Cretaceous

achieve our goal within the originally scheduled time. We were sitting there debating as the final core brought up what we wanted—a rock from the basement. To our great joy, the basement rock was exactly like the ophiolite gabbro I had seen on my way to the Allalinhorn.

Tethys and the Mediterranean

The sedimentary sequence at the Gorringe Bank is very similar to that drilled by the Leg 11 cruise northeast of the Bahama Islands and northwest of Bermuda. The oldest sediments consist of dark shales, white limestones, and radiolarian cherts deposited on the seafloor some 120 million years ago. The *Glomar Challenger* returned to the North Atlantic and penetrated early Cretaceous sediments again numerous times during DSDP Legs 14, 41, 43, 47, 50, 51, 52, 53, and so on, always finding this very sequence. Like similar rocks in the Alps, the radiolarian cherts consist almost exclusively of siliceous skeletons of radiolaria, and the white limestones contain countless skeletons of nannoplankton.

Why were we finding the ophiolites and the pelagic sediments of the same kind and of about the same age in both the Atlantic Ocean and the Alps? The mystery was solved by Walt Pitman and Manik Talwani of the Lamont Geological Observatory when they used the magnetic lineations data on the Atlantic seafloor to interpret the movements of the continents.

The drilling of Leg 3 had confirmed the theory of seafloor-spreading. The next step was to identify the geological age of all the stripes of magnetic anomalies on the seafloor. By 1970, Legs 2, 4, 11, 14, and 15 had come up with much data; the magnetic stripes were dated by paleontological determinations of the ages of the oldest sediments above the basalt seafloor, as we did in Leg 3. Since the geological epochs have been dated radiometrically on land, the ages of the magnetic anomalies could be expressed also in terms of millions of years (m.y.). For example, Magnetic Anomaly 5 has been dated as 9 m.y. of age, and Magnetic Anomaly 13 about 38 m.y., and so on.

What Pitman and Talwani did was restore the positions of the continents before the seafloor-spreading started. Africa and North America were a part of the supercontinent Pangea at the beginning of the Mesozoic, and Tangier in Africa should have been located far west of Gibraltar (Figure 6.3). Then, the continent was subjected to extensional stress, and Pangea broke up, before the seafloor-spreading started.

In the early Jurassic, 155 million years ago, the African continent made the first move to separate itself from Pangea. In a narrow crack, which must have looked somewhat like the Red

Sea of today, submarine lavas were poured out to form the oldest Atlantic seafloor. With the reversal of the magnetic poles, the lavas acquired alternately positive or negative magnetization, giving rise to a series of magnetic stripes, which are now called the M-series, from M-0 to M-29. Africa then started to move eastward relative to North America.

Pitman and Talwani found this series in the North-Central Atlantic between America and Africa, but not in the northern North Atlantic between Europe and Greenland or between Greenland and North America. This observation led them to conclude that Europe was still joined to North America and remained stationary when Africa first moved away. Tangier was thus moved to a position east of Gibraltar during the 74 million years during which Africa was moving eastward by itself (Figure 6.3). Then Europe also started to move eastward relative to North America 81 million years ago, when the North Atlantic had its first ocean crust.

6.3. Evolution of the Atlantic Ocean. Adopting the position of North America as the reference, the positions of Africa and Europe can be calculated as shown by this figure. Africa moved away from Europe before the late Cretaceous, when Europe was still attached to North America. Africa and Europe have been coming close together again, because Europe has been moving faster away from North America since that time. Numbers 155, 81, 25, 21, 13, and 5 refer to millions of years before present.

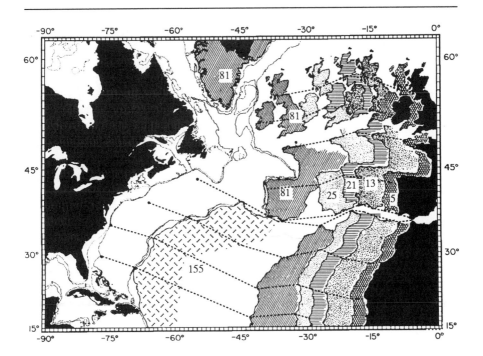

This timetable for the opening of the North Atlantic has also been manifested in geology. The early history of the Atlantic is recorded by the geology of the North American and African margins and is verified by the deep-sea drilling (see chapter 19). Rifting of Pangea became the dominant structure in the late Triassic, 180 million years ago. Depressions, similar in structure to the present Rhein-Graben, were formed, and Triassic New Red Sandstone formations were deposited in those grabens on both sides of the Atlantic.

The first ocean crust was formed under the present North-Central Atlantic 155 million years ago, when Africa split away from North America. The Alpine Tethys Ocean has the same age; at the time, Africa was also moving away from Europe. The Atlantic and the Tethys constituted then one ocean! No wonder the gabbro we obtained from the Gorringe Bank looked exactly the same as the gabbro Ruth and I saw on the cliffs of the Allalinhorn. No wonder the oldest ocean sediments obtained by the JOIDES drilling on both margins of the Atlantic, by Leg 11 in the region near the Bahamas, and by our Leg 13 west of Portugal, looked exactly the same as the radiolarian cherts and white limestones of the Swiss Alps. The Atlantic and the Tethys were twins!

But why should one twin remain submerged while the other was raised to form lofty peaks? Pitman and Talwani's timetable again provided the answer. Yes, Europe did not leave North America until late in the Mesozoic era, when Africa was long gone (Figure 6.3). However, the younger sister was much more dynamic, and she ran away from North America at a much faster rate than Africa did when splitting away from North America. The ocean between Europe and Africa, i.e., the Tethys, became narrower and narrower, because the distance between the two continents was being reduced by Europe's faster eastward movement. In other words, we can imagine that the traffic signal between Africa and Europe changed from green to red some 81 million years ago. Prior to that, the sediments were being deposited in an ever widening and ever deepening Tethys Ocean. After that, the ocean began to narrow. Finally, Europe caught up completely and collided with Africa during the Eocene epoch some 40 million years ago. The game was up. The collision of the two continents gave rise to the Alps. The Tethys met her death.

While Europe moved toward Africa, both continents became more distant from North America. The North-Central Atlantic Ocean thus continued to spread, increased in width, and grew up to become a full ocean. Now the ocean crust under the Atlantic is 6,000 kilometers wide, produced by 155 million years of seafloor-spreading!

The JOIDES drilling solved the mystery of the origin of geosynclines. The so-called Alpine geosyncline was the Tethys Ocean, created when Africa moved away from Europe. The JOIDES drilling also verified the age of magnetic lineations and permitted the reconstruction of the kinematics of the Alpine deformation. Still we did not know how the Alps formed.

7 Arc and Trench in the Mediterranean

Alfred Wegener was mistaken when he thought that continental crust was floating on a liquid substratum. Airy was also mistaken, but his mid-nineteenth-century notion persisted for more than half a century after it was proven wrong. The theory of isostasy was rescued when the idea of crust was replaced by that of lithosphere. Continental crust did not drift, but continental lithosphere did move, on top of ocean lithosphere first before its eventual collision with another continental lithosphere. The Challenger *went to drill the Hellenic Trench to verify a prediction.*

Did Continents Drift?

We missed each other, when I went to meet Dan McKenzie at Zürich's Kloten Airport in the winter of 1969. I was then the chairman of a committee searching for a new director for the Institute of Geophysics at the ETH. McKenzie had been recommended to us as a leading candidate, and he was invited to come and give a talk. I was informed of his arrival time and waited for him outside customs. When he walked out, we saw each other momentarily, but both dismissed the notion that the other could be the person we had come to meet. He was misled by the "umlaut" on the German romanization of my Chinese name, and expected a native-born Swiss! I was deceived by his youth; McKenzie was only 27 years old then!

McKenzie was not interested in the possibility of a professorial appointment at Zürich. He had several good reasons, and one alone was enough for me: he liked Cambridge and the academic atmosphere there.

The Geophysics Department at Madingly Rise on the edge of town was chaired for many years by Sir Harold Jeffreys, and was at one time a bulwark of resistance against the Wegenerian heresy. Cambridge in the sixties became, however, a center of excitement, where revolutionary ideas were born. Maurice Hills and Teddy Bullard had taken over, and they gave high priority to their efforts to explore the seafloor. Tension was in the air.

Orthodox doctrines had not adequately explained geology on land, and failed altogether to interpret the geology of the ocean, which, as we are often reminded, covers three quarters of the earth's surface! Then Drum Matthews came back in 1962 from the Indian Ocean expedition with a set of data on seafloor lineations. Fred Vine, inspired by a talk by Harry Hess, proposed the theory of seafloor-spreading to explain Matthews's data. J. Tuzo Wilson came to visit in 1965, talked to the young rebels, and was inspired to think up the idea of transform faulting. Then Hess came to Madingly Rise for his sabbatical and told Vine that he had a "fantastic idea." Encouraged, Vine worked with Wilson on "the magnetic anomalies over a young ocean ridge off Vancouver Island" (the Juan de Fuca Ridge), and they came up with their 1965 paper that did much to gain acceptance for the seafloor-spreading theory.

The theory of continental drift had always been controversial. Jeffreys did not ignore Wegener, but he went too far when he denied that the two sides of the Atlantic coastlines could be fitted together. Perhaps he inherited this prejudice from the famous Cambridge physicist Lord Kelvin, who believed that what cannot be expressed in numbers is not science! Matching coastlines did not require numbers then. It was child's play, not science. Sir Edward Bullard, Jeffreys's successor at Cambridge, decided to take a crack at the problem with numbers; he was helped by two of his students, J. E. Everett and Alan Smith. They used a computer to fit together the continents on both sides of the Atlantic, and published their amazing result in 1965. The fit, since known as the "Bullard fit," was too good to be coincidental (see Figure 6.1); Bullard himself was no longer in doubt that the continents had moved, and his students were taught the theory of continental drift.

How could continents drift? What is the mechanism? Vine was to discuss with me years later the preoccupation of scientists with mechanisms, before they have even defined the change of states. This is, of course, particularly true with earth scientists. Chemists, used to working with the first law of thermodynamics, are taught that change of state is independent of path, and that one needs to define a change first before seeking a mechanism to explain the path of the change. Wegener would have been right if he had emphasized that the position of continents had shifted, and his theory would have gained ready acceptance if he had not gotten carried away by the drift mechanism.

Drift involves the displacement of a lighter object above a heavy fluid. Wegener had good geological evidence indicating the relative displacement between continents, but his one piece of geodetic evidence that continents are drifting was easily proven wrong (see chapter 4). On the contrary, geophysicists had good physical evidence that continents are not floating on a heavier fluid. Continents that consist of granitic rocks will melt at a temperature much lower than the melting point of basalt or other mantle rocks; the continental crust could not float on a liquid substratum and drift. It was as simple as that.

Geophysicists did not present any evidence, however, that the relative positions of the continents had to remain fixed. Wegener insisted that continents moved because they drifted. In the heat of argument, one forgot that continents might have moved, even if they had not drifted. This is exactly what the theory of plate-tectonics is all about. But, first all of, where did Wegener get his fatal idea that continents could drift in the first place?

Earth's

Gravity and

Isostasy

The story goes back to the early part of the last century. After the British colonized India, they carried out land surveys and made maps. An indispensable item for a surveyor is a lead weight; hanging on the end of a string, the weight makes a plumb line, directed vertically down toward the center of the earth. Geodetic surveys measure angles and distances. The line joining two points lying north-south on the surface of the earth describes an arc. The angle of this arc is the latitudinal difference between the two points, and the difference is determined by the angle between the directions of the plumb line at the two stations. For example, the latitudinal difference between Kaliana and Kalianpur in India was determined by surveying to be 5°23′42.29″ (Figure 7.1).

One way to check survey results is to make astronomical observations at those two stations. To the discomfort of the surveyors, they found that the latitudinal difference measured astronomically was 5°23′37.06″. The discrepancy of 5.23 seconds of arc, corresponding to a distance of about 150 meters between the two stations, seemed harmless enough, and might have been ignored by surveyors less pedantic than the proud British.

There was an unavoidable source of error. Ever since the time of Isaac Newton, everyone knew that there is a gravitational pull between two objects on the earth. When one object is as massive as the Himalaya Mountains, its pull on the other can be considerable. It seemed probable indeed that the plumb line at Kaliana, sitting practically at the foot of the Himalayas, was not pointing straight down; the lead weight should have been deflected toward the mountain, however slightly, and so the plumb line was not exactly vertical.

7.1. Gravitational pull of the Himalayas. The effect of the Himalaya Mountains on the computed distance between Kaliana and Kalianpur is to make the observed angle of the arc (angle 2) smaller than the latitudinal arc (angle 1). The angles in this diagram are greatly exaggerated.

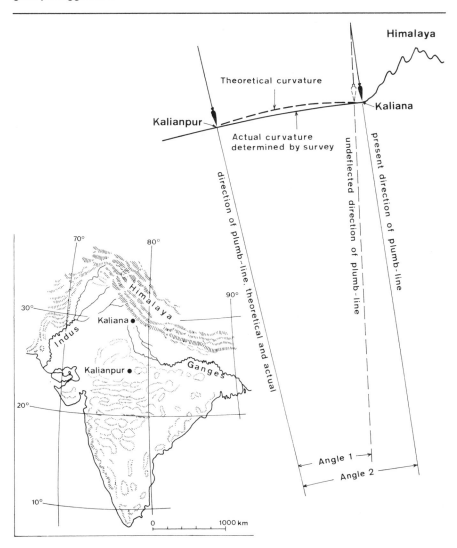

Archdeacon J. H. Pratt, a British clergyman who was also an amateur scientist, like many of his contemporary colleagues, became interested in the problem. Not having a computer, of course, Pratt had to make very laborious calculations. His results showed that the plumb line has been deflected both at Kaliana and at Kalianpur, but much more so at the former, which lies close to the Himalayas. The error caused by the deflections of the plumb line is theoretically 15.885 seconds, about three times the discrepancy of 5.23 seconds actually determined by the land surveying, which is shown in Figure 7.1. In other words, the gravitational pull by the Himalayas determined by the surveying was far less than that calculated by Pratt. Since neither the surveyors nor the archdeacon could find any fault with their calculations, an explanation had to be found.

When Pratt made his computations, he assumed that the material in the outer shell of the earth has exactly the same density everywhere. If he assumed that the material lying under the Himalayas has a slightly lower density, the mass of the mountain would be smaller, and the attraction on the lead weight of the plumb line would be smaller. In fact, he could accept the data by the surveyors and calculate what the average density of the material lying under the Himalayas should be.

Why should the density of rock materials under the mountains be lower than that under the plains of India? Pratt assumed volume expansions or contractions. If the earth started out with an outer shell of uniform thickness and uniform density, Pratt reasoned, the regions where the crust has expanded have risen as plateaus or mountains. In contrast, the regions where the shell has contracted and become denser are now underlain by ocean basins.

Sir George Airy, the Astronomer Royal of the British Court, was probably asked by the editor of the *Transactions of the Royal Society* to comment on the interesting idea by Pratt. Airy's short commentary was published in the same (1855) issue of the *Transactions* as Pratt's massive opus. Influenced by the observations made by early-nineteenth-century naturalists of lavas pouring out of volcanoes, Airy thought that the earth's interior was liquid; only its thin outer shell was solid. This outer shell was called crust, and rocks in the earth's crust, according to Airy, should have a lighter density and much greater strength than the underlying liquid substratum. Airy disagreed with Pratt's idea that the reason the Himalaya Mountains have a mass smaller than the assumed mass was that the crust under the mountains is less dense than that under the plains; the mountains are lighter than they should be because the lighter crust under the mountains is much thicker than the crust under the plains, according to Airy. Why should the mountains be where the crust is thick? Airy offered a very simple theory of flotation equilibrium, a theory that was to beguile the geologic profession for more than a century: solid crust floats on liquid substratum like icebergs in the sea; thicker crustal segments, like bigger icebergs, stand higher (than their thinner neighbors) to make the mountains (Figure 7.2).

The idea of flotation equilibrium can actually be, and has been, applied to Pratt's model for the outer shell of the earth. Segments with different densities can also float on a liquid substratum, and the highest mountains are where the shell segments consist of the least dense material (Figure 7.3). A major difference is that the surface separating the solid outer shell from the liquid substratum, according to this model, should be a level (also called the level of isostatic compensation by some scientists). In contrast, the surface separating the two, according to Airy, is a magnified mirror image of the surface relief; the higher the mountains, the deeper their "roots." The level of isostatic compensation would thus be the level of the deepest "root," as the weight of crustal segments plus the weight of liquid substratum above this level should add up to be exactly the same everywhere.

7.2. Airy's model of isostasy. Airy assumed the flotation of lighter crust segments of the same density, but different thickness, on a denser fluid substratum. The densities shown in the lower diagram were not specified by Airy, but are based on modern estimates.

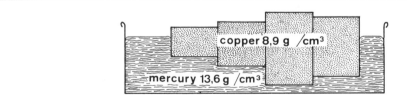

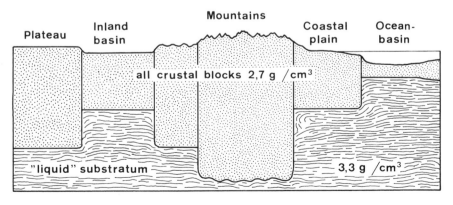

The gravitational pull of a mountain can also be determined by the gravitational acceleration of a fallen object there. We all know from middle-school physics that the acceleration is "constant" at a rate of 9.81 meters per second per second. Yes, the acceleration would be constant if you do not worry about a millionth of a meter per second per second. If you do, the acceleration is different at each spot on the surface of the earth. By the turn of the century, instrumentation had been devised to measure those very minute differences in gravitational acceleration; this kind of work is now called gravity surveying. The results of such surveys could be used to check the flotation equilibrium of the earth's outer shell.

The surveys carried out by the United States government during the early part of this century did indeed support the idea that the mass under mountains is apparently "deficient," either because of a lower density of the crustal material, as Pratt assumed, or because of a thicker crust, as Airy assumed. It was difficult to tell for sure from the survey data who was correct.

Joseph Barrell preferred the model by Pratt. By then, studies of the transmission of earthquake waves had already invali-

7.3. Pratt's model of isostasy. A model of isostasy according to Pratt's concept assumed the flotation of lighter crust segments of different density and different thickness on a denser fluid substratum. The densities shown in the lower diagram were not specified by Pratt, but are based on modern estimates.

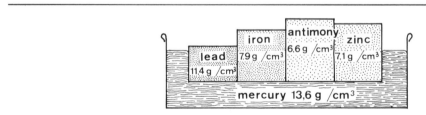

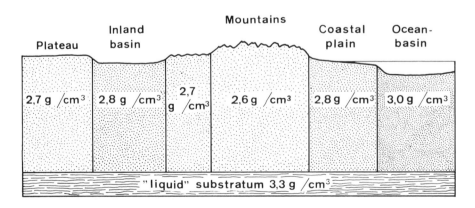

dated the idea of a liquid substratum. Therefore, Barrell used the term *asthenosphere* as a substitute. The material within the asthenosphere is a solid, not a liquid, but it is very weak, and may yield under very small stress like a liquid if the stress is applied slowly. The solid, however weak, has an elastic response to rapidly applied stresses associated with earthquake waves. The outer shell above the asthenosphere is called the *lithosphere*, and behaves as a solid in every aspect.

On the basis of the geodetic survey data, Barrell thought that the base of the lithosphere should be about 100 km or more below the surface, and certainly less than 400 km. Assuming that the weight of rock column above the weak asthenosphere is at some depth (i.e., the depth of isostatic compensation) everywhere the same, we could use Pratt's model to interpret the data of the geodetic survey: mountains are underlain by a lithosphere of lower density. Barrell further found that the earth's crust is in a state of regional isostatic equilibrium.

Seismology and Crustal Structure

Studies of the transmission of earthquake waves gave us another glimpse into the earth's interior, because the waves moved with different speeds through materials of different density. Mohorovicic found a surface of discontinuity that separates rock materials of different density, and this discontinuity is called the Moho (see chapter 1). The part of the earth above the Moho is the crust, and the part below is the mantle. Data on wave propagation indicates that the crustal material is lighter, and the mantle is heavier. Furthermore, as Airy had postulated, the crust is thicker under mountains, where the Moho is deeper, and thinner under oceans. It seemed that the mountains with their thicker crust are sticking out higher, like big icebergs floating in the Arctic Ocean. Airy's prediction was confirmed by seismologists; Airy must have been right! Or was he?

A very serious mistake, which may have delayed the revolution in the earth sciences for many decades, was the assumption that the crustal segments above the Moho, as defined by seismologists, are in a state of flotation equilibrium above the mantle, like Airy's crustal segments above a liquid substratum. To use Airy's model to interpret seismic results we must assume that the mantle below the Moho has practically no strength. In fact, this is not so. Numerous experiments to determine the strength of rocks were carried out during the 1930s, and the pioneering work was mostly done by David Griggs, who worked under Percy Bridgeman at Harvard.

The mantle rocks consist mainly of iron and magnesium silicates, and they have been found to be strong at the relatively low temperature prevailing below the shallow Moho under the oceans. Only when heated to temperatures close to their melting point, and when subjected to very slowly applied stress, could these rocks flow like a fluid. The experiments did not bring out great surprises. The surprises were that many earth scientists continued to think, like Airy did in the 1840s, that the mantle rocks just below the Moho were the liquid substratum below a floating crust. It was this misconception that led Wegener to postulate continental drift. Knowledgeable geophysicists and geologists knew that crust cannot float and cannot drift, and, therefore, Wegener must have been wrong.

Yes, Wegener was wrong; it was even more apparent after Airy was proven wrong by seismic surveys of the 1950s!

Continents Cannot Drift

Seismic waves can be produced artificially by exploding dynamite, and many studies to determine the depth of the Moho through the use of manmade explosions were carried out during the fifties. The general results were encouraging. When I attended the 1957 meeting of the American Geophysical Union, I saw prominently displayed a profile diagramming the latest results of studies of the earth's crust under North America. Like Airy had predicted, the crust was shown as thicker under the Sierra Nevada, the Rockies, the High Plains, and the Appalachians, and the crust is thin on the continental margins and very thin under the Atlantic and Pacific Oceans. The depth of the Moho, as determined by seismic studies, could be everywhere the same as what Airy predicted on the basis of a theory to explain the mass distribution of the earth's outer shell. There was one eyesore, however. The Colorado Plateau, which stands some 2 kilometers above sea level, should have a crust of about 40 kilometers or so, but H. E. Tatel and M. A. Tuve came up with the anomalous result of a 30-kilometer Moho depth (Figure 7.4).

The eyesore did not go away when the experiment was repeated. Finally, more than one person, including myself, came to the realization that the crustal segments above the Moho are not floating above a dense liquid-like mantle. We had to take a second look at Pratt's model, which assumed differences in the density of the materials that constitute the outer shell of the earth. The other shell above the depth of isostatic compensation is not Airy's crust, or the light crust above the Moho, but the lithosphere of Barrell, or the strong outer shell above the asthenosphere.

Two wrong assumptions had been made—that the upper mantle beneath the Moho is weak, and that the upper mantle is homogeneous. Now we had to go back to Barrell's old assumption that a strong and rigid outer shell extends down to a depth

7.4. Lithospheric isostasy. The upper diagram shows the Airy model of isostasy. The model has been disproven by seismic investigations that show that the crust of western North America is thinner than predicted by the model. The lithospheric model of isostasy (lower diagram) assumes a level of isostatic compensation considerably deeper than the Moho, and also assumes that the density of the upper mantle within the lithosphere is laterally variable. This explains the apparently anomalous thin crust of western North America.

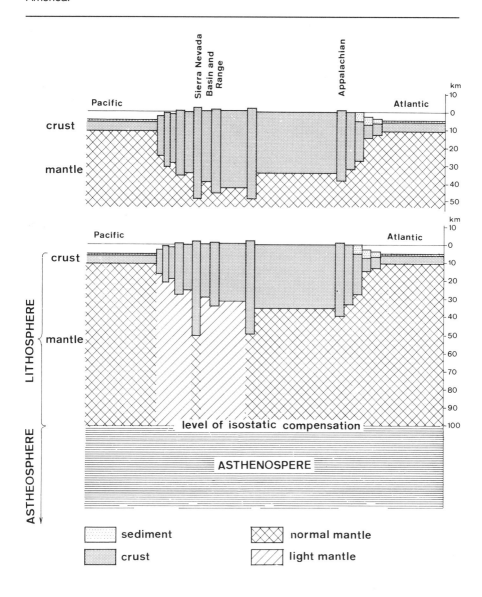

of 100 km or more; such a lithosphere includes both the strong crust and an uppermost part of the mantle, which is also strong. Furthermore, the mantle materials in the lithosphere may not have the same density.

The assumption of heterogeneity of mantle density provided new opportunities to explain the vertical movements on the earth's surface. One of those fundamental problems was the origin of continental margins.

North American geologists had known for more than a century that high mountains should have stood where the Atlantic Coast is; they called the vanished highland *Appalachia.* In 1963, Harry Hess and Robert Dietz were invoking continental drift and seafloor-spreading to explain the disappearance of Appalachia: the highland Appalachia was Africa before it drifted away, so we now have the Atlantic Ocean where Appalachia was.

Not accepting continental drift, I invoked vertical movement to explain the disappearance of Appalachia. In 1963, in a talk I gave at Princeton University, I postulated an expansion of the earth's lithosphere when Appalachia stood there as highland. Appalachia later sank to become the basement under the Atlantic coastal plain because of the contraction of the lithosphere. Hess had just published his theory of seafloor-spreading, and he was in the audience. He was, however, too polite to criticize my feeble attempt to ignore the theory of continental drift.

Wegener's theory had been proven wrong. But if continents did not drift, how did they move? Vine and Matthews's theory of seafloor-spreading implied that continents must have shifted in their positions, because they have to be pushed aside when new ocean crust is formed. Continents can move because they are a part of the lithosphere, and the lithosphere has moved.

Lithosphere and Lithospheric Plates

The lithosphere is not a perfect spherical shell without flaws. What are the flaws? They are the great fault or great fracture zones on the surface of he globe. Shallow earthquakes, with epicenters 10 or 20 kilometers down, have occurred mainly in regions of mid-ocean ridges; the median rift-valley is mainly a zone of graben-faulting. A map of the epicenters of those shallow earthquakes gave us, as a matter of fact, the first portrayal of the geographical distribution of the Mid-Atlantic Ridge and of the mid-ocean ridges of the Indian Ocean—a distribution subsequently confirmed by an analysis of bathymetric data (see Figure 4.4).

Not all the earthquakes are confined to shallow depths. Around the margins of the Pacific Ocean, deep earthquakes are common, and may extend down to a maximum depth of 700 kilometers. Hugo Benioff of Caltech observed that the epicenters of those were largely confined to narrow zones, plunging down from the surface at various angles from the horizontal. An inclined zone in the earth's interior where deep-focused earthquakes have been occurring is now called a Benioff zone (Figure 7.5) in his honor.

In studying the deep earthquakes of the Tonga region, Jack Oliver and Bryan Isacks noted in 1967 that there exists in the mantle an anomalous zone whose thickness is on the order of 100 km and whose upper surface is approximately defined by a highly active seismic zone that dips to the west beneath the island arc of Tonga and extends to a depth of about 700 km. The zone is anomalous because seismic waves propagating in the zone are subjected to much less attenuation than waves propagating in deeper mantle. Correlating wave attenuation to strength, Oliver and Isacks compared their "anomalous" mantle to the lithosphere in geological literature. The expression anomalous zone was soon replaced by lithospheric plate. Used in the modern sense, the surface separating the anomalous zone of wave attenuation is a discontinuity in strength: the lithosphere is strong above and can be displaced as rigid plates; the material is weak below and can creep like a fluid, under small stress differences (slowly applied) such as that induced by convection currents in the mantle (Figure 7.5).

Where could a lithospheric plate go and where is it going? Oliver and Isacks's data show that the lithospheric plate under the Pacific Ocean east of the Tonga Arc has been pulled down, or underthrust, westward under the Tonga Islands and the Tonga Basin. This process in today's jargon is called *subduction*.

7.5. A model of plate-tectonics. This block by Isacks, Oliver, and Sykes illustrates the movement of lithosphere in plate-tectonics. Arrows on the lithosphere indicate relative movement of adjoining blocks. The Benioff zone is where the lithosphere plunges down into the asthenosphere.

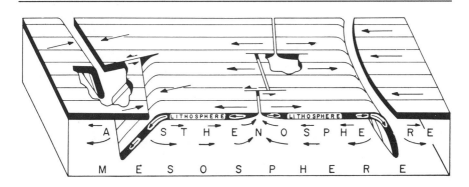

In addition to shallow earthquakes in the region of mid-ocean ridges, and deep earthquakes plunging into the subduction zones, there is a third class, those that are associated with the great fracture zones of the oceans.

East-west trending fracture zones were first discovered in the eastern Pacific by Bill Menard in the early 1950s, by echo-sounding of the seafloor. He found steep, mainly east-west trending escarpments that could be explained only by assuming a large fault, or fracture. Several years later, Victor Vacquier noted that magnetic lineations seemed to have been offset thousands of kilometers along such fracture zones. Similar fractures, offsetting mid-ocean ridges, or other seafloor features have since then been identified. The oceanic fractures are now called transform faults, thanks to Tuzo Wilson's insight; they are considered the tracks left behind by the movement of segments of the earth's lithosphere. I shall come back to a discussion of these curious features in chapter 12.

Theory of Plate-Tectonics

The distribution pattern of earthquakes suggested to Jason Morgan in 1967 that the earth's lithosphere has been broken into numerous plates (Figure 7.6). Those plates are moving either toward, away from, or sideways past each other.

Just as Bullard realized a few years back, when he tried to put the continents on both sides of the Atlantic together, Morgan at

7.6. Lithospheric plates. This map by Jason Morgan in 1968 shows that the earth's crust is divided into units that move as rigid blocks. The boundaries between blocks are rises, trenches (or young fold mountains), and faults.

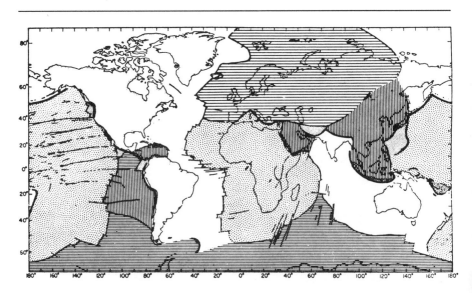

Princeton came up with the idea that displacements of spherical segments are governed by the Euler theorem: the motion of any point on a spherical surface can be defined as a rotation about an axis, with the motion describing a circular arc or a small circle on the globe. Almost simultaneously, McKenzie and Parker at Scripps were also using Euler's theorem for their computations of the movement of lithospheric plates. The theory of plate-tectonics was born.

Where the plates are moving away from each other, the shocks of breaking apart rigid crustal segments that have been welded together are the earthquakes registered from the regions of the mid-ocean ridges (Figure 4.4). The axis of seafloor-spreading is the place where new seafloor is created, acquiring magnetic lineations that record the successive motions of the departing plates. This is now called an accretionary margin—the margin where new seafloor is accreted to old seafloor. Continued accretion of new ocean crust onto "spreading" plates would widen the ocean, just like the continued addition of water to be frozen in the crack of a rock would widen that crevice.

When the plates move toward each other, deep-sea trenches or mountains are formed at their junction. This is now called an active plate-margin—the margin where the pressing together of plates is causing active building of mountain chains.

The return to Barrell's concept of lithosphere thus broke the impasse imposed by the Wegenerian theory of continental drift. The continents cannot have drifted, but they did move. Now we know that the continents, perched on top of lithospheric plates like containers on freight trains, did move. They moved because plates have moved; the plates are the rolling freight trains for the containers!

If a computer can put continents together, it can also tear them apart. Xavier Le Pichon did just that in 1968, when he made use of the vast amount of oceanographic data stored at Lamont, along with the theory of plate-tectonics, to compute the displacements of six lithospheric plates and the rates of their motions. His results enabled him to predict where and when new seafloors were created, where and when new mountains had risen. Now there are numbers to quantify our predictions. If Lord Kelvin were alive today, he would have to admit that geology has become a physical science; it is no longer "stamp collecting."

At active plate-margins, oceanic plate is thrust under continental plate. This is the situation along the Pacific Rim. The North American continent, for example, is carried westward on the

North American plate as the North Atlantic Ocean is widened by seafloor-spreading. Off the Pacific Coast, where the continent meets the Pacific, the ocean plate plunges under the North American plate. The downward movement, or subduction, causes friction and induces stress; the release of the stress gives shocks along a Benioff zone that extends all the way down to 700 kilometers. Beyond that depth the plate is so heated up that it becomes merged with the asthenosphere material and no longer maintains its identity. Through such subduction processes, the floor of an ocean thousands of kilometers wide can be swallowed up or consumed in the "bowels" of the earth.

The melting of the lithospheric material of a down-plunged plate yields molten liquid, which should rise, on account of its lower density. The magma intrudes into rock formations near the surface of the earth, and cools down as great bodies of granite. Or molten liquid could find its way to the surface and come out as volcanic lava, such as *andesite*. Both kinds of rocks are found in the Andes Mountains of South America and other Pacific coast ranges.

Subduction of ocean plate leads eventually to continental collision, when one continental plate meets another at the active margin. After the Tethys Ocean was swallowed up by the northward march of Africa toward Europe, their final collision caused the Alps to rise. Yet there is the Mediterranean Sea between the two colliding bodies. Is the Mediterranean a relic of the great Tethys? That was the burning question in 1969 when we were planning the Leg 13 drilling.

Plate Tectonics of the Mediterranean Sea

Emile Argand wrote in 1924 that all of the Tethys was eliminated when Africa collided with Europe during the late Oligocence. The Mediterranean was a young inland sea, and had its birth after the collision: it came into existence when Corsica and Sardinia drifted away from France and Spain, when Italy split off from those islands, and when Africa moved from southern Europe. All that, according to Argand, happened some 25 million years ago.

The results of the deep-sea drilling during Leg 13 in 1970 and subsequently during Leg 42A in 1975 gave considerable support to Argand's postulate. The oldest sediments above the oceanic basement of the western Mediterranean (Balearic and Tyrrhenian) basins are Miocene, some 20 or 10 million years of age, as Argand had predicted. The western Mediterranean basins were created in the manner described by Argand. As continents moved away from each other, the cracks were floored by submarine lavas to make new oceanic crust. The Balearic Basin ceased to expand during the last few million years, and the basin has been floored by a blanket of sediments that hid the old and extinct submarine volcanoes. The Tyrrhenian Basin is still growing, and its deep-sea floor is dotted with old and new volcanoes, some of which rise above sea level to make islands, such as the famous Stromboli off the Italian coast (see Plate IX).

The eastern Mediterranean seemed, however, to be quite a different story. Computations by Le Pichon showed that the eastern Mediterranean Sea has been there for a long time. Europe and Africa are not being split apart; they are being compressed and pressed together. Ryan's geophysical data indicated that the sediments of the easten Mediterranean are very thick, reaching more than 10 kilometers in some parts. We could never hope to get to the bottom of the sediment pile to date the age of the ocean floor there. We had to rely upon our knowledge of Cypriot geology to formulate a working hypothesis to be tested by the deep-sea drilling.

The oldest sediments on the island of Cyprus are radiolarian cherts and white limestones, deposited on top of the ophiolite basement. The sequence is similar to that in the Alps, although the Cypriot rocks are somewhat younger. The eastern Mediterranean Sea could thus be considered a younger sister of the Tethys. This part of the ocean floor has not yet been smashed in the jaw between Africa and Europe. On the other hand, its destiny is clear. The Mediterranean Ridge, with an elevation of some two or three thousand meters above the abyssal plains of the eastern Mediterranean, is a giant submarine mountain chain dimensionally comparable to the Alps; Cyprus is the highest peak of this mountain chain in the making. Continuous seismic profiling gave indications that the sedimentary formations on the ridge have been folded and faulted like those in the Alps. Apparently, the last remnant of ocean floor on the northward-marching Africa plate is being dragged down the subduction and consumed, and the Mediterranean Ridge will be the future Alps between the Balkans and the lands of Northeast Africa.

There is an island arc in the eastern Mediterranean—the Cretan Arc. Rhodes and Crete are the largest islands on the arc, which extends westward to the Peloponnesian Peninsula. South of the arc is a deep-sea trench, the Hellenic Trench; the deepest spot in the Mediterranean, over 5,000 m deep, is in the trench. South of the trench is the Mediterranean Ridge. Observers in a deep-sea submarine have found vertical cliffs on the inner trench walls, although the average slope is only about 35°.

Vening Meinesz was among the first to study an island arc.

He was a very tall man; he had to adjust himself to the narrow confines of a submarine to make gravity surveys under sea. The measured results were pretty much what had been expected. His one big surprise was that the gravitational acceleration is smaller, much smaller, on the inner wall of the Java Trench south of the island arc of Indonesia. This negative gravity-anomaly indicates a mass deficiency. Now the study of earthquakes by Oliver and Isacks in Tonga told us that an ocean trench is indeed where an oceanic plate is dragged, or *underthrust*, down the subduction zone. The "mass deficiency" corroborated their postulate that the place of denser mantle is taken up by lighter crust underthrust by subduction. One should find rocks from the underthrust lithosphere under the inner wall of an ocean trench.

Bill Ryan and his associates at Lamont gathered considerable geophysical data in four oceanographic cruises during the 1960s. They found a belt of negative gravity-anomaly on the inner wall of the Hellenic Trench. There is also the Benioff zone defined by seismicity, dipping some 30 to 40 degrees north from the Hellenic Trench. There are furthermore several volcanic islands in the Aegean Sea, Santorini being the best known. The source of volcanic lavas, according to the new theory, should have been the remelting of Mediterranean sediments that had been brought down to the Benioff zone through the subduction of the African Plate. Therefore, all indications were that the Hellenic Trench marks an active margin; the eastern Mediterranean on the north fringe of the African Plate is plunging under the European Plate (Figure 7.7). One of our goals on Leg 13 was to drill a hole into the northern inner wall of the Hellenic Trench to verify the prediction that young trench sediments of the African Plate are tucked under older rocks of the European Plate.

Subduction under the Hellenic Trench

Now I resume my narrative of the voyage of *Glomar Challenger* during our Leg 13 expedition in 1970. We departed our drill site over the Gorringe Bank on 17 August, and entered the Mediterranean at dawn on the eighteenth. The vessel was parked in the strait south of Malaga to drill one site, before we went up the Spanish coast to drill a couple more in the Valencia Channel. The drill results verified Argand's theory that the western Mediterranean was a young ocean. From there we went around the Balearic Islands and drilled JOIDES Site 124. It was a historical moment on the morning of 27 August when a core, since then nicknamed the "Pillar of Atlantis," gave us the first proof of the desiccation of the Mediterranean Sea. The story is told in chapter 15.

The _Challenger_ entered the eastern Mediterranean after passing through the Strait of Sicily at midnight starting the month of September. After drilling two sites on the Mediterranean Ridge, the _Challenger_ reached a spot above the southern edge of the Hellenic Trench in the early morning of 6 September 1970. Soon echo-sounding sent back signals telling us that we were above the trench, which is floored by a thick blanket of flat-lying sediments (Figure 7.7). Our plan was to spud our hole near the northern edge of the trench, and first drill through the youngest trench sediment, then penetrate the older rock of the European Plate, and finally reach the underthrust trench sediments of the African Plate.

The drilling capacities of the _Challenger_ were limited, as was our time. We had to reach our objectives with a hole less than 1,000 meters deep. To do this we had to position our site very accurately, and precise navigation was thus a must!

7.7. The Hellenic Trench, the surface expression of an active margin, where the Mediterranean seafloor (African Plate) plunges under Crete (European Plate). Site 127 was drilled at the foot of the inner trench wall and Site 128 at the foot of the Mediterranean Ridge.

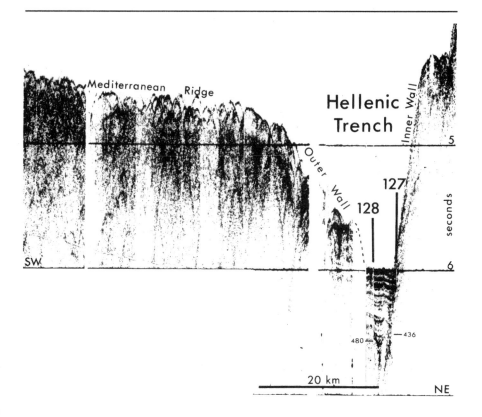

After four weeks at sea, Ryan and I had both become experienced in this game. Our target lay near the edge of the trench, so near that echoes bouncing off the steep sides of the trench wall should begin to show up on our precision depth recorder when our vessel arrived at a spot above the desired location. There we should drop the acoustic beacon immediately; the beacon, a guide for dynamic positioning, would then mark the exact spot of the drill site.

We stayed in the navigation room all morning on the sixth of September. The first side echoes from the landward slope appeared at 11:13 a.m., but the flat-echo trace continued beneath the side echoes. The vessel was still directly over the trench. It was still too early; we could never drill through where the trench sediments are too thick. We had to go near the edge, where the sedimentary cover is thinner. Tension mounted. Minutes went by. The second mate came twice from the bridge and checked with us. With infinite patience and confidence we waited for four more minutes while the side echoes from the steep wall became stronger and stronger. Finally, the signal was given, and the beacon was dropped. Thirty seconds later the echo sounders told us that we had reached the edge of the trench. We had found the perfect spot; later computations revealed that the drill site was less than 500 m from the base of the rock wall (Figure 7.8). We could use the American expression that we had stopped our 11,000-ton vessel "on a dime." Years later, we were considered lucky to have found the spot. Yes, there was some luck involved, but precise navigation was the secret of our success!

After the nerve-racking experience in navigation, the drilling was almost routine. Very rapidly we penetrated the trench sediments, which are less than 500 meters thick here. They are all very young. The top sediments include some coarse debris, slumped down from the steep northern wall of the trench. The older sediments are marls, a mixture of terrigenous mud and dead skeletons of nannofossils.

At noon on 8 September, the drill bit hit something hard; the rate of penetration dropped sharply. We asked the driller to cut a core, because we knew we must have reached the buried rocky wall of the trench. The core told us that the trench wall here is underlain by a dolomite, a carbonate of calcium and magnesium. The rock is fragmented and recemented, as one might expect at the spot where Africa is plunging down under Europe.

Cita and Stradner, our micropaleontologists, were used to working with soft sediments; microfossils are sieved out, and a tiny speck of ooze smeared on a glass slide will reveal, under a microscope, the presence of thousands of nannofossils. They could not work with this rock. We had to cut a thin slice of the rock and grind it down into a thin film (0.03 millimeter in thickness) so that the thin section could be studied under a microscope.

All evening I sat in the laboratory, peering through a microscope. Patiently I searched, in vain, for traces of fossils. Cita had sympathy for my plight. "Few fossils are ever found in dolomite rocks," she said. "Fossil skeletons are gone when lime mud is changed into dolomite." Nevertheless I persevered.

Toward midnight, I finally saw something, tiny round shells of some kind. I rushed down to the paleo lab and found our micropaleontologists. Neither Cita nor Stradner could help me, but they thought that Wolf Maync might be able to do some-

7.8. Sediments of the Hellenic Trench. The active plate-margin is overlain by very young sediments of the trench, more than 450 m thick. The deep-sea drilling during Leg 13 penetrated the Quaternary sedimentary cover and discovered underthrusting at this margin.

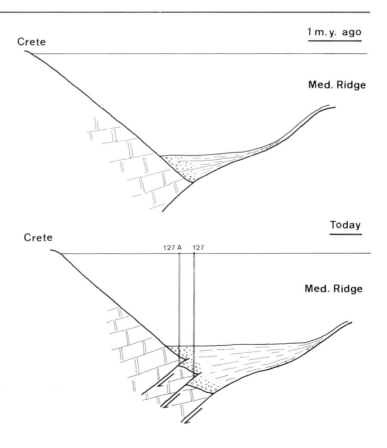

thing with that creature. But more rock slices had to be cut, and more thin sections made. Ryan had to do that, because our marine technicians could only handle oozes; the co-chief had to be a jack of all trades.

I went to wake up Maync. His specialty was identifying the single-celled foraminifers that live on the deep-sea bottom. We had not come across many of those samples since we got into the Mediterranean. Now he was eager to go to work.

When I got up the next morning, Maync told me the good news—he had found many foraminifers in the thin sections provided by Ryan. There was no question about it: the dolomite was a Cretaceous rock, about 100 million years of age. In fact, he knew that kind of dolomite very well, because it is quite common in the Alps, the Carpathians, and the Hellenides, the mountains of Greece.

This Cretaceous sediment was originally made up of fossil skeletons of organisms that lived in shallow seas, near a coral shoal perhaps. The so-called *Urgonian* facies of shallow marine carbonates had long been interpreted by geologists as the sediments deposited near the northern shores of the Tethys Ocean. Having been lithified into limestones and dolomites, they form the cliffs of many mountain peaks. Every tourist coming back from skiing holidays and stuck in the traffic jam near the Lake of Walenstadt in Switzerland, for example, can glance out the car window and see these massive carbonate rocks on the Alpstein Mountain north of the lake.

The Cretaceous dolomite was deposited on the European margin. After the Tethys was eliminated by subduction, Africa continued to move forward, and rode on top of Europe. The marginal sediments were scraped off from their basement and pushed forward, deformed as they were displaced, to form the beautiful folds of the High Limestone Alps of Switzerland. But what was such a Cretaceous rock doing down there in the Hellenic Trench?

The plate-tectonics theory provided the answer: Africa got on top of Europe in the western Alps, but the African Plate got under the European Plate in the eastern Mediterranean. The Mediterranean seafloor, a part of the African Plate, is thrust under the southern margins of Europe. Such an underthrust has managed to drag the outer European margin down to its present abyssal depth. The shallow carbonate deposit on the former Tethyan shore was thus taken down to 5,000 m. Someday, many millions of years from now, after the island of Crete and Egypt have collided, the Cretaceous dolomite under the trench could again be raised up, and could perhaps even form cliffs on the sides of high mountains.

Our penetration through the dolomite rock was slow, but we had learned patience. Suddenly, in the early hours of the ninth, the drilling speeded up. We took a core then and found a white ooze. Cita and Stradner could soon give us a middle Pliocene date as the age of the sedimentation of the ooze. It was fantastic! We had found Pliocene ooze under the trench wall!

We now had the evidence that we had come here in search of. We had a core with the older Cretaceous rock of the European Plate above the younger trench sediment of the African Plate. The elation was great. The new theory had again triumphed. The _Glomar Challenger_ had not failed the challenge.

Our other drill sites on the Mediterranean Ridge were the icing on the cake (Figure 7.8). We dug out information to indicate that the sediments on the Mediterranean Ridge had been laid down on an abyssal plain before they had been heaved a couple thousand meters to their present subsea elevations. This is an example of an embryonic mountain caught in the collision of Europe and Africa—an "infant" under the sea now, but destined to rise higher than the Alps!

8 Swallowing Up of the Ocean Floor

William Smith gave a new dimension to geology, but his follower Murchison did much to ensure that the geology in the century after him, like the geology of the era in which he had learned the science, would require no formal training. Geologists were ending up in blind alleys. The revolution of the sixties was not only conceptual but also methodological, and the new geology was oriented less toward map geometry and more toward understanding processes.

Smithian Stratigraphy

We all have our heroes in science. My hero is William Smith. He gave us order in timing—he gave us stratigraphy.

Layered rocks are beds or strata, and the study of strata is called stratigraphy. Unlike the natural phenomena described in physics or chemistry, the earth's history is too complicated to be expressed by formulas or digital models. Stratigraphy is an analog model of the earth's history, and "laws" in stratigraphy are expressions of common sense. Nicolas Steno (1631–1687), a bishop of Florence, stated the obvious when he formulated the "law of superposition": the oldest sedimentary layer is found at the bottom of a pile.

Roadcuts are rarely more than a dozen meters high, and even the Grand Canyon is only a few thousand meters deep. Nowhere do we find a sequence of strata, or a stratigraphic section, that encompasses the oldest and the youngest ever deposited on earth. The second common-sense observation, that a sedimentary layer has a certain horizontal extent, is now called the "law of lateral stratal continuity"; a layer cropping out at one locality should be sufficiently continuous and be found at another outcrop not too far away. A limestone containing mollusk fossils, for example, may be the highest (or youngest) bed at one roadcut, but the lowest at the next. Stratigraphical correlation is the methodology of certifying that the limestone exposures at the two roadcuts belong to one and the same bed. Through correlation, we could link up segments of local succession to patch together a sequence.

One of the earliest such stratigraphic sequences was recognized in the eighteenth century by Johann Gottlob Lehmann and George Christian Füchsel. They noted three rock units, always in the same descending order, at different places in Thuringen, Germany, namely:

variegated shales, marls, limestone, and gypsum

limestone with abundant mollusk remains

red sandstone

The gypsum was always near the top, and the sandstone was always at the bottom. Keuper, Muschelkalk, and Buntsandstein were the names eventually given to designate those units, and the whole group was called the Triassic, referring to its threefold division.

Lehmann and Füchsel recognized the value of rocks as a record of the ancient history of the earth. Füchsel was one of the first to propose that continuous series of strata of the same composition constitute a formation, which is now defined as a rock unit that can be put on a geologic map. He did not worry about intercontinental stratigraphical correlation. Naturalists of his time did not travel far, and they thought sedimentary formations were universal and everywhere the same. When traveling became less hazardous, geologists realized that the three formations of the Triassic are not present everywhere. Leopold von Bush went to the Jura Mountains at the beginning of the nineteenth century and was able to find a semblance of the German Triassic there, but no similar succession of formations was found south of the Alps. Are there rocks elsewhere that were deposited during the same time interval as the Triassic of Germany?

As a young man just finished with his apprenticeship, William Smith was hired to do surveying for a network of inland waterways to be built by the British government for the transport of coal. He joined the project in 1794 and for the next six years surveyed every segment of the course and supervised construction. Excavation laid bare outcrops on both sides of the newly dug channel, but almost no two sections showed exactly identical sequences of strata. Limestones, shales, sandstones, chalks, and clays, like strange faces in a crowd, seemed to possess no distinctive features by which they could be identified. Yet, after six years of growing familiarity, Smith realized that each stratum contained its own peculiar fossil remains, and that the sequential relation of the fossil faunas along the excavated channel was everywhere the same. Using fossils as a guide, Smith published in 1815 a geological map of England and Wales, and a year later an article entitled *Strata Identified by Organized Fossils*. Thus was born the science of *stratigraphy*, the study of strata, and Smith gave us the third law of stratigraphy: the law of paleontological dating.

King of Siluria

Every revolution was preceded by a tyrant. There was Ptolemy before Copernicus, and there was the biblical authority before Darwin. I often wondered who had the *après nous le déluge* mentality that eventually unleashed the earth-science revolution of the 1960s. Certainly not James Hutton, nor Lyell; the young rebels of the 1960s fought the dogma and brought forth a renaissance of the Huttonian uniformitarianism. I recently chanced to read a biography of Sir Roderick Impey Murchison, Bart., K.C.B., 1792–1871, by Robert Stafford. Now I think I know who the culprit was!

I did some work in 1971 investigating the history of research on *Flysch,* which is a rather unusual type of sandstone first described in the Alps. I came across Sir Roderick and his 1848 monograph *The geological structure of the Alps, Apennines, and Carpathians, more especially to prove a transition from Secondary to Tertiary rocks, and the development of Eocene deposits in Southern Europe*, written after a summer "invasion" of central Europe. The title page was not big enough to record in full all the honors received by this great man, and he still had another 23 years to add to his harvest of medals and titles. It was no wonder that a Swiss schoolteacher, Bernhard Studer, should have discarded his correct interpretation and accepted Murchison's authoritarian opinion, and this mistake was to hinder the progress of Flysch research for a century. Murchison's biographer tells us that the great man was not only vain and authoritative; he was also more a self-serving soldier-politician than a scientist.

I shall use a technique of Chinese poetry, and string key words and phrases together to convey an impression of Murchison's life. Imperial heritage; performing poorly in the classroom; deficiency in mathematics; military career; riding, hunting, drinking bouts; debt; gay life; shooting partridges with Sir Humphry Davy (1823); study of science for the rising bourgeois (1823); admission as a Fellow of the Geological Society (secured by Davy, 1825); beginning of independent field work; election as a Fellow of the Royal Society (secured by Davy, 1826); reliance on paleontological dating; striding at such a relentless pace that he walked the legs off the younger Lyell; frontier skirmish; campaign of pacification and annexation for the "invasion of Graywacke"; bitterest controversies; inductive accumulation of data; against theory; president of the Geological Society (1842, 1869); vice president of the Royal Society (1849); president of the Royal Geographical Society (1843, 1851, 1862–71); Silurian; Devonian; Permian; Grand Cross of St. Stanislaus; Imperial Academy of Science (Russia); alliance between science and national interest; director-general of the Geological Survey (1855–71); founding of colonial surveys; conquests on a scale commensurate with his ego; K.C.B. (1863); baronetcy (1866); mapping; discovery of exploitable minerals; arch-patriot and consummate promoter; the Geological Survey as a global arena for self-fulfillment; general in science like Napoleon; glory in the number and power of his clan; ability to make or mar careers; authority to enforce conformity; exercising his omnipotence; utilitarian quackery of the age; "far wider reputations than the real men of science"; influence on honors, publication privileges, and so on; patronage of survey and academic appointments, etc.; bargain between science and the forces of expansion (of the empire); his mind comparable to "a Silurian matrix, impervious and resistant"; British natural science as a gigantic looting operation that helped maintain British ascendancy; desire for new data, careers, classificatory conquests, and power in administrative affairs.

I could go on and on, but that is not the point. Suffice it to say, if William Smith is my hero in geology, then Sir Roderick would be the antipode; he was everything that William Smith was not.

Murchison had a fatal influence on the development of geology. "Instead of promoting the development of the new increasingly 'scientific' facets of geology which he found so bewildering," his biographer tells us, "Murchison ensured that the geology in the century after him, like the geology in the era in which he had learned the science, would require no formal training." I know, because I grew up in the next century, when the best geologists were those who could walk the legs off their younger assistants, when data gathering was considered fact-finding, and when the formulation of testable theories was dismissed as idle speculation.

Sir Roderick did, of course, have some redeeming features. His friendship with Sir Humphry Davy would not have carried him that far if he had not jumped on the bandwagon invented by William Smith; his reliance on paleontological dating was the key to his success. We should recognize that William Smith's discovery of biostratigraphy was a key contribution to the revolution in science that established geology at the beginning of the nineteenth century. Murchison was perceptive enough to mop up with the new paradigm. Abraham Gottlob Werner was wrong. Universal formations cannot be defined by lithology, or by the kinds of rocks making up a formation; universal formations had to be identified by fossils. Fossil organisms lived and died, and they evolved with time, just like kings and queens came and went. Universal formations are dynasties in the history of the earth. Trilobites, ammonites, and mammals appeared and disappeared like the Plantagenets, the Tudors, and the Stuarts; their ages give the age of their epoch.

Quitting his life as a soldier-playboy, Murchison at age 33 threw all his energy into conquering new kingdoms and establishing new dynasties. Murchison went to Wales and founded the "kingdom of Siluria." Having recognized the characteristic faunal assemblages of the Silurian rocks of the British Isles, Murchison then went to Russia and central Europe, and the "kingdom of Siluria" expanded. His sent his disciples to Africa, Australia, Asia, the Americas, and Antarctica, and the flags of Siluria began to flutter on every continent. Murchison had in the Silurian a truly universal formation.

The great conqueror did not, however, have enough. He was not able to subjugate his friend Sedgwick's "kingdom of Cambria," but he was quick to establish another: the Devonian. Like the Silurian, the Devonian was to become global. Global "formations" acquired then the dignity of being called systems; thus the Silurian System, the Devonian System, and so on were established.

Murchison still did not have enough. He went to the Urals and founded the Permian. Rocks yielding Permian fossils used to be called upper Carboniferous, but Murchison wanted another kingdom, another universal formation, another system. Independent-minded North Americans resisted the new aggression; they did not want to cede the Permian to Murchison, not immediately, anyway. Only after the turn of the century was the Permian System officially recognized by the United States Geological Survey.

There are other systems that were not named by Murchison. There is the Ordovician, a "buffer state" between Murchison's Siluria and Sedgwick's Cambria. There is the Carboniferous, referring to this system's richness in coal. There is the Cretaceous, a system of chalk formations. There are, of course, the Triassic and Jurassic, originating in central Europe. Finally, there are the two eighteenth century relics from Italy, Tertiary and Quaternary—leftovers from a pre-Murchisonian world before the "kingdoms of the Primary and Secondary" were occupied by Siluria, Devon, and others. Rocks older than those of the Cambrian System are called Precambrian. Now the names are codified in every geology textbook, and their age given in millions of years (Table 5.1).

A different system of terminology was proposed by another Englishman, J. Phillips, in an article for the 1840 *Penny Cyclopaedia*. "Kingdoms" (systems) now were assigned to various "empires" (eras), Paleozoic, Mesozoic, and Cenozoic, signifying the ages of ancient, middle, and new life forms. The Paleozoic oceans were populated mainly by invertebrate animals and by fish, while land plants began to evolve on continents. The Mesozoic Era is well known as the Age of Reptiles, when dinosaurs dominated the earth. The Cenozoic Era is the Age of Mammals, in which evolution has proceeded to its most advanced state.

For the most recent era, Charles Lyell suggested a subdivision of the Tertiary and Quaternary on the basis of comparing fossil mollusk assemblages with their living counterparts: older Cenozoic rocks have few fossil species that are still extant, whereas the youngest rocks have mostly modern species. Lyell's terms Eocene, Miocene, and Pliocene designate the epochs of the dawn of, the less recent, and the more recent life forms, respectively. Later, the Paleocene, Pleistocene, and Holocene epochs were introduced to designate the ancient (more ancient than Eocene), the most recent, and the wholly recent life forms (Table 5.1).

William Smith and Murchison gave us order in geology. Their method of mapping became the basic—and for many the only—methodology in geology. We go to the field and look for outcrops. We correlate rocks at different outcrops, guided by the laws of superposition, lateral continuity, and paleontological dating.

Non-Smithian Stratigraphy

After we were married in 1958, my late wife, Ruth, was transplanted to the flat plains of Texas, and she felt alienated; she liked oceans and mountains. I was a research scientist for Shell Oil then, but I maneuvered to start a project on the Franciscan rocks of the California Coast Ranges.

The Franciscan used to be called basement, the foundation of sedimentary strata. Basement rocks are "hard" rocks that rarely have the porosity to store hydrocarbons. Petroleum geologists do not study basement. After successes with my first two projects, however, my credibility with the management was such that I was able to convince them that Franciscan geology was relevant to oil exploration. The family moved to Morro Bay, California, and we were able to spend the spring and summer of 1963 on the coast and have some idyllic days there with our three preschool children a year before Ruth's death.

I had been fascinated by the Franciscan as a graduate student in California, and I had taken a number of excursions to the Coast Ranges. The mountains are full of blocks of different rocks, green, gray, red. A number of them seemed to be strange companions indeed, and brought together by some magic! For several decades geologists skirted the problem by calling those rocks "basement," and walked past them when they should have taken a closer look. Some of those "basement" rocks are layered like the sedimentary formations mapped by William Smith. Occasionally a Boy Scout or an amateur collector will find a fossil in the Franciscan. Now, William Smith would be puzzled, because the fossils in the "basement," more often than not, are younger than those of the immediately overlying sedimentary formation. The law of superposition does not seem to apply.

"Paleontologists don't know what they are doing," my favorite professor said, dismissing the problem lightly.

It is easy to dismiss troubling observations if you are not looking for solutions. Now I had to work with the problem. I went first to San Francisco, because such rocks were first described there by Andy Lawson, one of the most famous geologists of the first half of the twentieth century. What did he say?

Using the methodology of Smithian stratigraphy, Lawson divided the group into five formations. He thought the Franciscan rocks, in their normal stratigraphic superposition, were folded. The basalt and serpentinites were considered igneous extrusives and intrusives.

This time-tested Smithian methodology seemed, however, inapplicable to the study of the Franciscan rocks. Lawson's formations cannot be recognized outside of the San Francisco Peninsula area. Some sedimentary formations do have some lateral stratal continuity, but they still cannot not be traced for any distance. Paleontological correlation was hardly possible, because those strata rarely yield fossils. When fossils were found now and then, the normal superposition assumed by Lawson was proven wrong again and again.

Worse still, in many places Franciscan rocks do not seem to have much lateral continuity at all. Even a casual observer visiting San Francisco could notice that the green rocks on Nob Hill look quite different from the red strata on Twin Peaks. The green rocks are ophiolites, the red are radiolarite, and they are all a part of the Franciscan. We now know they are ocean crust and ocean sediments, which should have covered a large area. Yet we found flysch sandstones and shales on the way to Nob Hill and Twin Peaks. The same kinds of green and red oceanic rocks are present up and down the coast, but their outcrops at any one place are very limited in lateral continuity.

Deprived of the standard working tool of stratigraphical field mapping, and facing contradictory paleontological evidence, I felt helpless. But the die was cast. The family was on the coast, and I had to turn in a report to the company.

The discovery I was to make belies the common wisdom that hard work produces success. Initially I followed the dictum and plunged into a tedious fight against sagebrush on steep mountainsides under the California sun. I could not make sense of the chaos. I would find a green ophiolite at one roadcut, red chert at the next, and a buff-brown sandstone at the third. After weeks of hardship, I was rewarded only with more confusion.

Discouraged and disgusted with my inability to crack the problem, I took a working day off and went with the family to the seaside near San Simeon. I took a seat on a Franciscan sandstone while we had our picnic. Looking across, I saw Ruth sitting on a green "pillow lava"—a special kind of submarine volcanic rock—and the children climbing the steep face of a red-chert cliff. It suddenly dawned on me that I was looking at rocks that had undergone a traumatic experience: they had all been broken up, badly broken up (Plate IX).

The oceanic rocks were once regular layers. They are solidified basalt lavas, lithified sands, muds, and ocean oozes, and they should have originally been superposed in an orderly fashion. Somehow, they got caught in the jaws of a colossal rock-crushing machine and were torn into millions of pieces, ranging in size from centimeters to kilometers. The broken fragments of different sorts and various sizes were then chaotically mixed together in a matrix of rock flour, and shoved over one another, pretty much in the fashion of rock debris and mud scraped under and mixed together by a moving bulldozer!

That was the secret of the Franciscan rocks. A big part of the Franciscan no longer consists of regularly layered strata; nor is the Franciscan made of basement. Rocks in the Franciscan constitute a mélange (this was the name I adopted)—a mixture of many different kinds, all sheared, broken, and deformed.

We cannot study mélanges, or map mélanges like William Smith taught us to do, because the shearing, fracturing, and mixing have rendered the Smithian stratigraphy irrelevant; all three laws of stratigraphy are not applicable. One has to accept these facts and search for a different methodology in order to study mélanges. One should also ask different sets of questions.

**Missed
Opportunity**

I was very excited about my findings then. I wrote up my conclusion in a short contribution, which, as I mentioned previously, was an unsuccessful candidate for a limelight spot at the 1966 meeting of the Geological Society of America. I had no chance when Vine was giving his epoch-making talk on the seafloor-spreading theory.

Neither Vine nor I realized then that we were looking at two sides of the same coin. He was talking about the spreading of the Pacific Ocean, at a rate of hundreds of kilometers per million years. Ocean crust is created anew year after year, and the addition, or _accretion_ as we say in geology, of new crust should make the surface area of the earth greater and greater. Could that be possible?

Yes, it is possible; the surface area will increase if the earth has been expanding.

Like the theory of continental drift, the basic assumption is charmingly simple and elegant. We know that only one fourth of the earth's surface is covered by continental crust; the other three fourths are covered by oceanic crust. We could postulate a primeval earth enveloped entirely in continental crust. Ocean was produced when the earth expanded and increased its surface area.

The theoretical basis was a postulate in 1937 by Nobel Prize–winning physicist P. A. M. Dirac that the universal gravitational constant (G) may have decreased since the beginning of time. If so, another physicist, R. H. Dicke, reasoned in 1957, the radius of the earth should have increased at a rate of 0.1 mm per year. A year earlier, L. Egyed had estimated that the earth expanded at a rate of 0.5 mm/yr.

One prominent geologist well known for his advocacy of an expanding earth was S. Warren Carey. Carey was a supporter of the theory of continental drift, but he was bothered because Wegener could not come up with an acceptable mechanism. Now, in the early 1950s, Carey learned of Heezen's discovery of mid-ocean ridges (see chapter 9). He may have remembered that Walter Bucher had compared large global fractures to expansion cracks. Combining new observations with the old theory, Carey formulated his extreme theory of an expanding earth. The earth had expanded twofold, Carey believed. The radius of the earth had become 1,600 km longer during the last 200 million years of accelerated expansion!

The theory had a fascination, for the lay public especially. A classmate of my daughter's in high school wrote a term paper on the theory of an expanding earth. She read Carey's 1956 book _Continental Drift_ and was impressed. Since she could not understand the geological arguments against Carey, she thought I was just another reactionary obstructing the progress of science.

"Carey was right about the theory of continental drift," she argued. "You people finally got around to admitting that he was right. Now you still can't bring yourself to accept his expanding earth."

"Continents did not drift. Wegener was wrong. Carey was wrong. Continents did shift their positions as the plates moved."

"You are indulging in semantic triviality," the strongminded young lady said, dismissing my argument.

Carey's belief in an expanding earth might be attributed to the fact that he never did work in the California Coast Ranges, or any other terrain underlain by a mélange. The origin of mélanges can best be explained if the earth has a finite size.

Carey asked what had happened to all the seafloor that had been created during the last 200 million years. If I had gone to Vine's talk on seafloor-spreading, I could have said in 1966 that the crust had been swallowed up.

No, the earth did not expand. New seafloor is created on a spreading ridge while the old is disappearing down into the earth's mantle to make mélanges.

Benioff Zone and Subduction

The Pacific does not have a mid-ocean ridge, unlike the Atlantic and the Indian Ocean. The Pacific seafloor is being ripped apart at the crest of the East Pacific Rise, which lies closer to the Americas. This lack of symmetry of the Pacific seafloor is now explained by the theory that the east side of the Pacific has disappeared under the continents of North and South America. In the jargon of plate-tectonic theory, we say that the Pacific Plate has been subducted under the North American Plate along a Benioff zone.

That was the explanation of Morgan, McKenzie, and other theoreticians of the new school. Earthquakes along the Benioff zone under the western margin of North and South America is the geophysical evidence. The Franciscan mélanges of the California Coast Ranges is the geological evidence. However, back in 1966, I was not aware of the fact that my discovery of the coastal mélanges was a corollary of the theories of seafloor-spreading and plate-tectonics.

In December 1969, the Geological Society of America sponsored an interdisciplinary conference to discuss the new plate-tectonics theory in Monterey, California. Some 90 geologists, geophysicists, and oceanographers from all over the world were brought together. I was invited to come back from Switzerland to lead a field excursion to the beaches where I had once spent many happy days. This time, I met with spontaneous acceptance, and the term mélange is now a "household word" in geological literature. Mélanges are found in coast ranges, where the Pacific Ocean is subducted under continents. Mélanges are found in the Alps, and in the Himalayas, where an ancient ocean was swallowed up between colliding continents.

At the Monterey conference, I met Roland von Huene again. He was a student at UCLA when I was a teaching assistant there. We met again in 1957 in Europe and went skiing together in Salzburg. After he finished his degree, he joined the Geophysics Branch of the U.S. Geological Survey and made a career studying ocean trenches, particularly those bordering the American continents. Like myself, von Huene had also been skeptical of the new theory. He had had the same professors as I. Furthermore, he also had some data that seemed to argue against the assumptions of plate-tectonics: trenches are supposed to be compressed and thrust under continents, but von Huene's seismic profiles showed that the trench sediments are not deformed.

At Monterey, we spent a few quiet moments together away from the madding crowd. By then, I was already a convert. I told him, like others had probably also told him, that the trench sediments are not deformed because they are very young, too young to be deformed by subduction. The subducted trench sediments and their underlying ocean floor would be deformed; they would make mélanges under the steep trench wall, where they cannot be detected by seismic profiling.

Von Huene agreed to such a possibility, but he was bothered. Subduction is supposed to be occurring off the west coast of North America. Leg 5, led by Dean McManus of the University of Washington, was out there drilling in April and May. They drilled into Miocene rocks, but the sediments were mostly flat-lying. Von Huene could not understand why the soft sediments of Miocene and Pliocene age are not folded or faulted if the Pacific Ocean floor is being thrust under the North American continent.

Deformation takes place only after ocean sediments are dragged into the Benioff zone, I told him, using the argument of the plate-tectonic theory. Von Huene understood the argument. Still, he wanted to see some proof; he wanted to drill into the deformed sediments under the continental slope or inner wall of a trench.

Half a year later, Ryan and I took the _Challenger_ close to a margin of plate subduction. We did find, as I described in the previous chapter, the underthrust that brought young trench sediments under the older rocks of the inner trench wall, but we did not have enough shiptime to drill into the underthrust sediments. Ryan and I were convinced that the Mediterranean seafloor has been tucked under the island of Crete. Others, however, continued to be skeptical. Additional tests of the theory were necessary.

Von Huene's chance came in 1971, when he and Laverne Kulm of Oregon State were co–chief scientists of DSDP Leg 18. The _Challenger_ left Honolulu, Hawaii, on 29 May 1971 and headed toward the coast of the Pacific Northwest. The East Pacific Rise offshore from Washington and Oregon, whose local name was the Juan de Fuca Ridge, was where Raff and Mason first discovered magnetic lineations, and where Fred Vine and Tuzo Wilson first correlated the width of those lineations with the duration of epochs of magnetic polarity reversal. According to the plate-tectonic theory, the Pacific seafloor on the east side of the ridge should plunge under and be swallowed under the North American continent (Figure 8.1) Several indications supported the theoretical predictions: the Benioff zone of earthquakes had been recognized in the right place; mélanges of deformed rocks in an ancient Benioff zone, very similar to the Franciscan rocks, are exposed on the coast of Washington; finally, the lavas that made Mount Rainier are the type one expects from the remelting of a subducted lithosphere. The only problem was the absence of a trench, which is almost always present where an ocean floor is thrust under a continent. Instead there is a pile of marine sediments, spreading out from a submarine canyon as a fan of sediments—the Astoria Fan. This is explained by the fact that the Columbia River has carried so much detritus to the ocean that it has filled up the hole that would otherwise be there; we thus have a submarine fan—the Astoria Fan—instead of an oceanic trench (Figure 8.1).

Seismic profiling had shown that the Pacific seafloor plunges eastward under the thick fan sediments. Yet the fan sediments themselves are flat-lying, not disturbed by deformation. How could those soft sediments escape the fate of being smashed and

8.1. Astoria Fan. No ocean trench, but a submarine fan, is present at the Pacific margin west of Washington State. Leg 18 drilling was scheduled to investigate why the fan sediments are not deformed.

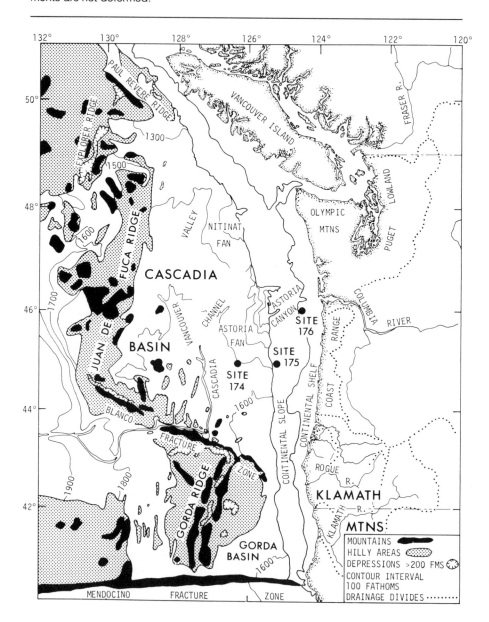

smeared into a chaotic mélange if the Pacific has been thrust under North America? That bothered von Huene, as he told me during the Monterey meeting.

The plate-tectonic theory predicted that the flat-lying sediments are so young that they have not yet been deformed. Only the sediments on an older Astoria Fan were indeed crumpled when their underlying ocean crust plunged into the Benioff zone. If this explanation was right, the rate of accumulation of the fan sediments must have been fantastically high. Also, the disturbed sediments at the foot of the continental slope should be the deposits of an older Astoria Fan. Leg 18 drilled several sites to test the prediction.

At Site 174, the drilling penetrated sediments as old as Pliocene. As the theorists had predicted, the oldest sediments were deposited on abyssal seafloor above newly created ocean crust. As the crust moved eastward toward North America, it was bent down to form a trench. But then, terrigenous detritus from the Pacific Northwest poured in and filled up the potential depression. Paleontological dating of drill samples indicated that sedimentation on the fan began about a million years ago. The accumulation rate was indeed, as predicted by the theory,

8.2. Deformation of the Astoria Fan. Leg 18 drilling found that the youngest sediments of the Astoria Fan at Site 174 are not yet deformed, but older fan sediments at Site 175 are deformed, where the ocean lithosphere (acoustic basement) is thrust under the North American continent.

LITHOLOGIES

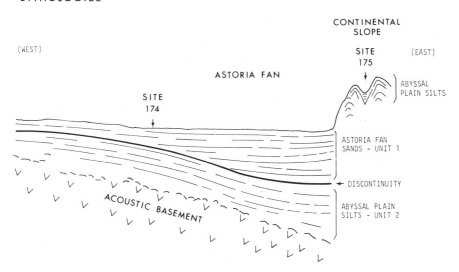

fantastically high, reaching a maximum of 1,000 m per million years, or about 100 times the normal rate of ocean sedimentation. The sediments under the Astoria Fan are flat, because they are too young to have suffered deformation. However, the fate of those sediments is sealed: in the not too distant future, after they are dragged down by subduction, those sediments will be chaotic, like the Franciscan rocks of the California Coast Range!

Hole 175 was drilled at a site near the base of the continental slope, some 800 m above the abyssal plain of the Astoria Fan. Von Huene found his deformed sediments; those originally abyssal-plain deposits have been heaved up (Figure 8.2). The deformation is consistent with the postulate that they were scraped off and compressed as their underlying crust plunged under the North American continent some one to two million years ago. The sediments are mildly folded, but they are not yet mélanges; they have not yet plunged into the jaw of the Benioff zone.

Glomar Challenger next sailed northward, to drill the Alaskan Trench. A deep trench was there, instead of a submarine fan of sediments, because the Aleutian Islands could not deliver enough detritus to fill it up. Four sites were drilled. Again, the same story: the floor of the trench was oceanic crust created at a distant spreading ridge. Like a conveyor belt, the seafloor was moved steadily northward at a rate of 6 cm per year, while receiving a rain of skeletons of foraminifera, nannoplankton, and radiolaria to cover the basalt basement. Finally, after some 50 million years of travel, the crust under Site 180 reached the outer edge of the Aleutian Islands and began to be filled up rapidly by detritus from continents at a rate of 100–200 m per

8.3. Accretionary prism of Alaska. Leg 18 drilling penetrated turbidite sediments at Site 181. They were originally deposited in the Alaskan Trench, but are now uplifted and form an accretionary prism under the inner wall of the Alaskan Trench.

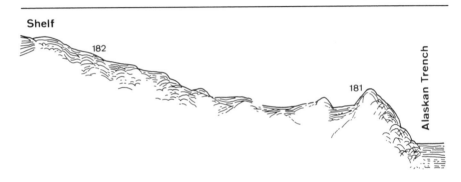

million years. All the time, the trench sediments on the inner edge were being crumpled as they were being detached from their underlying crust because of the subduction processes. At Site 181, on the inner slope of the trench, the drill string of *Glomar Challenger* penetrated deformed sediments that had been heaved up a few thousand meters from the depth of the trench floor, at a place and in the manner predicted by the theory (Figure 8.3).

Von Huene, Kulm, and their associates returned to Kodiak, Alaska, on 20 July. They had their baptism on *Glomar Challenger* and became new converts to the plate-tectonic theory. Von Huene was to undergo a radical change in his scientific philosophy, as I had two years earlier after my first tour of duty on the *Challenger*, in the South Atlantic. He became an ardent supporter of the plate-tectonic theory, and he would chair the JOIDES Active Margin Panel for many years, and participate actively in the drilling of Pacific margins during the IPOD, International Phase of Ocean Drilling, which will be discussed in chapter 18.

9 Marginal Seas

*Marginal seas owed their origin to oceanization. So geologists
thought until Dan Karig pointed out the obvious: small oceans, like
big ones, owe their genesis to seafloor-spreading. The predictions
were first verified by the "hit-and-run" Leg 6, led by co-chiefs Al
Fischer and Bruce Heezen, "a Mayor of Casterbridge."*

**Marginal
Seas and
Back-Arc
Basins**

"Earth scientists have long been fascinated by the complex
marginal seas, volcanic island areas and deep-sea trenches. In
fact, each generation of geologists since the turn of the century
has repeatedly recast the trinity of features into various models
of the mountain-building process." So reads the introductory
statement of the cruise report of DSDP Leg 31, coauthored by
Dan Karig and Jim Ingle, co–chief scientists, and their scientific staff.

I have described the triumph of the plate-tectonic theory,
with assistance from *Glomar Challenger*, in explaining the origin of the trenches and arcs. Now the drill vessel was called
upon to tackle the third of the "trinity of features."

The origin of basins beneath the marginal seas has indeed
been a favorite subject for speculation. A prominent Soviet
geologist, V. V. Beloussov, seemed to have an emotional attachment to the dogma that the earth's crust could mostly
move only up and down, and not very much sideways. Beloussov borrowed the idea, first postulated by the Dutch geologist van Bemmelen, that continental crust could somehow be
oceanized, and in the early 1960s he started to make marginal
basins by playing with the word oceanization. He was convinced that those basins were originally continental margins,
but were oceanized when granite in the crust was somehow
converted into basalt; the continent then sank like the mythical
Atlantis!

I knew Beloussov's lesson all too well, because in those
years I was groping my way down the same blind alley. I

even gave that 1963 talk at Princeton at which Hess was too polite to tackle me. I also wrote a venomous critique of Robert Dietz's paper on seafloor-spreading, and this was my claim to fame during the initial scrimmages of the earth-science revolution—my only citation in Bill Menard's *Ocean of Truth*.

I was, in fact, not exactly wrong, but my scenario was incomplete and misleading. I discussed the possibility that ancient highlands could be raised at a time when the underlying mantle is heated up. Erosion could then go to work and cause the continental crust of the region to become unusually thin, such as that under the Great Basin of western North America. Sooner or later, the heat source is gone, the upper mantle cools down, and its density returns to normal. The worn-down mountain would eventually sink below sea level (Figure 9.1).

I postulated this mechanism to explain the subsidence of the Atlantic coastal margin of North America. I postulated this mechanism to explain the Aegean Sea, a marginal sea. I still think it is a possible mechanism, but I do not think it is the only mechanism. It is almost certainly not the most significant cause of the thinning of continental crust that converts land into marginal sea. That the crust under marginal sea is thin is not so much a consequence of removing material by erosion; more significantly, it is related to extensional stress in the lithosphere.

Marginal seas are commonly underlain by deep basins behind island arcs. The Aegean Sea is an example close to home; it is situated behind the arc that stretches from the Peloponnese to the islands of Crete and Rhodes—the Hellenic Arc. The Caribbean Sea is another example, separated from the Atlantic by the arc of the islands of the Antilles. Marginal seas are more common in the Pacific: the Bering Sea behind the Aleutian Islands, the Sea of Okhotsk behind Kamchatka, the Sea of Japan behind the islands of Japan, the East China Sea behind the Ryukyu Archipelago, the Philippine Sea behind the Mariana Arc, and so on. The current extensional theory of the origin of marginal seas, like many other new ideas that ushered in the revolution in earth sciences, was the brainchild of a graduate student—Dan Karig, then studying at the Scripps Institution of Oceanography.

Karig was an unassuming young man and a wizard with tools. My wife, Christine, was particularly appreciative of that talent of his. In 1972, Jerry Winterer and I both took a sabbatical leave. He and his family came to Zürich and stayed in our house; we went to live in theirs in La Jolla. I am notoriously inept at fixing things around the house. After six months it seemed that we had wrecked everything. The radiator of the car was always boiling, and no garage mechanic could cure the malady. The dishwasher had stopped functioning. The refrigerator had burnt a fuse. The garden door had come off its hinges, and so on. We felt very bad about having to leave things in such a mess before going home. Just then, a young man showed up and introduced himself as Dan Karig. He was to stay in the Winterers' house for the two weeks after our departure.

9.1. Subsidence of continental margin. Supracrustal thinning by erosion and mantle-density increase could be a cause of the subsidence of continental margin. Note that block no. 3, which was uplifted and eroded during the initial stage of rifting, would subside eventually down to a depth of 3 to 5 km to form the outer continental margin of continental block no. 1. Note that the mantle density is 3.27 g/cm^3 prior to the rifting (*a*) and at the end of the subsidence (*d*), and that the density is 3.11 g/cm^3 during rifting (*b*) and seafloor-spreading stages (*c*).

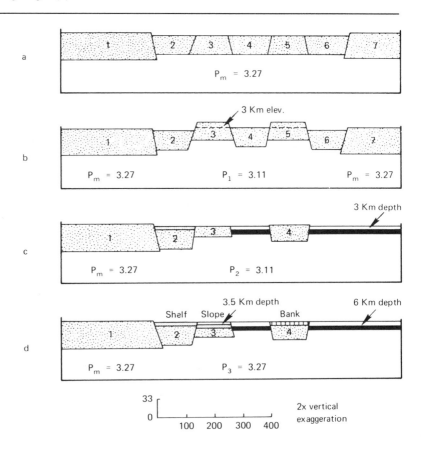

Christine pounced on him with our woes. Karig looked, searched, and worked. All the repairs were done in two hours, and we were soon able to leave with everything in the best of order and a clean conscience!

Karig's ingenuity with tools must have helped when he went on oceanographic cruises. However, he needed more than his hands to solve the puzzle of marginal basins. I asked Karig years later how he got involved in that problem.

"It was Bill Menard," Karig said. "He had a seminar and I had to give a presentation on the origin of marginal basins."

In 1967, when geologists were ready to concede to Vine and Matthews and accept their theory explaining the origin of ocean basins, they still clung to Beloussov and his oceanization hypothesis to account for the genesis of marginal seas. Karig did not like Beloussov's idea, because the hypothesis does not explain the fact that a marginal sea is where land used to be. The oceanization hypothesis is circular reasoning: The earth's crust must have been oceanized, because there is an ocean there. There is an ocean there, because the earth's crust there has been oceanized.

One of the things we did as students in chemistry laboratory was to heat up a glass tube on a flame, and give it a pull. Under the tension of pulling, the tube becomes thinner and thinner, and then it is split in two. This thinning process is called necking. Karig invoked the concept of necking and postulated that continental crust under tension is stretched and becomes thin, very thin, before the crust is split. New ocean crust forming in the crack starts the process of seafloor-spreading. Where the initial crack is in the interior of the continent, an open ocean comes into existence. This is the case with the Atlantic, Indian, and Southern Oceans. Where the initial crack is on the margin of a continent, a strip of continental lithosphere is split away to form an island arc, and the seafloor-spreading behind the arc causes a back-arc basin to form (Figure 9.2). Marginal seas are inundated back-arc basins.

To verify the idea of back-arc spreading, Karig was given a rather preliminary bathymetrical chart of the Tonga-Kermadec region. The very rough seafloor topography of the Lau Basin reminded him of the topography of the Mid-Atlantic Ridge, where seafloor-spreading has been taking place. That superficial resemblance was not going to convince anybody; Karig very much wanted to go to Tonga to get some more data.

Menard had the *Nova* expedition to the South Pacific in 1967, and Karig went along. I almost met Karig that year, because I too had wanted to get on the cruise. I was then interested in the mélanges on the Solomon Islands. I could fly to Guadalcanal, but I needed a ship to get me to other islands. Menard kindly made my request a rider on his 1.4-million-dollar project. He got the grant, minus the 0.007 million-dollar rider, so I missed the *Nova.*

Karig did go, and he was mainly busy with seismic profiling to determine bottom topography and sediment thickness. There were other things to be done, too: depth of the Moho had to be worked out, heat-flow measurements had to be made, and magnetic anomalies had to be delineated. When the work was done, everybody was anxious to fly back home, but the slow boat had to be taken back to La Jolla. Karig, the graduate student, volunteered for the chore. In exchange Menard gave him a week of shiptime to make a reconnaissance survey of the Tonga-Kermadec region while he was under way.

During the late 1960s excitement was in the air. The new theory of subduction of ocean floor had just provided the explanation for the origin of ocean trenches and island arcs. Karig took the next logical step, and recognized that marginal basins on the concave side of the arcs are the third member of a trinity produced by such a subduction.

9.2. Back-arc basin. The subduction of cold ocean lithosphere causes local thermal convection in the upper mantle under continental margin. The rise of mantle material causes the rifting of continental lithosphere, and the flooring of the crack by ocean basalt. The hypothesis concerning the origin of back-arc basins was verified by the drilling of the *Glomar Challenger.*

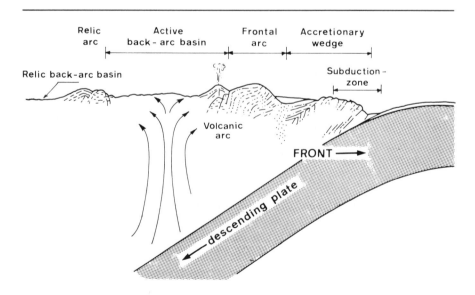

Karig noted the presence of a thin crust under the floor of the Lau Basin behind the Tonga Arc, as revealed by seismic refraction studies done by Karig's teacher George Shor; the basin is underlain by ocean crust. Karig found also that the heat flow in the area is high; there could be upward transfer of heat by convection. Finally, the bottom topography of the basin is similar to that of mid-ocean ridges. All these features are typical of spreading ridges. Karig had verified his prediction that back-arc basins are formed by seafloor-spreading.

The Lau Basin is situated west of the Tonga Arc. While the opening of back-arc basins was going on, the ocean floor on the front side of the arcs plunged under the arc to make a deep-sea trench. The subduction of a lithospheric plate produces compression, but the opening of back-arc basins implies tension.

Geologists were not used to thinking that a geological feature formed by tension could be a contemporaneous neighbor of one formed by compression. They made this mistake because they were misled by the old idea that the earth's crust, at any one time, is subjected either to global compression or to global tension.

Geophysicists deal with actual models. The down-plunge of cold oceanic lithosphere disturbs the thermal state of the earth's mantle in the region behind island arcs, forming a local convection cell. Rising current under the continental margin spreads out laterally and causes the tension and the seafloor-spreading of the back-arc basin (Figure 9.2).

I was in the audience when Karig gave his talk at the Washington, D.C., meeting of the American Geophysical Union in 1968. His ideas generated much interest and a lot of discussion, and he was under pressure because he had not had time to collect enough data to convince the skeptics. He had failed to find well-defined magnetic stripes—a typical signature of seafloor-spreading in the Lau Basin. He could not even identify a central rift-valley comparable to that on a mid-ocean ridge. Yet his idea of extensional back-arc basins was an alternative to oceanization. At least his theory had predictions, and again we could call on *Glomar Challenger*.

Karig predicted that we should find an ocean crust in back-arc basins, formed by submarine volcanism, not by a sunken continent. Karig's theory also required that the crust of a back-arc basin should be different from the crust on the front side of the island arc; the age of the basin floor should be younger, or much younger, than the seafloor being subducted under the oceanic trench. Finally, a back-arc basin created by seafloor-spreading should have symmetry about an axis; the age of the seafloor and overlying sediments should be oldest on the periphery of the basin, farthest away from the center of seafloor-spreading.

"The Mayor of Caster- bridge"

Not long after Karig presented his ideas in Washington, D.C., the first chance to test the idea came up, during the Leg 6 cruise to the Pacific in the summer of 1969. Al Fischer of Princeton and Bruce Heezen of Lamont were the co-chiefs.

Bruce Charles Heezen was a tragic giant, and I always compared him to Thomas Hardy's Mayor of Casterbridge. His energy, intellect, and driving ambition were the personal traits responsible for his success, and their excess led to his downfall. Heezen died of a heart attack in 1977, shortly after his "rehabilitation."

I first met Heezen in the early 1960s, when he came to Shell Research to give a talk on turbidity-current deposition. Deep-sea environments used to be considered stagnant. Sands had been sampled from the deep-sea floor ever since the HMS *Challenger* expedition. Thanks mainly to the work of Ph. Kuenen in the 1950s, many of us came to accept the theory of turbidity-current sedimentation. The subaqueous streams are flows of suspension laden with sand and mud.

Heezen had been studying the deposits of turbidity currents on Atlantic abyssal plains. He made a name for himself as a graduate student, investigating the 1929 Grand Banks Earth-quake. The earth tremor had triggered an underwater mud avalanche, which in turn was converted into a turbidity current. The transatlantic telegraph cables were broken off successively by the current. Since the exact timing of the interruptions had been recorded, Heezen and his mentor, Ewing, were able to determine the speed of the current and the distance it had traveled. Piston coring at the Atlantic abyssal plain more than 700 km south of the Grand Banks verified the deposition of deep-sea sand by this turbidity current.

In the early 1960s Heezen went a step further and discovered sand-transporting and sand-depositing ocean currents. I was finding evidence for the same kind of phenomena while working on older rocks of the Ventura Basin. Wanting to know more about what oceanographers were finding out in modern oceans, I had Heezen invited down to Houston for a lecture and for consultation.

He told me that he was an Iowa farm boy studying geology when Doc came to Iowa City on a lecture tour. Heezen was so fascinated by Doc's talk and by the wonders of the seafloor that he indicated a wish to come to Lamont, and Doc was happy to have him. Those were the good old days when Heezen was one of only two geologists on the staff; the other was Dave Ericson.

Heezen went to sea, with Marie Tharp, in the early 1950s. On their first cruise, they made three transatlantic profiles. The Mid-Atlantic ridge was obvious. Tharp noticed a notch in the ridge; she thought it might be a rift valley. Heezen did not think very much of it.

While working on the Grand Banks problem, Heezen had become interested in underwater earthquakes of the Atlantic, and the American Telephone and Telegraph Company had provided him with a plot of such occurrences. There are a lot of earthquakes on the Mid-Atlantic Ridge, and one of them lay close to the rift valley recognized by Tharp. Now Heezen was excited. There was not only the rift valley Tharp had found on their profile; there had to be a rift on top of ocean ridges everywhere, Heezen reasoned. Furthermore, the process of rifting, of splitting the ocean floor on mid-ocean ridges, must have been the cause of the earthquakes.

After studying numerous bathymetric profiles, Heezen was soon convinced of world-girdling mid-ocean valleys, even though Bill Menard told him that there is no rift on the East Pacific Rise. In 1965, Heezen gave a talk in which he pointed out the similarity, in seismicity and topography, between rifts on mid-ocean ridges and rift valleys in East Africa. Heezen was thus the first to recognize the significance of mid-ocean rift valleys and their extensional origin.

Heezen had published 44 papers with Doc Ewing by 1963, including his most significant contributions on turbidity currents and on mid-ocean rifts. They ceased active collaboration sometime in the early 1960s. Then there was a rift.

Heezen was not the only person who complained that Ewing insisted on putting his name on publications to which he had made little contribution. Tsuni Saito, my shipmate on Leg 3, asked Art Maxwell once: "Could you tell me, Art, why Doc would want his name on all Lamont publications?"

"I'll tell you a secret, Tsuni. It is every baseball star's dream to have a slugging average of 1.000. Well, Doc will get there, but now he is only slugging .693. He has to work hard before he retires."

As Bill Menard commented, Heezen had the same weakness. We probably all do. At least, who does not overestimate the contribution he has made to a coauthored paper? Without the organization, the finance, and in numerous cases the idea by the leader of a team, most coauthored papers would not have been written. Also, the offender is not necessarily the senior member of the team. We once had a project drilling a hole into the 143-meter-deep Lake Zürich, and some 15–20 scientists were involved in the half-million-dollar project. One graduate student was so inexperienced, he was planning to submit a single-author talk to a scientific meeting, skimming the cream off the team discovery. Fortunately he checked with me before he sent in the abstract. More often than not, of course, it is the professor who gets all the glory; at least, many young scientists feel that way. The sensitivity of a scientist for his due recognition, as substantiated in the question of coauthorship between a teacher and student, is the source of many a friendship gone sour. This was probably the basic cause of the rift between Heezen and Ewing.

In the mid-1960s, when I was at Riverside, my La Jolla friends told me that Heezen had gone into "desktop publishing." He sent mimeographed newsletters to his colleagues to establish priority in discoveries. I thought it was a little funny, but I understood; he must have been nursing a deep injury.

The final crash came in 1966, when Heezen rushed to the news media to announce, prematurely, a sensational discovery. The episode started in 1964 when Doc persuaded Neil Opdyke to set up a paleomagnetics laboratory at Lamont. Opdyke did, but he himself was less interested in working with sediments than his students. One, John Foster, built a spinner magnetometer to "run" deep-sea cores and in 1965 found negative magnetization in a sediment. Foster told his colleagues about the discovery. Opdyke was not excited, but Heezen was, and he jumped in.

Heezen had the idea of relating cosmic radiation to gene mutation and to biologic extinction. It was an attractive idea, and many of us were talking about it, but we knew of no way to determine the intensity of cosmic radiation. Now theorists told us that the cosmic radiation should have been more intense during times of weak magnetic field, such as times of polarity reversal. If one could find a correlation between such reversals and extinction events, he had it made. Well, Heezen was lucky (or unlucky, as it turned out)—the first samples of radiolarian oozes provided by a colleague, Jim Hays, yielded the desired correlation. Opdyke, meanwhile, was not happy with Heezen's intrusion; he went to Ewing and asked him to intervene. It was agreed by all parties concerned that Heezen should not be the first author of the first paper to publicize work done in Opdyke's laboratory.

But ambition was to get the worst of Heezen. On 2 June 1966 the *New York Times* published an interview conducted in Moscow headlined "Evolution Linked to Magnetic Field," containing the following sentences: "The theory was advanced by Dr. Bruce Heezen . . . Other members of the research team with Dr. Heezen were Dr. Neil Opdyke, Dr. Dragoslav Ninkovic, Dr. James Hays, . . ."

That was the bomb! Someone initiated the process, and the university administration acted. At age 42 the internationally famous oceanographer Bruce Charles Heezen was suspended by Columbia University. He could not get access to Lamont data, use Lamont facilities, or go on Lamont ships. Leg 6 sailed in 1969, 3 years after the bomb had exploded; Heezen was able to come out of his hermitage then, because JOIDES was not a Lamont project, and the *Challenger* was not a Lamont ship.

In the 1970s, I saw Heezen from time to time when I was in New York. His student Bill Ryan and I were close friends, working together on a Mediterranean project. We three used to go to a Chinese restaurant near Lamont for lunch. Heezen was not exactly subdued—he was almost garrulous at times—but we never talked about the sea or about Lamont. Eventually Doc left Lamont and passed away in 1974; Heezen was "rehabilitated" in 1977. He went down in a submersible (deep-diving submarine) on his first Lamont cruise to the Atlantic, and died of a heart attack while getting reacquainted with his beloved mid-ocean ridge.

Heezen had been too impatient. Magnetostratigraphy, developed by Foster, Opdyke, and others at Lamont, was eventually to provide the precision that enabled Luis and Walter Alvarez, 15 years after Heezen's press conference, to link evolution to a cosmic cause. If Heezen had been less zealous in 1965, he might have been a co-discoverer in 1980 of that important new theory. But then, if he had been less zealous, he would still have been down on a farm in Iowa.

Hit and Run across the Pacific

Leg 6 was known to us in the JOIDES community as the "hit-and-run" cruise across the Pacific. Heezen and Fischer's crew never stayed very long or drilled very deep at any site, because of repeated equipment failures.

During the initial phase of the JOIDES project, we were not provided with the right kinds of drill bits. On the very first cruise, Doc Ewing found out to his dismay that the drill string of *Glomar Challenger* could not penetrate a hard chert layer. Chert is a lithified ocean-sediment. Originally the sediment was an ooze, made up exclusively of dead skeletons of one-celled radiolaria (Plate VII). The skeletons consist of amorphous silica. In some places, the amorphous silica underwent changes and became opal or quartz. With such a recrystallization of amorphous skeletons, soft oozes were converted into hard, compact cherts.

During our Leg 3 cruise to the Atlantic, we ran into the same problem when we drilled into chert at Site 13 west of Sierra Leone. Commonly one might run into a thin layer of chert, while drilling into many meters of soft sediments. Yet it would take longer to drill through 1 m of chert than through 100 m of oozes. Furthermore, hard grinding on chert quickly wore out the drill bit. Our effort at Site 13 was given up because we could not penetrate a chert layer. Many other earlier holes had also bottomed in chert because of drill-bit failures.

To drill through a hard rock such as chert, a heavy pressure or weight must be exerted onto the drill bit for effective deepening; otherwise, the bit would just spin around and around, like the wheels of an automobile stuck in sand or snow. The pressure on the drill bit, transmitted by the drill pipes, is normally balanced by the counter-pressure exerted by the wall of the borehole. However, if the segment of drill pipe called the bottom-hole assembly is not yet buried in mud, the ocean water will provide little counter-pressure, and the drill pipes can be bent or broken. Eventually we learned that we should not try to spud a hole into ocean floor if the drill bit was to encounter a rock layer at a depth of less than 100 m beneath the seafloor. If we tried to drill our way through chert at a shallower depth, before the bottom-hole assembly was safely buried in mud, we would be courting trouble. We knew, because Heezen and Fischer had tried.

The primary objective of the Leg 6 cruise was to determine the age of the seafloor of the North Pacific. The ocean crust is in many places buried under chert, and the chert, as a rule, is covered only by a thin veneer of soft sediments. The drill string was punched into the soft stuff without difficulty. However, when weight was applied to the drill bit to drill through a chert layer, the stress was too great, and the bottom-hole assembly that was not yet buried would be broken and twisted off. Leg 6 lost 9 assemblies this way while drilling 34 holes at 17 sites. They drilled in so many places that the ship's store ran out of the beacons needed for dynamic-positioning. A fresh supply of bottom-hole assemblies and beacons had to be sent to *Glomar Challenger* via a vessel out of Saipan.

While chasing magnetic lineations across the Pacific, Heezen and Fischer confirmed the general pattern of the age of the seafloor as predicted by the new theory. Steaming west, they encountered successively older crust in the West Pacific. At Site 52, in deep waters northwest of the Mariana Arc, they drilled into chert of the Cretaceous period, more than 100 million years old. Now they had a chance to resolve the controversy of the origin of back-arc basins. If Beloussov was right, the seafloor west of the Mariana Arc should be underlain by older rocks of continental crust. However, if the young Karig had the correct idea, the boreholes in the Philippine Sea should bottom in young oceanic crust. Left with two last beacons before a fresh supply was to arrive, those on *Glomar Challenger* decided to cross the Mariana Trench and conduct this critical experiment.

This is an excellent illustration of the informal way science was done in the early days of *Challenger* drilling. Later, during the late 1970s and early 1980s, when I was active on JOIDES panels, every hole that was to be drilled had to be included in a preliminary proposal, which was sent in to a JOIDES committee; the committee decided if the proposal had sufficient merit to be included as an item in a research program to be presented to the U.S. National Science Foundation for funding support. With the stiff competition for use of the drill vessel, a proposal from an outsider (someone from an institution or a country not contributing financially to JOIDES:) would rarely receive very favorable consideration. After the funding was received, the proposals that had been included in the program would be referred to JOIDES panels for planning. Of course, each panel lobbied and fought for its proposal against the other planning panels. The JOIDES planning committee was to decide. I had no quarrel with the decisions by JOIDES committees and panels, although science politics did play a role, at times, in the panels' decisions. After the drilling program was approved and organized into drilling cruises, a JOIDES Safety Panel for the Prevention of Hydrocarbon Pollution would study the drill sites carefully to screen out those that were not completely "safe" bets. The final program naturally had to be reported to the National Science Foundation. When the drilling cruises were finally organized, not only the geographic coordinates of the drill sites, but also the projected depth, the coring schedule, and so on, were all specified. When the co-chief scientists of a cruise were appointed, they would find that they had little opportunity to alter the program.

Of course, actual conditions at sea might make last-minute changes unavoidable. I made several requests for such changes during my tours of duty. This required sending telegrams to the Deep Sea Drilling Project, and the chief scientist of the Project would then have to make emergency calls to various key people to have each minor change approved. Therefore, as a rule, the co-chief scientists at sea were hardly expected to change things, and they were certainly not permitted to make *ad hoc* decisions to drill unauthorized sites, especially sites that had not yet been thoroughly surveyed by geophysical investigations.

The bureaucracy was at a minimum when the project was young. Heezen and Fischer were thus able to make a quick decision. They got in touch with Dan Karig, who was chief scientist on a Scripps vessel, the R/V *Argo*, in the area. *Glomar Challenger* had very little geological information; the *Argo* had to lead the way and provide the last-minute survey of some potential drill sites. Karig transmitted part of his data by radio, with sketches sent by facsimile and key information by word of mouth. Two sites were picked out.

The improvisation was vindicated. Two boreholes in the region west of the Mariana Arc penetrated Oligocene volcanic rocks about 30 million years of age. There was no indication of an old sunken continent. Karig was right! The back-arc basin is a young basin. The basin first came into existence when the Mariana Arc was ripped away from Asia; the basin floor was paved by lavas from submarine volcanoes filling the crack behind the arc.

The scientific staff of Leg 6, however, remained cautious. Their drill string had barely scratched the top of a volcanic layer. Perhaps that was only a lava flow extruded on an old seafloor. Perhaps the basin was underlain by a continental crust, deeply buried beneath the volcanic flow and a thick pile of older sediments. Perhaps the Oligocene volcanism only did a "remodeling" job on a much older Pacific crust. The reconnaissance cruise did give very interesting indications, but the other options had to be kept open. The final proof had to wait until later cruises.

Seafloor-Spreading in Back-Arc Basins

Geophysical surveys of the 1970s indicated that the geological history of the Philippine Sea region is far more complicated than can be explained simply by assuming one back-arc basin. In fact, the region bounded by the Mariana and Bonin Troughs on the east and by the Ryukyu and Philippine Trenches on the west seems to be underlain by three pairs of arcs and back-arc basins (Figure 9.3).

The island of Guam is sitting on the Mariana Ridge, fringed on the convex side by the Mariana Trench and on the concave side by the Mariana Trough. Steaming west from Guam, as *Glomar Challenger* did during the start of Leg 31, one first crosses the Mariana Ridge and Trough; next, the West Mariana Ridge and the Parece-Vela Basin; and finally, the Palau-Kyushu Ridge and the West Philippine Basin. The ridges are submerged arcs; the troughs or basins are back-arc basins.

9.3. Evolution of the Philippine Sea. The West Philippine Basin (horizontal lines) owes its origin to the rifting of the Bonin-Guam Arc from Asia 30 million years ago. The arc was then split in two when the Parece-Vela Basin (vertical lines) came into existence between the frontal and relic arc. Another split led to the genesis of the Mariana Trough (oblique lines) between the West Mariana Ridge and the Mariana Ridge.

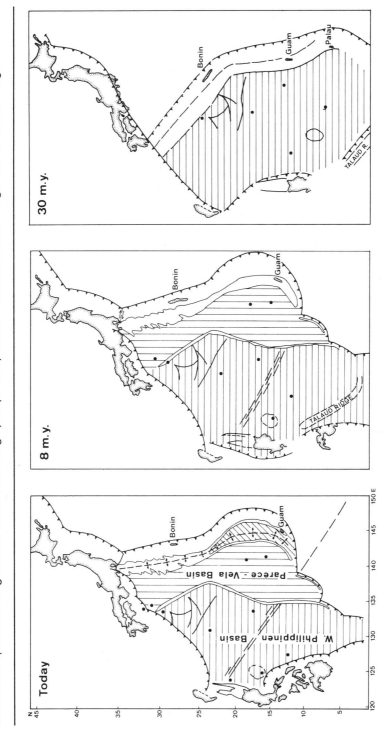

After the reconnaissance cruise of the first phase of deep-sea drilling, the origin of the back-arc basins was no longer being disputed by the experts. Leg 31 scientists could now concentrate their efforts and attempt to work out the geological history of island arcs and back-arc basins under the Philippine Sea. The theory states that basins formed when arcs were ripped apart and when volcanic rocks were erupted to floor the cracks behind the arcs. When were the arcs ripped apart? What was the age of volcanism? Those answers can be provided only by obtaining samples from those arcs and basins.

Karig's Leg 31 cruise drilled eight holes in the region, mostly into seafloor of the West Philippine Basin. At Site 292, the hole was bottomed in an upper Eocene basalt, which is believed to be the oldest seafloor of the West Philippine Basin. Synthesizing the geological data of the region, Karig came to the conclusion that the three basins west of the Mariana Arc are different in age. The West Philippine Basin is the oldest; this back-arc basin was created 40 or 45 million years ago when the Pacific seafloor plunged down in front of an old Mariana Arc. With the continuing subduction of the seafloor, the old Mariana Arc was split in two, the Palau-Kyushu Ridge on the west and a newer Mariana Arc on the east. Between those two ridges, the Parece-Vela Basin came into existence some 28 million years ago. The Mariana Basin is the youngest; it was created 10 million years ago when the newer Mariana Arc was again split in two, forming the West Mariana Ridge and the Mariana Ridge. The latter is the active island arc. Guam and the Mariana Islands sit on top of the arc, while the subduction of the Pacific seafloor is going on beneath the inner wall of the Mariana Trench. Meanwhile, the Palau-Kyushu and the West Mariana Ridges are now submerged and are referred to as remnant arcs.

A final clarification of the geology of the Marianas had to wait for an international effort during the IPOD phase of deep-sea drilling. DSDP Legs 59 and 60 were scheduled by von Huene's panel to drill holes along an east-west transect crossing the Philippine Sea. Several deep holes were drilled into the ridges and basins behind the arc. The submarine volcanism paving the new ocean floor of a back-arc basin is not fundamentally different from that at a mid-ocean ridge, except for its more explosive character. The basalt lavas in back-arc basins seem to have a chemical composition similar to that of mid-ocean-ridge basalt.

The 1978 drilling of the Parece-Vela Basin and the West Mariana Ridge answered the question raised by the Leg 6 scientists of whether or not the early drilling had reached basement. In several holes, several hundred meters of volcanic deposits were penetrated; they were laid down during the submarine volcanism that gave the new seafloor to the Parece-Vela Basin behind the frontal arc. There was no older seafloor.

A topographic depression was found in the Parece-Vela Basin, and was called the IPOD Trough by the shipboard staff in honor of the international program that scheduled the cruise. This depression is the equivalent of the rift valley on mid-ocean ridges. The age of the seafloor is older on both sides of—and oldest farthest away from—the IPOD Trough, as Karig's theory had predicted some 10 years earlier. The half-rate of seafloor-spreading has been determined to be about 2 cm per year during an interval between 30 and 10 million years before present. In more recent times, the center of seafloor-spreading has been positioned within the Mariana Basin between the Mariana and the West Mariana Ridges.

The drilling by *Glomar Challenger* has settled once and for all the controversy of the back-arc basins. Today few people bother to read the fictitious "oceanization" story by Academician Beloussov. Instead, the doctoral dissertation of student Karig has become one of the most cited references in geological literature.

10 Hope and Frustration in Nauru

This is a brief narrative of the life of Seymour Oscar Schlanger and his Pacific adventures on Glomar Challenger.

Introduction to a Distant View

I have always been uneasy with excessive praise. When I am praised, I am embarrassed; when someone else is praised, I am irritated, because nobody is perfect.

Seymour Oscar Schlanger, to whom this opus is dedicated, was at one time a victim of my prejudice, because everyone liked him—almost too much, it seemed. First my friend Max Carman at the University of Houston and his wife, Libby, talked only of Sy after they came back from Brazil; they had worked together there for Petrobras, training petroleum geologists for the Brazilian government. Then my old professor Cordell Durrell and my former classmate Mike Murphy also returned from Brazil, and during my occasional visits to Los Angeles, those friends of Sy would incessantly sing the man's praises. Finally, after I joined the faculty of the University of California at Riverside, Frank Dickson did not miss a chance to tell me how great Sy was, who was at that time taking a sabbatical leave from Riverside to visit Japan. After hearing all this, I decided that I was going to dislike the person. I was mistaken!

Sy was born in 1927, the son of a Brooklyn grocer who had emigrated from Austria. His Brooklyn was the "Brooklyn in which a tree grew," the small town described by Betty Smith in her best-selling novel. Sy was 18, finishing high school, when the Second World War ended. He entered the U.S. Navy, and was honorably discharged in 1946. Before starting college, he spent some time in Honolulu, working for a local newspaper. He became a great swimmer and received an athletic scholarship to join the swim team at Ohio State. His teammates were all international stars and Olympic medal winners.

Sy was restless. He wanted to go to the Holy Land, and was

in Europe looking for contacts to join the "Exodus" when the State of Israel was established. Coming back home, he elected to go across the Hudson to study at Rutgers.

Last November, half a year before he was to take his own life, we talked about the fractal geometry of fate, the inevitability of the improbable. He told me about the incident that was to determine the course of his life.

In the summer of 1951, he got his master's degree at Rutgers. One day he went to the city to look for a job. It was not very difficult to find something in industry; oil companies were in the midst of their postwar expansion. Sy got a job with Texaco to work in the Caribbean. He came back one evening to the geology building to clean out his desk. He was loading his belongings into his old car when a colleague, an instructor in the department, passed by. Sy told him of the new job. The colleague asked if Sy had ever thought of working for the U.S. Geological Survey. Yes, Sy had; he had even passed the civil-service examination, but there were no openings.

"Josh Tracy was here today, looking for someone to work on their Pacific project," his friend told him. "Perhaps you could call him up."

It was a casual remark, and Sy made a casual telephone call. He got an interview in Washington and was hired on the spot. Thus he joined the famous group of the "Pacific Islanders," which included Harry Ladd, Josh Tracy, and Dave Doan. Sy's life, private and professional, was to be intricately tied to the Pacific.

Sy talked often about his life on Guam; those were his happiest days. He did geological mapping on the island, and went to look for water on Ulithi Atoll in the Carolines. In the evenings there was the usual partying, often in the governor's mansion. Sy met Sally Wimsatt, the daughter of the commanding general at Anderson Air Force Base on Guam. They were married before coming back to the States.

The Survey had an atoll-drilling project during the late 1940s, in connection with the bomb tests at Bikini and Eniwetok. Geologists since Darwin's time had been fascinated by the origin of coral reefs. The government got a lot of cores from the atolls, and Sy was able to study them for his dissertation at Johns Hopkins University. The life of a graduate student was never "plentiful." Although the daughter of well-to-do parents, Sally did not want subsidies from her family; she worked in an office to help with household expenses. Yet the young married couple was happy in Washington. Sy could not imagine doing something more fun than looking through a microscope at his limestones in the office before coming home to Sally and their two infant daughters.

Sy went to Brazil in 1959, and rejoined the survey two years later, just when the organization was going through retrenchment. Someone high up thought the survey was overstaffed for fancy research, so numerous geologists were sent down to Kentucky on loan to do something practical. Sy was one of these victims of federal politics; dissatisfied, he quit the federal government and was invited by Murphy to be an acting lecturer at the University of California, Riverside. When I joined the Riverside campus in 1964, Sy had not yet been granted tenure. I thought all the praises I heard were merely propaganda paving the way for his promotion. I was wrong. Sy came back from Japan in the autumn of 1965, and he was very much what people had said, and more.

We did not go home for lunch in those days, and we spent many noon hours together. Sy was fascinated by the Orient and by Buddhism, and what impressed him most was the Oriental distant view. Yet he was still too steeped in Judaism and Christianity to rise above the ambition, the involvement, or the pressure of achieving.

Isostasy and Oceanic Rises

After crossing the Pacific in 1948 on my way from China to the United States, I did not have much chance to sail on this great ocean to learn its geology. The nearest I got was to lose myself in the Coast Range mélanges that included pieces of the ancient Pacific seafloor. I was, however, always curious what others were doing. Talking to Sy Schlanger every day at Riverside, I was step by step initiated into the geology of the Pacific Ocean.

The Pacific bottom is not featureless. Islands like Hawaii are volcanic mountains that stand above the deep-ocean floor. Seamounts are either new or old submarine volcanoes that are submerged. Atolls are dead volcanic islands fringed by a ring of coral reefs. Guyots are dead volcanoes capped by a flat top.

As I have already mentioned, the idea that volcanic islands should sink under their own weight was first advanced by Charles Darwin. Corals colonizing a sinking island built fringing reefs or atolls. Where coral growth could not keep up with subsidence, an island would sink out of sight into the abyss, and become a seamount or a guyot.

In my Riverside years, we were too cautious to follow Hess's "geopoetry" of seafloor-spreading. Schlanger and I focused our attention at the time on the history of vertical movements of the

Pacific floor, which could be easily verified by geological evidence. We wrote a paper together on the origin of oceanic rises or submarine plateaus. At Prague in 1968, during that fateful Geological Congress in August, the two of us presented a talk on Pacific volcanism before the Soviet tanks drove in. In fact, the original idea was embodied in the letter that I wrote to Hess after I first heard his talk on the sunken flat-top mountains of the Pacific (see chapter 1). Our postulate explained the vertical movements of extinct volcanoes on the Pacific floor.

We pointed out that active volcanoes are present on the East Pacific Rise, where the geothermal gradient is steep. Lava piling up around a vent makes a seamount. Where lava is poured out in great quantities, it flows into depressions between seamounts and eventually buries the seamounts under a thick pile of volcanic rocks. A net result of submarine volcanism is that the ocean crust there is thicker than normal.

Then we made use of the well-established geological principle of isostasy. In essence, at some depth beneath the surface of the earth, somewhere in the asthenosphere, the mantle is so weak that it cannot sustain big differences in weight between adjacent columns of lithosphere. As a consequence, each column of rocks above the level of isostatic compensation should have about the same weight per unit cross-sectional area. We suggested that when submarine volcanism is active, the mantle is heated up, and the thermal expansion should cause the overlying seafloor to rise. If the accumulation of volcanic materials on the seafloor is negligible, the seafloor should sink back down to its original depth when the volcanism ceases and when the mantle cools down to its original density. However, if the ocean crust is thickened a kilometer or two by volcanic rocks poured out of vents or fissures, the lithospheric column will then have an excess of lighter materials. This excess weight will push the seafloor down, but the conditions of the isostatic equilibrium dictate that the top of the thickened crust will never sink back to its original depth, because of the excess of lighter materials. Oceanic rises, underlain by thickened crust, stand higher than the surrounding ocean floor, which is underlain by thin-crust, just like big icebergs stand taller than small icebergs (Figure 10.1).

Invoking this principle to interpret the geology of the Pacific, Schlanger and I suggested that the submarine plateaus now present in the Pacific owe their elevation to extensive volcanism, which has thickened the ocean crust there. Seismic studies had indicated that submarine elevations are commonly underlain by a crust 10–15 kilometers thick, or some 5–10 kilometers thicker than the normal ocean crust (which is produced by seafloor-spreading). This fact gave credibility to our postulate of crustal thickening by volcanism.

Submarine volcanoes seldom occur in isolation. In some regions, a whole string of sunken volcanoes occurs, forming what is called a seamount chain, as we shall discuss later (chapter 11). At the time, when neither Schlanger nor I was much impressed by the new theory of seafloor-spreading, we talked only of rise or subsidence of the ocean floor; we did not bother to explore the implications of the seafloor-spreading theory.

Sy and Sally stayed with us after Prague, and they were fond of Switzerland. Sy decided to spend his next sabbatical leave

10.1. Stages in the evolution of oceanic plateaus: (I) seamount on an oceanic rise, caused by mid-plate volcanism; (II) emergence of atolls and volcanic islands above sea level; (III) subsidence of atolls and islands to form guyots. The dashed lines show the successive stage in crustal thickening due largely to the addition of volcanic material as a consequence of mid-plate volcanism.

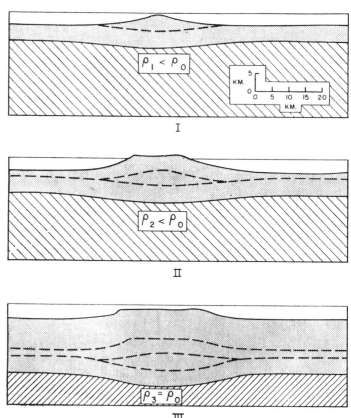

$\rho_1 < \rho_0$

5 KM.
0
0   5   10   15   20
KM.

I

$\rho_2 < \rho_0$

II

$\rho_3 = \rho_0$

III

with us. When he came back a year later, the intellectual climate had become entirely different. I had acquired a new outlook. Schlanger, who had been more open-minded all along, now found little difficulty accepting the new theory. After my South Atlantic adventures my enthusiasm for JOIDES-DSDP was contagious. Schlanger also began to see that *Glomar Challenger* might be exploited to give answers to his puzzles over the Pacific. When he went back to the States, he contacted Jerry Winterer, our friend at Scripps, and was invited to join the Leg 17 cruise of the Deep Sea Drilling Project.

Sunken

Volcanoes

and Guyots

With Jerry Winterer and John Ewing as co-chiefs, *Glomar Challenger* left Hawaii at the end of March 1971 to explore the Central Pacific. Holes were drilled into the Horizon Guyot, on the Magellan Rise, and on a submarine plateau in the region of the Line Islands.

The Horizon Guyot was an objective during the "hit-and-run" campaign of Heezen and Fischer. In fact, the very first hole of that cruise was drilled there. Chert was encountered at 62 meters, and the bottom-hole assembly was broken and twisted off at 76 meters. The Leg 6 co-chiefs were having their first hint that the failure may have been induced "by encountering chert at such shallow depth below the seafloor."

The survey of the guyot by R/V *Argo* and by D/V *Glomar Challenger* prior to the Leg 17 drilling revealed that the Horizon Guyot is in fact a 3,000-m-high submarine ridge, with steep sides rising up from 5,000 m to 2,000 m subsea (Figure 10.2). The top has a rough relief, and one row of volcanic peaks along the ridge axis stands at about 900 m above the low spots on the "flat top." Hole 44 drilled during Leg 6 was positioned near the top of one such peak.

The Horizon Guyot (Site 171) was the last site drilled during Leg 17, when the *Challenger* was returning to Honolulu. There were about three days left in which to drill a shallow hole *en route*. The drilling was to determine whether or not the guyot had been submerged in shallow sea at some time.

Hole 171 was positioned in a saddle on the "flat top" at 2,290 m subsea. Seismic profiling had indicated that chert is buried under more than 100 meters of soft sediments in this depression. Winterer and Ewing were able to benefit from the experiences of Leg 6 in choosing this site. A layer of black chert was encountered at about 150 m subbottom, but the bottom-hole assembly had already been buried in mud; adding weight to the drill bit did not break the drill pipes. The hole could thus penetrate chert layers, and was bottomed in basalt basement, almost 500 m beneath the seafloor.

The record at Site 171 indicated to Winterer and Ewing the following history:

1. Volcanic rocks erupted to build the foundation of the Horizon Guyot, which rose as an island some 3,000 meters above the adjacent ocean floor.

2. The last volcanic rocks were erupted about 100 million years ago, while the last basalt lavas were being poured out.

3. Lime muds were laid down in a shallow lagoon, which was then enclosed by coral reefs rimming the shoulders of the Horizon Guyot.

4. Subsidence started soon after the last volcanic eruption. Sandy debris from volcanoes was laid down in the lagoon. The tip of an extinct volcano still remained above sea level as late as 85 million years ago. Vegetation flourished on the island, and dead plant-debris was transported and buried in the lagoonal sediments.

5. Subsidence continued since then. Dead skeletons of floating and swimming microorganisms fell like snowfall onto the drowned lagoon, burying the last volcanic peaks with a thick blanket of ocean oozes.

10.2. Seismic profiling record of Horizon Guyot. The vertical ordinate is the two-way travel time of acoustic waves bouncing back from the ocean floor. The 7-second travel time is equivalent to a depth of about 5 km. The ship passed over the guyot on 19 May 1971 between 7 a.m. (0700) and 1 p.m. (1300), at a speed of 8–10 knots and over a distance of about 100 km.

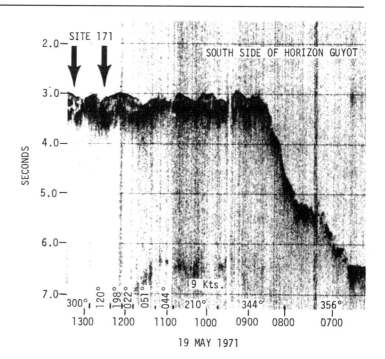

19 MAY 1971

The drilling of the Horizon Guyot thus corrected the misconception Hess made when he first talked about those "flat-top" mountains. Not all guyots of the Central Pacific have a very flat top. Some volcanic peaks stood more than a thousand meters above adjacent valleys or lagoons. Where the relief was reduced, rough topography was covered by (i.e., buried under) flat-lying sediments. There is no need to postulate, as Hess did, that the volcanoes had their tops cut off by wave erosion, or to assume a long pause of stillstand at sea level to let erosion do the work of flattening.

The drilling of the Horizon Guyot verified that its origin is linked to volcanic activity on the Pacific seafloor. The next questions to be asked were, Why should there be such volcanism? Why should the volcanoes be where they were? Why and when did the volcanism stop? Finding answers to those questions, not the origin of guyots, was the primary objective of drilling cruises to the Central Pacific.

Mid-plate Volcanism

We have discussed so far two kinds of volcanic activity within the framework of the plate-tectonic theory: volcanism on an active plate-margin, and volcanism on an accretionary plate-margin. What Schlanger and I proposed at Prague is a third kind, the so-called mid-plate volcanism.

Active-margin volcanism prevailed on continents landward of a subducted (swallowed) ocean trench. The volcanism typically produces andesite lavas such as those in the Andes or on the Island of Santorini in the Aegean Sea. Andesites contain more silica than basalts.

Accretionary-margin volcanism produces basalt lava, pouring out of rift valleys on mid-ocean ridges or out of cracks in back-arc basins. It is cooled immediately by ocean water. The viscous melt is then congealed into oval-shaped, bulbous lumps, of sizes comparable to bed pillows. The "pillows" at first had a thin solid crust; then the liquid interior also froze and became solidified. This type of basaltic rock has been called *pillow-basalt*. These rocks have been photographed by cameras lowered down to the ocean bottom and observed directly by scientists in submersibles. Pillow basalts have been seen in the Alps; they are an element of the ophiolites. Their presence is evidence that ocean floor existed where mountains now stand.

The chemical composition of fresh and unaltered basalts coming out of mid-ocean ridges is about the same everywhere around the world, and has been given a strange name, *tholeiitic*. The nearly uniform composition of tholeiitic basalt has been attributed to the fact that this type of basalt has come directly, without contamination or fractionation, from the partial melting of the earth's mantle. There are, of course, minute differences in the composition of tholeiitic basalts, especially in the trace-element chemistry, and many deep-sea drilling cruises have been sent to sample those rocks, which make up the bulk of the ocean crust (chapter 17).

Geologists have known for many years that thick basalts, commonly also tholeiitic, may cover areas many thousands of square kilometers in size in the continental interior. This is mid-plate volcanism on land; the Deccan Trap in India is one example. There is also mid-plate volcanism in the ocean, forming guyots, seamounts, and oceanic rises. If our idea is correct, drilling on those features should encounter volcanic rocks considerably younger than the "normal" ocean crust.

Drilling during Leg 17 did not give clear-cut answers. At Site 167 on the Magellan Rise and at Site 166 in the Central Pacific Basin to the south, the boreholes were bottomed in tholeiitic basalts, which have an age just about what the geophysicists predicted prior to the cruise on the basis of the seafloor-spreading theory. On the other hand, basalt samples from the Horizon Guyot of the Mid-Pacific Mountains and from Site 165 on the Line Islands Rise are younger than predicted (Figure 10.3). Also, those basalts contain more sodium than a normal tholeiitic basalt from a mid-ocean ridge; such sodium-rich varieties are called *alkali basalts*. Schlanger and Winterer had their postulate confirmed that the guyots and seamount chains owed their origins to mid-plate volcanism, which, as petrologists tell us, commonly ends with an eruption of alkali basalt.

Seamount Chains and the Theory of Hot-spots

Why should there be mid-plate volcanism? A very imaginative idea was published in 1963 by J. Tuzo Wilson of the University of Toronto, Canada. Wilson belonged to a generation that learned from its teachers the doctrine of the permanence of continents and ocean basins. A continent may have "grown" a little and enlarged itself after some ocean sediments on a continental margin were lithified, metamorphosed, and added on to the edge of the continent. But continents must have been fixed in their positions and have always been where they are now. I am old enough that I actually attended a talk, at the 1954 meeting of the Geological Society of America, given by Tuzo Wilson in support of the "fixistic" doctrine.

Wilson's scientific perspective changed when he started to study the volcanic islands of the Pacific in the early sixties. Noting that those islands are older the farther away they are from an ocean ridge, he became an advocate of the seafloor-spreading theory in 1963, as Hess was before, and Vine and

10.3. Line Islands Seamount Chain. Bathymetric contours in meters below sea level. Locations of DSDP sites are indicated by dots and numbers.

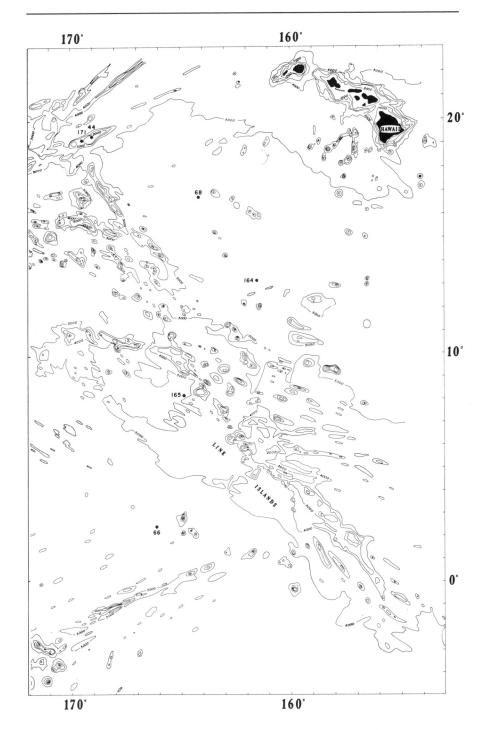

Matthews were after him. Wilson further took notice of the fact that many seamounts and volcanic islands in the Pacific have a linear arrangement. The concept of a *seamount chain* was proposed.

Geologists had known that the islands and seamounts in a chain were not produced by volcanic activities of the same age. The Hawaiian Seamount Chain, for example, has active volcanoes on the islands at the southeast end of the chain, whereas the oldest extinct volcanoes are found at the northwest end (see chapter 11). This age difference inspired Wilson. He hit upon the very simple idea that those islands came about because of a "hot-spot" in the earth's mantle. According to the seafloor-spreading theory, an oceanic lithospheric plate has moved horizontally over long distances during the last few hundred million years. The "hot-spot" in the mantle is stationary, but as a lithospheric plate moves, different spots on the moving plate are heated up successively by the stationary "hot-spot" in the mantle. The heating can cause partial melting of the material in the earth's mantle of the moving lithosphere. Rock melts generated in the upper mantle rise to erupt as lava under the sea, forming seamounts and volcanic islands.

The "hot-spot" that gave rise to the Hawaiian Seamount Chain, according to Wilson, is still there, sitting under Kilauea and Mauna Loa in Hawaii, and feeding fire to the active volcanoes. Older extinct volcanoes, now sunken as seamounts at the northwestern end of the Seamount Chain, had their day when their underlying lithosphere was being cooked by the "hot-spot."

The Wilsonian "hot-spot" stayed stationary as the Pacific Plate marched northwest. Older volcanoes successively moved away from the "hot-spot." Volcanism faded away and ceased. Volcanic islands sank, when the underlying mantle cooled down, and became submerged seamounts and guyots. Wilson's idea thus not only explained the rise and fall of oceanic volcanoes; it also elegantly explained why such volcanoes should form seamount chains. His publication actually appeared five years before Schlanger and I gave our talk. Still bound by our traditional training to ignore postulates of large horizontal displacement of the earth's crust, neither of us paid any more attention to Wilson's "hot-spot" than to Hess's "geopoetry." Nor could the planners of the Leg 17 cruise give much credence to Wilson's ingenious idea, which was publicized after the completion of the drill proposal.

While *Glomar Challenger* was out drilling in the Central Pacific, there was considerable talk on the "beach" about the latest "hit" by Jason Morgan, the bright young man who had first formulated the revolutionary new theory of plate-tectonics. Morgan's new "hit" was to provide a theoretical basis for Wilson's speculative idea. In a 1971 article published in *Nature*, Morgan portrayed the Wilsonian "hot-spot" as an expression of a plume-like rise of hot rock-material through the motion of convection currents in the earth's mantle. Morgan pointed out that the seamounts on the Line Islands Rise form a chain (see Figure 10.3), and suggested that this chain, like the Hawaiian chain, also owed its existence to the movement of the Pacific Plate over an ancient hot-spot.

As they were writing their cruise report for Leg 17, Schlanger and Winterer regretted that they did not have enough shiptime to drill more than one hole on the Line Islands Rise. To remedy the situation, they quickly prepared a proposal for a Line Islands transect and sent it to the JOIDES Pacific Advisory Panel in 1972. The proposed project was to provide an experimental test of the Wilson-Morgan theory. In due time, the proposal was accepted and was scheduled as DSDP Leg 33.

For the 1973 cruise Schlanger and Winterer were joined by Dale Jackson of the U.S. Geological Survey. Jackson was an expert in igneous rocks and was thus a logical choice as a co-chief scientist of this expedition to study oceanic volcanism. Winterer, who had been co-chief during two previous cruises to the Central Pacific, gracefully bowed out, and went on board as a sedimentologist, so that Schlanger had his chance to lead the Line Islands cruise.

With one hole, at Site 165, already drilled at the northwestern end of the Line Islands Seamount Chain, only two more, at Sites 315 and 316, were planned for the northwest-southeast transect. Morgan's theory postulated that the movement of the Pacific Plate was then directed from the southeast toward the northwest. Sites 315 and 316 are located southeast of Site 165. If the hot-spot had remained stationary while the plate moved, the chronological order of riding over the hot-spot should be Site 165, Site 315, and Site 316, successively. At Site 165, the basalt basement beneath the ocean sediments had been dated; the volcanism ceased there about 81 million years before present. Using Morgan's theory, the prediction was that the volcanic edifices near Site 316 should be 16 million years younger than those near Site 165.

Leg 33 drilling contradicted Morgan's prediction. The volcanism at Site 315 had ceased 85 million years before present, a few million years before, not after, the cessation of volcanism at Site 165. At Site 316, the top of the basalt sequence ranges in age between 79 and 82 million years, almost exactly the same as that at Site 165, not 16 million years younger, as Morgan's theory required. The discrepancies are far larger than the margin of error allows.

Much as they wanted to jump on the bandwagon, Schlanger, Jackson, and Winterer could not invoke the "hot-spot" theory to explain the Line Islands volcanism. Perhaps there had been a hot-spot, but the present topography of the Line Islands is related to mid-plate volcanism. The outpouring of that lava has made the ocean crust thicker than normal. Buoyed by this thicker crust in isostatic equilibrium, the Line Islands Rise stands 1 or 2 kilometers above the surrounding ocean floor.

Leg 33 taught us a lesson. Simple theories are elegant, but the earth's history is surprisingly complex in some areas. Theories explaining large horizontal displacements are spectacular, but classic ideas on vertical movements should not be altogether forgotten. It helped boost my ego when I read the following passage in the cruise report of Leg 33:

"Hsü and Schlanger proposed in 1968 that the history of vertical motion of an area (like the Line Islands Rise) could be explained in terms of changes in the thermal state of the underlying crust and mantle that accompany the volcanic event. As they pointed out: an uplifted oceanic rise will not simply disappear when its underlying mantle reverts to its normal density state; the linear belt will become a guyot-atoll ridge which still stands high above surrounding ocean floors after a long subsidence history because of the isostatic effect of a thickened crust!"

Eternal Hope and Frustration in Nauru Basin

Mid-plate volcanism made interpretations difficult for the scientists of Leg 33. Its occurrence in the Caribbean had, in fact, surprised and puzzled the shipboard scientific staff of earlier cruises. During the Leg 15 drilling of the Caribbean Sea in 1970–71, co-chief scientists John Saunders and Terry Edgar were bothered; they had not come up with decisive evidence in favor of the seafloor-spreading theory. The Caribbean Sea had been expected to have ocean crust more than 100 million years old; however, at three widely spaced sites in the Caribbean region, basalts of about the same age, 80 million years old, were encountered. We now know that the apparently anomalous results can be interpreted in terms of mid-plate volcanism. In fact, the crustal structure of the Caribbean is very similar to that of the ocean rises of the Central Pacific. The ocean crust in both regions is about twice the normal thickness, having apparently been thickened by a volcanism that ended some 80 million years ago.

Mid-plate volcanism was to haunt Schlanger once again in 1978, when he led DSDP Leg 61 to search for the old seafloor of the Pacific. Based on the pattern of seafloor magnetic anomalies, Roger Larson, the other co-chief from Lamont, recognized that the ocean crust in the Nauru Basin, where Site 462 was drilled, should be 155 million years old. Here was the place where the oldest seafloor still extant could be reached—at least, Schlanger and Larson hoped so.

The Leg 61 shipboard scientists did not expect mid-plate volcanism. Magnetic lineations from ocean crust are obscured if the original crust of the area is buried under a thick pile of volcanic rock. Larson could, however, easily map magnetic

10.4. Seafloor topography of the Nauru Basin. The atolls of the region have been built on top of extinct and sunken volcanoes.

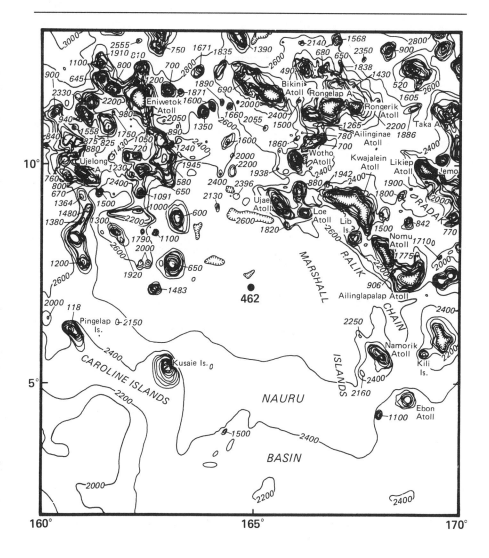

anomalies in the Nauru Basin. Also, the site lies in a basin, not on an oceanic rise, and the water depth is 5,190 m (Figure 10.4). Even if there had been mid-plate volcanism, the thickness of the basalt could not be very great; one should be able to penetrate the volcanic pile to sample the oldest crust, the co-chief scientists hoped.

The *Challenger* arrived on site in late May 1978. A pilot hole was drilled, and the first basalt was encountered at 558 m beneath the seafloor. The basalt was middle Cretaceous, about 100 million years old, more than 50 million years younger than the crust produced by seafloor-spreading. By now, Schlanger recognized that he had again encountered his nemesis—mid-plate volcanism.

The main effort was then made in Hole 462A nearby. The top of the basalt layers was encountered at about 560 m subbottom in this second hole on 11 June. After that it was a game of endurance and patience. Often it took 7 or 8 hours to cut through 2 meters of basalt. After penetrating 20 m through basalt rocks, the drill bit was worn out. The first reentry was tried. It took 24 hours to raise the drill string, change the drill bit, find the reentry cone again on the seafloor, and start cutting with the new drill bit. But the drill bit was gone again in less than 2 days. Once again, it took a day to raise the drill string, change the drill bit, and make the reentry. The drill bit was gone again in 2 days. Another reentry trial.

On the morning of 22 June, the drilling rate speeded up. Everyone started to smile again. Indeed, core number 41 sampled ocean sediments. The age was early Cretaceous, 110 million years old. There was still some distance to go, but it should be easy going if the hole had indeed gone through the mid-plate volcanics. So it was hoped.

These hopes were soon dashed. Toward evening, after two more cores containing sediments were hauled up, the drilling slowed down again. Core number 44 contained only hard rock, and the drill string was getting into another thick pile of basalt! The game of patience continued. Two more changes of drill bit and many days of slow cutting of hard basalt. Finally, the time was up. The last core came up from 953 m depth.

From 20 May to 4 July, more than six weeks, *Glomar Challenger* had stayed within a few hundred meters of a spot in the Pacific at 7°14′ N, 165°02′ E. One core barrel after another had been pulled out of the drill string, and almost invariably the core was a basalt. Just the same, the co-chief scientists had kept on hoping, like gamblers in a roulette game. They betted even, but the little ball always dropped into an odd compartment. Some less patient souls, like me, might have given up. But my friend Schlanger kept dreaming about the never-never land. If they could only get through the mid-plate volcanism, they would have smooth sailing in the "Jurassic Ocean." The Jurassic sediments buried beneath the basalt pile would give us so much information, telling us new and exciting stories about the Pacific 150 million years ago—information not available anywhere else on earth!

Schlanger and Larson had sent telegrams to the "beach" asking for permission to return to their drill site at the start of the next leg, which had been scheduled to investigate the North Pacific. There was always the possibility that the very next core would sample the Jurassic sediment. Their hope and enthusiasm were shared by their friends at home. The JOIDES planning committee made the rather exceptional decision to take 14 days from Leg 62 and invest them in the "bottomless" hole of the Nauru Basin.

After the drill pipes were hauled up on Independence Day, 1978, *Glomar Challenger* made a quick trip to Majuro in the Marshall Islands for an exchange of the ship's crew, and returned to resume drilling Hole 462A on 19 July. The drill string found its reentry cone again, after this absence of two weeks. Schlanger and Larson stayed with the new scientific team to supervise the continuation of drilling in their old hole.

The same game of patience was played again. Slow grinding, change of drill bit, reentry, and so on. Basalt, basalt, basalt! A ray of hope came in the early morning of 21 July, when core number 79 sampled a sedimentary layer at a depth just short of 1,000 m beneath the seafloor. But the sediment was only 20 cm thick, and optimism was snuffed out a few hours later, when the drill string started to cut through hard rocks again. That was the last sediment sampled from that hole, still less than 120 million years of age. The drill string would never be able to penetrate through the mid-plate volcanism! Not here.

Time was up again on 25 July. At noon, Schlanger and Larson were to have their last core from 1,068.5 m depth in the hole. As Jørn Thiede, who was a co-chief for the succeeding Leg 62 and who watched the agonies of the dashed hopes during the last two weeks of July at Site 462, told me later, Schlanger kept on hoping that the next attempt would break them out of the "traffic jam." Again and again, he was disappointed.

When I saw Schlanger two years later, at the International Geological Congress in Paris, he told me that they had been on the very verge of breaking through when they had to pull up and return to port. He and Larson still hoped for an opportunity to return, if the JOIDES organization could persuade the U.S.

National Science Foundation to extend the Deep Sea Drilling Project to 1983. As I said, deep-sea drilling reminded me of a roulette game. The rotary stem turned and turned. Every core coming up could bring in what you were looking for. If not, there was always the next core! Schlanger and Larson "lost their shirts" in the Nauru Basin, but they still retained the gambler's "eternal optimism."

Is the Caribbean a Twin of Nauru?

The Nauru Basin venture was not an unmitigated disaster. Drilling through a volcanic pile 500 m thick, an excellent record of mid-plate volcanism became available. Unlike those of the Central Pacific, the basalt rocks of Nauru are chemically similar to tholeiitic basalt lavas pouring into the rift valley of mid-ocean ridges. Like the mid-plate volcanism of the Caribbean, the voluminous outpouring of lavas did not produce any edifice such as seamounts or guyots in the basin. Shipboard scientists described their finding as "both unique and enigmatic"!

Drilling results from several cruises, Legs 17, 33, 61, and 62, gave evidence that the volcanic activities persisted for about 40 million years during the middle Cretaceous period, from about 110 million years to 70 million years before present. And this mid-plate volcanism of the Central Pacific covers an area millions of square kilometers in extent.

During the 40 million years of mid-plate volcanism, the mantle of the underlying lithosphere became less dense than normal. The ocean bottom of the Central Pacific oceanic rises was elevated and was about 2 km shallower than the surrounding ocean bottom. Volcanic eruptions built up seamounts and volcanic islands. Around the islands were fringing reefs or atolls, colonized by sedentary mollusks called rudists and by reef corals. In places, those reefs continued to grow as the volcanic foundation sank with the cessation of volcanism. Some of the atolls of the Marshall Islands northeast of the Nauru Basin, for example, have been building up since the Cretaceous (see Figure 10.4). Where the reefs were displaced by the northward movement of the Pacific Plate and moved out of the zone of tropic reef growth, the coral islands and atolls sank into the abyss. They became the guyots of the mid-Pacific mountains.

Schlanger and Premoli-Silva studied the faunas that lived during the Cretaceous in shallow waters near reefs and atolls around the Nauru Basin. They found a close similarity between the bottom-dwelling faunas and the Cretaceous faunas of the Caribbean region. In fact, the paleontologists of Leg 15 had already noted an apparent anomaly—the Cretaceous radiolarian faunas of the Caribbean are very similar to those of the

Pacific. That was not surprising, because radiolarians are floating organisms that can be carried far and wide by ocean currents. However, bottom-dwelling benthic organisms, living around reefs, usually migrate by crawling on shallow sea-bottom; even their swimming larvae cannot be expected to be dispersed across a deep ocean more than 10,000 km wide. The similarity of the benthic faunas led Schlanger and Premoli-Silva to conclude that the ocean crust under the Caribbean was originally Nauru's neighbor in the Central Pacific. They shared the same fate of mid-plate volcanism during the Cretaceous. Later, the Caribbean was pushed thousands of kilometers away to its present position by seafloor-spreading.

I have compared the Atlantic and the Tethys; the two twins did not go very far away from their home, but they encountered very different fortunes (chapter 6). The Central Pacific and the Caribbean are twin sisters who shared a common fate, but the younger sister has been deported to the realm of the Atlantic, and imprisoned between the Antilles and Central America. The Central Pacific has remained home.

This interpretation of the genesis of the Caribbean implies that its origin is quite different from that of the marginal seas of the West Pacific studied by Karig and others. The Antillean Arc was not ripped off the continent of Central America; there has been no back-arc spreading. The Antilles are a fragment of Atlantic seafloor that was jacked up to form an east-facing arc. The Central American coast range is a piece of Pacific seafloor that was jacked up to form a west-facing arc. Trapped between the arcs is the Caribbean Basin.

Vindication

and Nirvana

Sy Schlanger did not give up easily. In 1977 he joined the staff of the University of Hawaii, which had become a JOIDES institution at the start of the International Phase of Ocean Drilling. Schlanger served on various JOIDES panels to lobby for his objective, finding "old Pacific crust." He was a member of the JOIDES Ocean Paleoenvironment Panel and chairman of the Mesozoic Working Group, 1980–1982, chairman of the Science Advisory Committee, Joint Oceanographical Institutions, Inc., 1983–1986, and chairman of the JOIDES Central and East Pacific Panel, 1986–1990. Two more cruises were to be scheduled for the obstinate hunter.

Sy and his friend Ralph Moberly, also of Hawaii, were co-chief scientists of Leg 89. The *Glomar Challenger* left Yokohama, Japan, on 11 October 1982 to drill in the Mariana Basin before returning to Nauru. Their primary objective was again to sample the oldest sediments of the Pacific under the pile of mid-plate volcanism. Schlanger's disappointment after this cruise was great, although few readers could appreciate the depth of his frustration from the cruise report: "JOIDES planning for MZP-6 (Hole 585) had always been in terms of reentry. . . . As a result of drill-string losses within the past year . . . , however, DSDP engineers were loath to lower the weight of a reentry cone to the depth of the Mariana Basin. Their proposal, announced to the scientific party only after our arrival in Yokohama, was to make a single-bit attempt. . . . That program jeopardized not only the MZP Jurassic objectives but also all secondary leg objectives."

Added to the arbitrary executive order were other kinds of bad luck: stuck core barrel, bad weather again and again. After 20 days on Site 585, Schlanger and Moberly had this to report: "Faced with a delay of at least 48 hr. and no good chance of improvement even after that, we reluctantly decided to leave the site. . . . We could not waste additional JOIDES time waiting to attempt yet another futile single-bit hole at this site (after two were drilled). Thus we accomplished neither our primary leg objective of penetrating lower Cretaceous and Jurassic sedimentary rocks and oceanic crust in the Mariana Basin, nor one of our lesser objectives of logging there."

The *Glomar Challenger* then reoccupied Site 462, where the Nauru Basin hole was drilled 4 years earlier when the drilling had to be terminated at 1,068.5 m subbottom. Schlanger and Moberly were now given ten more days. The hole was reentered, the debris was cleaned out, and coring operations resumed on the evening of 9 November 1982.

The weather was not kind. "For a few hours on 13 November, . . . the weight indicator for the drill string (now some 6,350 m long) reached the new operating limits imposed for Leg 89 on two or three occasions." Fortunately, the winds and waves calmed down, and the hole was deepened another 137.3 m, to 1,209 m subbottom, penetrating mainly sheet-flow basalt with minor sedimentary rock. At 1654 hours on 16 November, when Sy Schlanger had to leave again with the *Challenger*, he was convinced that the "never-never land" could not be more than 50 m deeper than the bottom of the hole.

Neither Schlanger nor the *Challenger* ever returned, but Sy's friends Roger Larson and Yves Lancelot did. After the end of the Deep Sea Drilling Project in 1983, JOIDES secured funding to start, in 1985, a new Ocean Drilling Project with a new drill vessel, the *JOIDES Resolution*. As chairman of the JOIDES Central and East Pacific Panel, Schlanger was able to convince his peers of the importance of finding old Pacific crust. Instead of again trying a frontal assault at Nauru Basin, the planners favored the possibility of an "end run" in the region near the

Mariana Basin; seismic investigations had shown that the volcanic rock produced by mid-plate volcanism is thin or missing there.

This last attempt was scheduled as Leg 129 for the end of 1989; Larson and Lancelot were appointed co-chief scientists. "After 20 years' search the Jurassic Pacific has been found at last!" The news was publicized in the June 1990 *JOIDES Journal*. Schlanger was vindicated.

Hole 801 was drilled in the small Pigafetta Basin, some 600 km north of the Mariana Basin Site 585. The hole was spudded in 5,682 m of water, and penetrated some 300 m of red clay and Cretaceous radiolarian chert. The volcanic formation consisted mainly of turbidity-current deposits of volcanic debris, and was only a little more than 100 m thick. At 435 m subbottom, the sediments below the volcanic pile were sampled. The sedimentary sequence deposited on the "normal" ocean crust was only about 30 m thick. The middle Jurassic radiolarian chert above the basalt indicated that the crust was 170 million years of age. The drill string went more than 50 m into the basalt basement before Larson and Lancelot called it quits. They had finally found the oldest Pacific.

The scientific value of the final success cannot be overestimated. Not only were valuable samples obtained, which tell us the story of the old Pacific and its inhabitants. More significant was the verification of a theory: The volcanic rocks in the Nauru and Mariana Basins are not the ocean crust formed by seafloor-spreading; these rocks are the product of mid-plate volcanism, as Schlanger told me before we wrote that 1968 paper. The ocean crust formed by seafloor-spreading is Jurassic, as Roger Larson's magnetic lineations told us.

The satisfaction came too late to save Sy. He had lost Sally in an automobile accident in Riverside. Our friend Jerry Winterer called one morning in 1971 from La Jolla to tell me the sad news. Sy never recovered from the loss. I knew how he felt, because I too had lost my wife, Ruth, in an automobile accident, seven years before Sally died.

Sy remarried, and I saw him often in Europe during the late 1970s, when he was at Utrecht and we were both officers serving the International Association of Sedimentologists. He was not happy with the pomposity of European science, and elected to return to the States when Ralph Moberby offered him a post at Hawaii. Sy was happy there, but the isolation of the island was hard on his family, and he moved to Northwestern University in Evanston, Illinois, in 1981.

Sy had changed after Sally's death. He became increasingly distant and detached. His first disillusionment was national politics. He had been an engaged volunteer worker for the presidential campaign of Eugene McCarthy, and in 1986, prior to Ronald Reagan's reelection, I expected to see some fire of opposition in him. No, he was resigned to the inevitable.

The enthusiasm, the devotion to science, the spirit that had made him lose his shirt in the "roulette of the Nauru Basin," was also being eroded. He was still productive, and had in fact just completed, prior to his suicide, a grand synthesis on the nature, extent, and timing of mid-plate volcanism in the western Pacific; it will be published by JOIDES. Just the same, the distant view prevailed. Geological controversies, such as the current debates on the cause of catastrophic extinction at the end of the Cretaceous, seemed trivial to him.

Sy and I had dinner with Jerry and Jacqueline Winterer when we were all in Washington, D.C., to attend the International Geological Congress. That was July 1989, and Sy told me that he was living alone. I advised him to make a new start, but his eternal hopes of reconciliation reminded me of his frustration in Nauru Basin.

I went to Evanston after the Geological Society of America meeting in November, and Sy and I spent several evenings together. He cooked us a spaghetti dinner, remarking that he was "broke." With all their international travel, they had saved no money. In fact, they had spent all his pension fund. "I'd have to work until I'm 85 to get enough of a retirement salary for a decent living," he told me.

Through the years, we had become very close. All the praise that I had once heard was no exaggeration. He was an extraordinary friend: warm, perceptive, understanding. And he was generous, freely dispensing his time, his love; he could never be a miser.

We talked and talked, but we did not talk about mid-plate volcanism, even when Larson and Lancelot were out there to verify our theory—or, rather, his theory. We reminisced. Sy talked about his youth in Brooklyn, his student days at Rutgers, and his happy time on Guam. Yes, he mentioned Brazil. He remembered Riverside. That was all; his will to live left him when Sally was gone.

I tried to rekindle the fire that was once in him. I tried to persuade him to come to Switzerland. He could stay in Appenzell, and we could do field work in the Säntis Mountains. He could get back to reef limestones, the kind of sediment he first encountered on the island of Guam. For a moment or two, Sy was tempted. He would perhaps come in September, he said. But he did not.

Sy's health was failing. He would carry on as long as he could hold out, but . . . He was a proud person, and he would never be a burden.

On those November evenings we talked of life, of death, of the epistles of Paul the Apostle to the Corinthians. Life is service to God, and death a reward. Sally had served, and she had earned her reward. Did I not always feel the same way about Ruth's death?

In June 1990, Sy took a four-week leave to visit his two daughters in New Mexico. He went to a lawyer's office and secured forms for changing his will, but he left them blank; it seemed small from a distance. On the early morning of Sunday, 1 July, Seymour Oscar Schlanger took out the gun he had bought in Hawaii for his wife to guard against burglars, and fired one shot.

Sy Schlanger had given all, and he had earned an earlier reward.

11 Hawaiian Hot-Spot

Tuzo Wilson and Jason Morgan postulated that the movement of the Pacific Plate over hot-spots during the last 100 million years produced seamount chains. After the theory failed to explain the origin of the Line Islands Rise, Dale Jackson led Leg 55 to drill the Emperor-Hawaiian Seamount Chain. The shipboard operations summary told a "tale of horrors," but, except for the untimely passing of the co-chief, all's well that ends well.

Emperor Seamount Chain

The Pacific floor is dotted with innumerable submarine volcanoes, active and extinct. They are called seamounts if they have a conical summit, and guyots if they have a flat top. Seamounts are either active volcanoes building up from the seafloor or sunken, extinct volcanoes like guyots. More than a hundred years ago, the prominent American geologist R. D. Dana noted that seamounts tend to be arranged linearly, forming what we now call seamount chains. He also noted, by studying the chain northwest of the Hawaiian Islands, that the seamount farthest away from the active volcanoes of Hawaii seems to be the oldest and has sunk the deepest below sea level (Figure 11.1). This simple set of observations, as I discussed previously, prompted Tuzo Wilson to propose the hot-spot hypothesis to explain the origin of the Hawaiian Seamount Chain.

In the early 1950s Japanese oceanographers discovered a number of seamounts in the North Pacific; R. Tamaya named some of them after Japanese emperors. Two Soviet oceanographers, P. L. Bezrukov and Gleb Udintsev, recognized in 1955 that the Emperor Seamounts constitute the northern continuation of the Hawaiian Seamount Chain. A few years after Wilson published his "hot-spot" articles in 1963, E. Christofferson extended the idea to explain the genesis of the Emperor Seamounts. However, the Emperor Chain has a northward trend, in contrast to the northwest trend of the Hawaiian Chain. There is a kink, or bend, in the chain. Christofferson had to postulate a change in the direction of the Pacific Plate motion to explain the bend.

When Morgan in 1972 further extended the Wilsonian postulate to interpret other seamount chains of the Pacific, he was influenced by his observations of their parallelism. Besides the Emperor-Hawaiian Chain, Morgan identified three parallel chains: the Marshall-Austral, the Line-Tuamoto, and the Gulf of Alaska chains (Figure 11.2). All but the last have a peculiar bend. Morgan also noted that there is a site of present Pacific volcanism at the southern end of each of the four chains—on the Hawaiian Island, at MacDonald Seamount in the Austral

11.1. Emperor Seamounts. Linearly arranged extinct volcanoes on the ocean bottom constitute a seamount chain. The Emperor Seamount Chain was first investigated by Japanese scientists, and the individual seamounts are named after Japanese emperors. Locations of drill sites 430, 431, 432, and 433 are indicated.

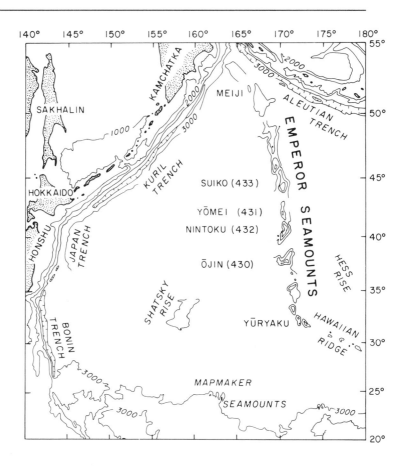

Chain, near Easter Island in the Tuamoto Chain, and near Cobb Seamount on the Juan de Fuca Rise. Those sites, according to Morgan, have been the four hot-spots in the earth's mantle, where convective plumes have been rising. The movement of the Pacific Plate over the hot-spots during the last 100 million years produced the four seamount chains. The speed of the motion should have been 7 or 8 centimeters per year; the plate made a turn from a northwestward to a northward direction some 40 million years ago, and this turn produced the kink on all the chains.

Morgan's idea was elegant in its simplicity. Unfortunately, Nature did not cooperate, and it was not very long before the hypothesis was shown to be an oversimplification. In fact, a year after Morgan's publication came out, two of Jerry Winterer's graduate students, David Clague and Richard Jarrard from Scripps, examined available geological data on the age of the Pacific seamount chains to test Morgan's hypothesis. The

11.2. Pacific seamount chains. Three seamount chains, the Emperor-Hawaiian Chain, the Marshall-Austral Chain, and the Line-Tuamotu Chain, are all distinguished by a kink. These three, and a fourth in the Gulf of Alaska, were thought by Jason Morgan to owe their genesis to four hot-spots in the earth's mantle. This idea has been verified only in part.

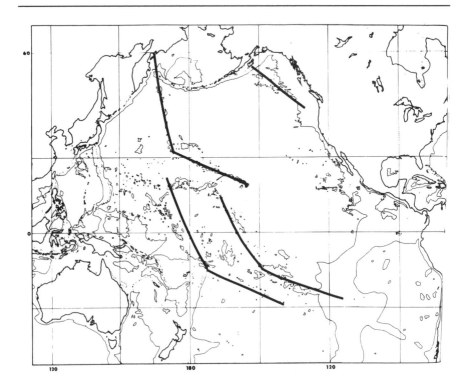

ages of the seamounts in the Emperor-Hawaiian Chain, which formed the basis of his kinematic analysis, naturally should be what the postulate stated. An extension of the idea to explain the genesis of South Pacific seamounts, however, ran into difficulties; the Line-Tuamoto and Marshall-Austral chains would have to be generated by some earlier motions of the Pacific Plate if they had anything to do with hot-spots at all.

As I related in the previous chapter, the results of drilling in the Line-Tuamoto Chain during Leg 33 and the investigations of the Marshall-Austral Chain during Legs 61 and 62 have contradicted Morgan's prediction: the volcanoes of those regions were produced by a mid-plate volcanism. However, the Wilson-Morgan idea was too good to be given up without first drilling the Emperor-Hawaiian Seamount Chain, the prototype of volcanoes generated by hot-spots. The hypothesis was to receive one final experimental test during Leg 55.

Portrait of a Chief Scientist

The *Glomar Challenger* left Honolulu, Hawaii, on 23 July and docked at Yokohama, Japan, on 6 September 1977. Dale Jackson and Itaru Koizumi of Osaka were the co-chiefs. Jason Morgan, the theorist, sailed with the cruise as a geophysicist, making routine measurements of physical properties of rocks on board ship. My assistant, Judy McKenzie, went as a sedimentologist to collect samples for geochemical analyses. She brought back numerous anecdotes of that memorable cruise—memorable because Jackson died of cancer shortly after he disembarked. The cruise report was dedicated to its late leader, and his shipmates pitched in and produced a masterful monograph, ensuring that Jackson will be remembered as the leader of the expedition that gave proof to the hot-spot theory.

The Emperor-Hawaiian Seamount Chain includes an array of 107 volcanoes that stretches nearly 6,000 kilometers across the floor of the Pacific. At the southeast end lie the eight principal islands of Hawaii, with the chain's only active volcanoes, Mauna Loa and Kilauea. Beyond the Island of Kure on the submarine Hawaiian Ridge, extinct volcanoes of the chain are entirely submerged beneath the sea. The chain of seamounts bends sharply to the north at a point some 3,500 kilometers northwest of Hawaii (Figure 11.2).

The lavas from the volcanoes of the chain were erupted onto the floor of the Pacific Ocean without any regard to the age or the preexisting structure of the ocean crust. The age of the crust of the Pacific Plate under the chain is 70 million years or older. The hot-spot was believed to have remained stationary at the southern end of the chain during the last 100 million years. Morgan predicted, therefore, that all the volcanoes of the

Emperor Seamounts were born at the latitude of Kilauea and Mauna Loa, or about 19° N. The lavas of the seamounts and guyots should thus have been covered by coral reefs, which grew at such a low latitude. Those tropical islands were eventually "shipped" out of the reef-building climate, and are now sunken under the stormy North Pacific, at latitudes as high as 50° N.

The JOIDES experiment was designed around drilling holes on four of the seamounts named after Japanese emperors—Suiko, Yomei, Nintoku, and Ojin. The ages of the seamounts, signifying the timing of the last volcanic eruptions at each site, could be determined by paleontological dating of the oldest sediments lying on top of the youngest lavas, and by radiometric dating of the lavas themselves. Also, the angle of inclination of the natural remnant magnetism of the lavas was to be determined, in order to test Morgan's prediction of the northward shift of the seamounts. If the theory was correct, there should be corals, as well as fossil skeletons of other tropical organisms, in the oldest sediments on top of each of the seamounts, even if their present latitudinal positions lie far north of the zone of coral growth.

Jackson and I were graduate students together at Los Angeles in the early fifties. He served in the Marines during the Second World War, and later studied with the financial support of the G.I. Bill, a subsidy provided by the U.S. Congress to help returning soldiers pursue higher education. Jackson fit the image of a battle-hardened veteran of the U.S. Marine Corps—he was dark haired and leather skinned, with awe-inspiring, penetrating eyes.

In my first year at Los Angeles, the geology department was housed in the old Chemistry Building. Ten graduate students, all male, had their desks in one big room, called the "bull pen." Jackson was an occupant of the bull pen, but being usually very quiet, and more mature, he seldom got involved in the trivial debates we younger students liked to indulge in. I was working with Dave Griggs then, and I saw Jackson in Griggs's laboratory, where Jackson worked occasionally on an hourly basis. He and I were also among the four graduate students who took Griggs's seminar on experimental geology.

In order to support his family, Jackson had to interrupt his graduate studies after a few years, and he went to work for the U.S. Geological Survey. He completed his doctorate some years later, after he had had a successful career studying volcanic rocks.

Jackson was a chain smoker, a heavy drinker, and a very intense person. He was a kind man, but he seldom smiled. Judy

McKenzie confided in me later that she was always a little afraid of doing something wrong in front of him, even though he never chided anyone.

Being a co-chief on *Glomar Challenger* could be a nerve-racking experience. I was sick with exhaustion for almost a week after I completed my duties as co-chief on Leg 42A. In the ensuing leg, I signed up as a member of the "troop" working as a sedimentologist "swimming" in oozes and mud. Many of us actually preferred the less demanding, routine activities of a shipboard scientist to the frustration, the responsibility, and the fear of defeat of a chief. I had many such moments of anguish as the leader of a cruise. However, I always had a co-chief who shared my angers and sorrows, and who also helped with decision-making. Jackson's co-chief was a friendly gentleman, but being Japanese, he felt more like a guest than a boss in this predominantly American venture. Jackson, the veteran, the marine, thus had to take charge on every occasion, and had to face crises alone. Leg 55 was Jackson's cruise, and the shipboard operations summary kept by him told a "tale of horrors."

Drilling into a thinly veneered hard basement in the shallow waters of the North Pacific is always an operational nightmare. The "hit-and-run" Leg 6 had nine bottom-hole assemblies twisted off and lost. Jackson was to get into similar difficulties. As I mentioned before, the soft-sediment cover had to be sufficiently thick at a drill site to support the drill string, so that pressure could be applied to the rotating drill bit to penetrate into hard basement. Seamounts are, however, rarely covered by a thick blanket of soft sediments. Bottom-currents flowing over the submarine elevations tend to remove the fine oozes. The residue of coarse sand and gravel can make further troubles, by slumping down into a drill hole and stopping the rotary motion of the drill pipe. Finally, the strong bottom-currents on top of a seamount may move the acoustic beacon around, preventing a stable positioning of the drill vessel.

At Site 430 over the Ojin Seamount, drilling at 1450 m water depth, Jackson had his first taste of things to come. The first hole had to be abandoned after 14 meters of penetration when the hole caved in; sand and gravel fell down into the hole and rendered drilling progress impossible. Jackson did better with the second hole at this site, but he ran into other problems—the drill bit was jammed, the core barrel was plugged, or the core catcher had failed.

After hours of waiting in anticipation, there would be calls: "Core is up!" Hopes were raised, but Jackson was to find again and again an empty barrel. Finally, the acoustic beacon used for dynamic-positioning also acted up; it drifted aimlessly around on the seafloor under the influence of a strong bottom-current. Not able to maintain a stable position for the drill vessel, Jackson had to abandon his second hole after only 118 meters of penetration, but he did get his basalt samples.

Jackson then tried to drill a third hole, but he got only a barrel of water for his effort; the core barrel had again been blocked by a piece of hard rock.

Jackson's trouble really came at the next site, over the Yomei Seamount. His first hole there was drilled down only to 29 meters beneath the sea bottom, into the sand and gravel above the seamount, when the "drill pipe began to vibrate on the drilling platform, broke free, and began to rotate independently. The drill string was pulled, and was found to be twisted off beneath the lowest bumper-sub. The remainder of the bottom-hole assembly, including the bit and the inner core barrel, were lost."

The quoted passage was entered into the operations log of the expedition at midnight, 6 August; such prosaic statements belied the frustration of the chief scientist.

Another trial at the same site was made. Fifteen hours later, the second hole managed to penetrate 4 meters deeper than the first, but again "the drill pipe at the drilling platform began to vibrate, sheared off." So reads Jackson's report. Left unsaid was the fact that another bottom-hole assembly was lost.

"The captain, the operations manager, the drilling superintendent, and the two co-chief scientists met to examine the cause of the failures. All agreed that the weak points in the bottom-hole assemblies were the small bottom subs, which had been machined in dry dock in Long Beach, and which had apparently been overshortened and weakened."

That was the published version of the discussion at the meeting. In fact, reading the shipboard reports, I realized that at the time, all but Jackson blamed the location. Impatient, they wanted to try their luck elsewhere. Jackson alone was certain that the problem lay with faulty equipment; they could not expect to do better at another location. However, he had to give up the site after losing two sets of expensive drill pipes, sacrificial offerings to Emperor Yomei. Judy McKenzie remembered the sad expression on Jackson's face as he sat defeated in the mess hall of the *Challenger* while the vessel departed, shortly after midnight on 7 August; they had gotten nothing but a handful of sand and gravel at this fateful location.

The *Glomar Challenger* now came to the Nintoku Seamount. Perhaps Jackson prayed to the ancient Japanese emperor in whose honor it was named. If so, in vain. The drill string broke even before the bottom-hole assembly was buried in the mud. The same emotionless passage appeared in the shipboard operations narrative by the chief scientist: "As the corer was being pushed in, the drill string at the platform began to vibrate, and a break occurred."

Those ominous words: The drill string at the platform began to vibrate. Another bottom-hole assembly was lost.

Jackson had only one bottom-hole assembly left, for one final trial. From this second borehole at Site 432 he managed to obtain a few pieces of basalt from the seamount before he again had to write the same sad expression: "The drill pipe at the platform began to vibrate."

Eventually a postmortem was done. Many different opinions were expressed. Some thought it was faulty planning; they condemned the very idea to drill into seamounts as downright foolishness. Others thought that the planning was all right, but that there had not been a thorough enough search for a suitable location where a thick sedimentary cover was present. Still others blamed the bottom-currents—perhaps no beacon could stay fixed on the seafloor above a seamount. The beacon moved and *Glomar Challenger* followed. A drill string cannot pull a 11,000-ton vessel back if the beacon has lured the giant away, they reasoned, so the drill string broke when the *Challenger* was lured away from the location. All these critics overlooked the fact that Jackson did all right during his second trial at the Ojin Seamount, when he drilled into hard-rock basement with less than 60 meters of sedimentary cover, and when he had plenty of time to terminate drilling when the bottom-current got too strong. The final verdict of a thorough investigation after the cruise vindicated Jackson. No, it was not the location, nor the currents. It was faulty equipment, as Jackson had insisted. The batch of bumper-subs used in the drilling of the first holes of Leg 55 did not meet specifications. Some "wise guy" had made the mistake of coming up with a new design, and the screw at the end of the drill pipes had been threaded too thin. The joint between the bumper-subs had been weakened, and broke easily under stress. When old bumper-subs were used, no more breakage occurred.

Meanwhile, Jackson had to give up Nintoku and ten days of his shiptime for a trip to Adak, Alaska, to pick up four more bumper-subs; losses had been so great, he had run out of equipment! The interruption may have been a joyful relief for the ship's crew, but it was sheer agony for the leader. On 18 August, *Glomar Challenger* returned from Adak with new supplies, and was positioned over the Suiko Seamount. Jackson's troubles, however, were not yet over. While drilling the first hole the stern-thrusters, which were needed to keep the vessel stationary on location, behaved "irregularly." The roughnecks on the derrick floor had to hurry and pull the drill string up out of the hole before it could be twisted off again. Thousands of meters of drill pipes had to be raised on deck and stacked up. Divers were sent down into the water. They found the culprit: "The aft stern-thruster was wallowed in a wad of nylon fishing-net, and polyethylene filament line!"

Who had been fishing at Suiko?

Fortunately, the thruster was not damaged, and the *Challenger* could return to her duty station.

At 1400 on 18 August, with less than ten days of the leg left to go, everything finally settled down. Dale Jackson was to be rewarded with a good bore hole and several good reentries to change the drill bit. Everything, in fact, went so perfectly that he let success go to his head—he started to discuss with his staff about sending in a request for a few days' extension of the leg, to get a few more meters of the valuable basalt of Suiko. He spoke too soon; someone must have overheard the conversation during the midnight-snack time. Soon things went wrong again! Core barrels came up empty, repeatedly, one after another. Jackson was told that the drill bit was plugged, and no further cores could be obtained. There was no choice but to raise the drill pipes back on deck.

The episode was reported by Jackson in his operations resume, with no humor intended: "At 0000Z 29 August (1100L), the pump pressure and drilling rate indicated still another bit plugged with basalt above the cones. We tried repeatedly to clear the bit with the chisel bit attached to the core barrel. Two short cores were pulled, the second one empty. By 0245 (1345L), the bit valves were hopelessly jammed open . . . The string was pulled and the bit was on deck at 1030Z (2130L). The top of the bit cones was jammed with basalt, and the center of the bit opening was jammed tight by a piece of rag (still recognizable as a pair of size 44 men's underpants) completely entwined about a 5-cm piece of basalt core. The bit teeth were still in a very serviceable condition (the underpants were not)."

Obviously, the sailors were anxious to get back to their families, and one of them had sacrificed his underpants to Emperor Suiko! They had been counting down the days left in the cruise; they were not about to make a new start if an extension should be granted.

At the time of the "bit failure," Jackson had 28 hours of drill time remaining. However, 26 hours would have been needed just to put the drill bit back down into the hole. He wanted to ask for an extension, to make up for the time lost due to defective equipment, and his wish would probably have been granted. Unfortunately, it was Sunday, San Diego time, an inconvenient time to make such a request. A good officer of the marines follows orders and would not think of taking unilateral action, however justifiable. Besides, Jackson knew that his gain would be someone else's loss. Taking inventory of the cores, Jackson realized that he had drilled into 114 lava flows, and had in fact accomplished all his objectives. Bowing to the inevitable, he gave the order for departure, and returned to Yokohama a day ahead of schedule. Having been a co-chief myself three times, I knew how every minute of shiptime was as precious as a jewel, and I could understand very well how heavily the pair of underpants, size 44, weighed on his mind.

Verification of the Hot-Spot Theory

So Jackson gave in, but his cruise was a great success, despite all the troubles. Reading the press release given out at the conclusion of the leg, one can barely detect traces of the hell that Jackson had gone through. The pretty words, peppered with bureaucratic pronouncements by the publicity department of the Deep Sea Drilling Project at Scripps, bear little resemblance to the shipboard operations narrative jotted down by Dale Jackson. You might have read a newspaper report in September 1978 like the following:

Emperor Seamounts in the North Pacific Were Once Coral Islands

Scientists aboard the Drilling Vessel *Glomar Challenger* for Leg 55 of the Deep Sea Drilling Project (DSDP) have discovered that the Emperor Seamounts in the North Pacific Ocean once rose above the sea as tropical islands. The evidence, gathered on a recent 45-day scientific drilling voyage, indicates that 50 to 60 million years ago these huge extinct undersea volcanoes were once islands rimmed with coral reefs and sandy beaches much like the Hawaiian Islands are today. The purpose of the scientific expedition was to test an important hypothesis about the origin of the Emperor Seamount Chain and the history of the motion of the crust of the Pacific Ocean during the past 70 million years.

The "hot-spot" hypothesis, first advanced in 1963 and refined during the following ten years, proposes that the Emperor Seamount Chain is a continuation of the Hawaiian volcanic chain. According to this idea, the Hawaiian and Emperor chains of volcanoes formed as the crust of the Pacific moved first north and then northwest over a fixed source of lava, called a "hot-spot," in the earth's mantle. The Hawaiian hot-spot is now presumed to lie beneath the active volcanoes of Kilauea and Mauna Loa on the Island of Hawaii. It was proposed that the Hawaiian-Emperor volcanoes become progressively older and sink deeper beneath the sea the farther away they are from

Hawaii. Scientists on Leg 55 say that the preliminary findings from four drilling sites on the Emperor Seamounts confirm this hypothesis.

The international team of scientists, led by Dr. E. D. Jackson of the United States Geological Survey, Menlo Park, California, and Dr. Itaru Koizumi of the College of General Education, Osaka University, included Dr. Gennady Avdeiko, Institute of Volcanology, Far East Science Center, Academy of Sciences, Petropavlovsk-Kamchatsky, USSR; Dr. Arif Butt, Institute for Paleontology and Geology, University of Tübingen; Dr. David A. Clague, Department of Geology, Middlebury College; Dr. G. Brent Dalrymple and Mr. H. Gary Greene, U.S. Geological Survey; Dr. Anne-Marie Karpoff, Institute of Geology, Strasbourg; Dr. R. James Kirkpatrick, Scripps Institution of Oceanography, University of California at San Diego, La Jolla, California; Dr. Masaru Kono, Geophysical Institute, University of Tokyo; Dr. Hsin Yi Ling, Department of Oceanography, University of Washington; Dr. Judith McKenzie, Geological Institute, Federal Institute of Technology, Zürich; Dr. Jason Morgan, Princeton University; and Dr. Toshiaki Takayama, College of Liberal Arts, Kanazawa University. The cruise drilling operations were supervised by Mr. Barry Robson of Scripps Institution of Oceanography.

Scripps Institution of Oceanography manages the Deep Sea Drilling Project, which is funded by the National Science Foundation through a contract with the University of California. The University of California subcontracts with Global Marine, Inc., of Houston, Texas, for drilling and coring using GMI's drilling vessel *Glomar Challenger*. Dr. Melvin N. A. Peterson is the project manager and Dr. David G. Moore is chief scientist.

Scientific advice is furnished to the Project by panels from the Joint Oceanographic Institutions for Deep Earth Sampling (JOIDES). JOIDES members are the Lamont-Doherty Geological Observatory of Columbia University; Rosensteil School of Marine and Atmospheric Sciences, University of Miami; Department of Oceanography, University of Washington; Woods Hole Oceanographic Institution; USSR Academy of Sciences, Moscow; Bundesanstalt für Geowissenschaften and Rohstoffe, West Germany; Ocean Research Institute of the University of Tokyo, Japan; Scripps Institution of Oceanography; Hawaii Institute of Geophysics; University of Rhode Island; Natural Environmental Research Council (NERC), United Kingdom; and National Center for the Exploitation of the Oceans (CNEXO), France.

The *Glomar Challenger*, under the command of Captain Joseph Clarke, departed from Honolulu on 23 July and arrived in Yokohama on 6 September 1977, with an intermediate stop at Adak Island in the Aleutians. During her 45-day voyage the *Glomar Challenger* traveled more than 5,000 miles and recovered rock samples from holes drilled into four seamounts of the Emperor Chain. The hole at Suiko

Seamount penetrated 550 meters below the ocean floor and sampled 385 meters of volcanic rock, making it the deepest hole ever drilled into basement rocks of the Pacific Ocean.

Jackson's summary of his successful cruise in the trade journal *Geotimes* had fewer words, but conveyed a more informative message. He told us that "in a general way the "hot-spot" origin of the Emperor Seamount Chain has been confirmed. The oldest fossils above basalt on Ojin, Nintoku, and Suiko seamounts became progressively younger. The ages of the seamounts indicate that the Pacific Plate moved at 9 centimeters per year. The chemistry of basalt from the seamounts is similar to that of Hawaii, suggestive of a common origin. The presence of coral reefs on seamounts proves that they once occupied more southerly sites, and the latitudinal changes are further indicated by the magnetic inclinations of the basalt flows."

This preliminary conclusion was drawn on the basis of scientific investigations carried out on board *Glomar Challenger*.

11.3. Age of volcanism in Emperor-Hawaiian Seamount Chain. The prediction by Jason Morgan's hot-spot theory that the volcanism is oldest at the northern end and is actually taking place at the southern end of the seamount chain is verified by sampling of the seafloor. The relation indicates that the Pacific Plate has been moving over the Hawaiian hot-spot at a rate of about 8 cm per year, during the last 70 million years at least.

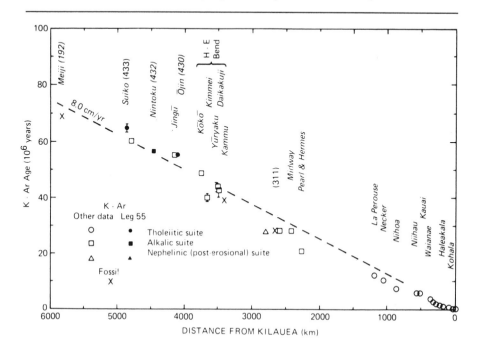

Detailed studies of samples in laboratories on land have confirmed the preliminary conclusion. The radiometric dating of basalt samples from Suiko, Nintoku, and Ojin seamounts all gave more accurate ages of volcanism than the paleontological dating on board the vessel. Suiko, the northernmost of the seamounts drilled, is also the oldest, having been formed 65 million years ago. Nintoku and Ojin are 56 and 55 million years of age, respectively. Combined with the radiometric ages of samples obtained by dredging from other seamounts, a remarkably linear relation between the ages of the seamounts and their distances from Hawaii has been obtained (Figure 11.3). The correlation indicates a speed of movement of the Pacific Plate of about 8 centimeters per year, and also that the shift of the movement's direction took place 43 million years ago, almost exactly what Morgan had predicted.

The latitude of the seamounts at the time of volcanic eruptions was determined by measuring the natural remnant magnetism of basalt samples. The analyses of the Ojin samples yielded results that agreed exactly with the predicted values, while the data from Nintoku and Suiko show some discrepancies, which might be related to small excursions of the earth's

11.4. Paleolatitudes of Emperor Seamounts. Crosses indicate the present latitude; circles, the paleolatitude at the time of active volcanism of seamounts on the Emperor-Hawaiian Chain. The predicted paleolatitude for all seamounts is 19° N, and the prediction is more or less verified.

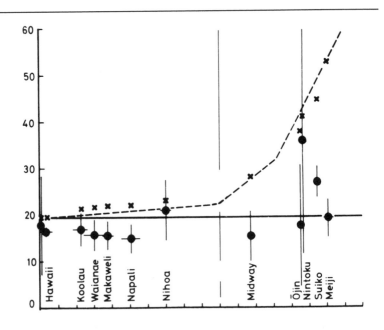

magnetic poles during that time. Combining the data with available results for other seamounts in the Emperor-Hawaiian Chain, it seems that during the last 70 million years the hot-spot has been more or less fixed at about 19° N, while the Pacific Plate moved several thousand kilometers (Figure 11.4).

Judy McKenzie brought back to our laboratory in Zürich samples of corals and bryozoans to determine the temperature of the ocean water in which these fossil organisms lived. Using the oxygen-isotope method first developed by Harold Urey (see chapter 2), she was able to ascertain that they must have grown in more southerly water, far warmer than the surface waters of the North Pacific above the sunken seamounts. The publicity people of Scripps did not exaggerate too much when they spoke of those undersea volcanoes as once being islands rimmed with coral reefs and sandy beaches like the Hawaiian Islands are today.

The story of the hot-spot had a happy ending, except Dale Jackson was no longer there to see the successful conclusion of his experiment. He died on 28 July 1978, 10 months after the end of the cruise. It would be overly sentimental to dramatize the story by saying that Jackson gave his life for the success of the "hot-spot" leg. The fact is, however, that he knew he was a dying man when he went on board *Glomar Challenger*, but he did not confide in anyone. He was completely exhausted after the cruise and disappeared for several months, somewhat mysteriously. Then he told his shipmates of his illness. He was getting better from treatments, he wrote, and expected to come to the post-cruise meeting. He did not make it; he died three days before his scientific staff gathered at La Jolla to prepare for the publication of the cruise report. Our profession lost a brilliant scientist, and we lost a great friend.

12 India's Long March

The earth's history since the end of the Paleozoic era is the saga of the breakup of the supercontinent Pangea. The early Mesozoic separation of Gondwanaland from North America and Eurasia produced the Atlantic and the Alpine Tethys, while a strip of continental hemisphere was split off the northern Gondwanaland margin to form the Himalayan Tethys. India was split off of the Gondwanaland later. Its long march northward is manifested by oceanic fractures, which were recognized as a new class of fault by Tuzo Wilson. Leg 22 drilling verified the geological and geophysical predictions.

Tethys and Gondwana-land

We all expect to see strange beasts and exotic plants in faraway lands; paleontologists also are used to finding fossil faunas of faraway places different from those at home. It was, therefore, quite a surprise to Eduard Suess when, during a visit to England in 1862, he noticed that the fossils brought back by Major General Stachey from Triassic formations in the Himalayas were almost identical to those found by his colleagues in the Dolomites of Austria. Suess, the foremost geologist of his day, was forcefully struck by this resemblance, but he had other, more urgent projects to pursue. Finally in 1892 an Austrian expedition to the Himalayas had an opportunity to investigate the matter more thoroughly. Many more fossils were collected and were examined by Suess's colleague E. Mojsisovics, who was able to confirm the faunal similarity. Meanwhile, Suess's son-in-law, Melchior Neumayr, was studying the fossils of the overlying Jurassic formation, and Neumayr was actually the first to propose an east-west ocean, extending from the Alpine region to the Himalayas. This Central Mediterranean (Zentrales Mittelmeer) was later given the more exotic name of Tethys by Suess, after the wife of Okeanus in Greek mythology. While

Africa's northward march eventually eliminated the west end of the Tethys through the elevation of the Alps (see chapter 6), a similar northward march by India closed off the eastern end and heaved up the mighty Himalayas. In between, the Tethys gave birth to the mountain ranges of southern Europe, Asia Minor, Iran, and Afghanistan. Only the eastern Mediterranean remains as the last relic of this mighty ocean.

Deep oceans are good barriers to the migration of animals and plants living on land. They can also prevent the dispersal of animals living on the bottom of shallow seas. The similarity between ancient faunas of India and Austria can be explained in part by the fact that many of the fossil species were pelagic, floating or swimming on the surface of deep oceans. Moreover, both the Dolomites of Austria and the High Himalayas were sediments laid down on the continental shelf on the south side of the Tethys, so animals crawling on the bottom could wander thousands of miles along the shallow shelf during the many million years of geologic time. Neumayr recognized an equatorial ocean because he could distinguish the different shelf faunas on the opposite shores of the Tethys. Later Suess gave the names Angaraland and Gondwanaland to the northern and southern continents, respectively.

Gondwanaland, as I mentioned before, was a giant southern continent of the Permo-Carboniferous age characterized by a distinctive group of plants called *Glossopteris* and by the widespread distribution of ancient glacial deposits (see chapter 4). Toward the end of the Paleozoic era, Gondwanaland and the northern continent were welded together to form Pangea, the one supercontinent of the earth. Some 160 million years ago, during the early Jurassic period, Pangea was again split into two. The Gondwanaland south of the Jurassic Tethys was thus a Neo-Gondwana, although it covered pretty much the same ground as the old. The Neo-Gondwana was separated from Angaraland when Africa moved away from North America to give birth to the Central Atlantic and to the Alpine Tethys (chapter 6). This southern continent was subsequently broken up into four continents (Africa, South America, Antarctica, Australia) and one subcontinent (India).

The drilling of the South Atlantic during Leg 3 had indicated that South America moved away from Africa some 130 million years ago; the South Atlantic was widened or spread apart as the two continents moved away from each other (chapter 5). Meanwhile, studies of magnetic lineations on the seafloor gave indications that India left Antarctica 20 million years later, and the process of seafloor-spreading between the two continents was responsible for the genesis of the Indian Ocean.

That the Himalayas owed their origin to the collision of India and Tibet had been accepted by many geologists even before the earth science revolution occurred. The controversy was centered on the distance traveled by India. If India had come from somewhere near the South Pole, where are the "footprints," the tracks left behind by the "long march"?

Augusto Gansser, my predecessor as Chairman of the Geological Institute at Zürich, had the answer. In a daring article written in 1966, a few years before the invention of the plate-tectonic theory, Gansser suggested that the rails for the "transport" of India are the two major fracture zones in the Indian Ocean—the Ninety-East Ridge to the east and the Owen fracture zone to the west.

Oceanic

Fracture

Zones

When one part of the earth's crust moves and an adjacent part stays stationary, there must be a break between the two to allow the motion. Such a break is called a fault, or a fracture zone. Geologists studying the motions of the earth's crust have traditionally recognized three major types of faults. Graben faults have a large vertical component; the Rhein-Graben, for example, was formed when a fragment of the earth's crust was lowered down between the Black Forest and the Vosges Mountains. Thrust faults describe the movement of one slice of the earth's crust on top of another; they are common in mountain ranges such as the Alps and the Himalayas, where the rock formations have been pressed together. The third type of fault, in which one piece of the ground moves past another, is called strike-slip. Strike, the intersection of the plane of fault motion with a horizontal surface, gives the direction of the slip of those faults—in other words, strike-slip displacements are horizontal.

Commonly, strike-slip faults can be recognized only when one sees a horizontal offset of linear fractures on the surface of the earth. The San Andreas Fault is the best-known example. The movement along the fault was first detected after the San Francisco earthquake of 1906, because water and gas pipes buried in shallow ground had been displaced horizontally 3 or 4 m by the earthquake's motion. We have no manmade objects to help us detect ancient strike-slip movements that took place a million years ago; we have to map linear features of the earth and decide if they have been offset horizontally.

A mountain chain is a linear feature. The Sierra Nevada of California ends rather abruptly in southern California north of the San Andreas Fault. Two California geologists, Mason Hill and Tom Dibblee, believed in 1953 that they had found the continuation of the Sierra Nevada on the other side of the faults, in the Coast Ranges of northern California, north of San Francisco. If they were right, there should have been a displacement of about 600 km along the San Andreas Fault during the last 150 million years. John Crowell, an instructor of mine at UCLA, later refined the study; his more reliable estimate was that the slip on the fault was about half Hill and Dibblee's value, and this all took place during the last 20 million years or so. At that rate, Los Angeles would merge with San Francisco in about 50 million years.

As a graduate student at UCLA in the early 1950s, I directed my antiauthoritarianism against those who disagreed with my dear professor. The simple matching of mountain ranges by Hill and Dibblee seemed to me more a child's game of jigsaw puzzles than real science. Crowell also seemed to have made debatable assumptions. I remained skeptical, because it was very difficult to prove that the linear features on different sides of the fault were once continuous and were indeed offset by the fault. My reluctance in believing in strike-slip faults was, of course, conditioned by my training in the tradition of a "fixist," who could not accept postulates of large horizontal displacements of the earth's crust.

Ten years later I was still an obstinate young man arguing against strike-slip faults when I was told of the horizontal offset of magnetic lineations on the seafloor. In 1952 Bill Menard and Bob Dietz discovered the Mendocino fracture zone through echo-sounding surveys of the depth of the Pacific bottom offshore western North America (Figures 8.1 and 12.1). The Mendocino is a spectacular topographic feature: the seafloor rises about 1,500 m from south to north across the fracture zone. Later Menard discovered several east-west trending fracture zones parallel to the Mendocino and called them the Murray, Clarion, and Clipperton fracture zones.

Menard first thought that the faults of his fracture zones had vertical displacements, as indicated by their submarine relief. However, when Raff and Mason went on Pioneer to map the lineations of the anomalous magnetic field of the Pacific floor (chapter 4), they found that the north-south trending lineations have been offset by the east-west trending fracture. Across the Murray fracture zone, a 200 km section of the anomaly pattern on the north side can be matched against a section on the south side (Figure 12.1). The match indicates a displacement of 155 kilometers, with the south side moving apparently westward. The match is indeed excellent, but I was still unable to overcome my prejudice and dismissed the evidence as an insignificant coincidence.

To my annoyance, those "coincidences" would not go away. Instead, offsets of magnetic lineations were soon found along other fracture zones. I did not let the discoveries bother me; I was working for Shell and my job was looking for oil.

One day in 1959 a friend, John Handin, came to my office

12.1. Pacific fracture zones. The ocean fracture zones of the Pacific were first discovered by Menard's echo-sounding during the early 1950s; they were then found to mark the displacement of seafloor magnetic lineations. The Mendocino fracture zone was considered an offshore continuation of the San Andreas fault, but the sense of movement seemed to be all wrong. Tuzo Wilson invented the concept of transform faulting to explain this paradox. The figure also shows the displacement across the Murray fracture zone by comparing magnetic anomalies; the anomalies in the boxed areas (shaded) can be matched across the fault.

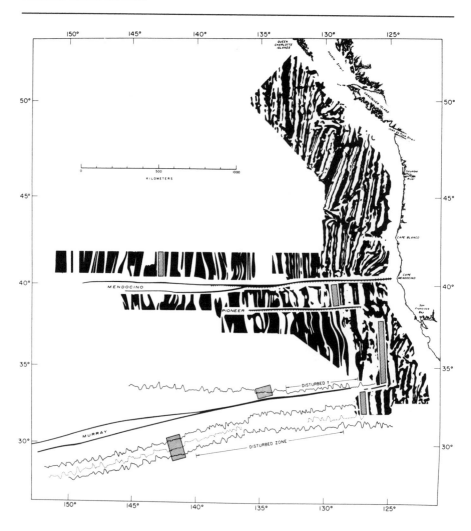

and told me of a *Nature* article by Victor Vacquier on ocean fractures. I looked up the paper. The new findings were even more incredible. Whereas the offset along the Pioneer Fault was a modest 265 kilometers, Vacquier claimed a displacement of more than 1,000 km along the Mendocino Fault!

The E-W trending Mendocino Fault and the famous NNW-SSE trending San Andreas Fault seem to come into each other somewhere offshore Cape Mendocino in northern California (Figure 12.1). One might thus assume that the Mendocino Fault was the offshore extension of the San Andreas Fault. But the sense of the movement was all wrong: the land southwest of the San Andreas was moving 600 kilometers westward, while the seafloor south of the Mendocino was moving 1,200 kilometers eastward.

"What kind of a hopeless traffic jam is this kind of movement leading to?" I commented with undisguised sarcasm. Handin, to my amazement, actually believed in Vacquier's sensational pronouncement.

Fracture zones are not confined to the Pacific. In 1965 Heezen and his associates from Lamont and Woods Hole were finding

12.2. Displacement of Mid-Atlantic Ridge. Arrows indicate first motion of the ocean floor during an earthquake. Lynn Sykes first found that motion is eastward for the seafloor east of the Mid-Atlantic Ridge, and westward for the seafloor west of the ridge.

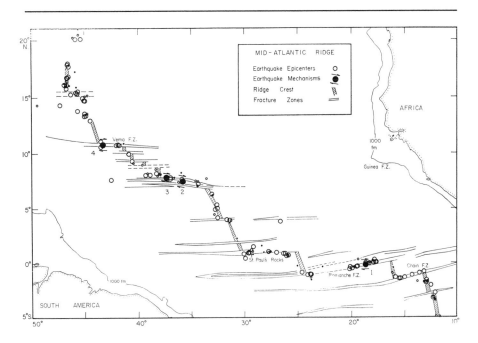

large fracture zones in the Equatorial Atlantic. The displacements are manifested by apparent offsets of the Mid-Atlantic Ridge (Figure 12.2). The Atlantic fracture zones have a steep escarpment rising above a narrow valley, which is underlain by very thick sediments of very young age. One of those is called the Vema fracture zone, named after Doc Ewing's famous Lamont ship.

In 1966, when I was still at Riverside, I was invited down to Scripps by Jerry Winterer to give a talk on the origin of oceanic plateaus, the idea I was then working on with Sy Schlanger (see chapter 10). During the visit I got into a heated debate with Jerry van Andel, who told me that not all oceanic features could be explained by vertical displacements: "How do you explain the Vema fracture zone?"

I did not know anything about the Vema fracture zone, but I dismissed it as just another graben fault, like the Rhein-Graben. Van Andel did not let me slip away; he bored in and gave me a lesson in marine geology.

The crest of the Mid-Atlantic Ridge between 15° N and 5° S is offset to the east nearly 4,000 kilometers through a series of fracture zones (Figure 12.2), of which the Vema fracture zone is one of the most prominent. This narrow east-west trending trough cuts through the ridge at latitude 11° N, and was first described and named by Heezen and others in 1964. Van Andel led a 1966 Scripps cruise to survey this fracture zone, and was just back when I wandered into his office.

One of the most amazing features of the Vema fracture is the thickness of the sediments in the trench. The reflection time was more than 0.9 second, representing a kilometer-thick layered sequence. "Whatever its origin," van Andel reasoned in his 1967 *Journal of Marine Research* article, "a sediment thickness of approximately one kilometer represents, at this distance from the continent and in this water depth, a very long time interval, probably in excess of several tens of millions of years." The Atlantic Ridge, which is offset by the Vema fracture zone, is, in contrast, a very young feature; its ridge crest is only a few million years old, according to Vine and Wilson's latest estimate.

"Your graben-faulting cannot explain the Vema fracture zone," van Andel argued. "You cannot dump old sediment in a young graben. You have to accept Tuzo Wilson's mechanism of transform faulting."

"Tuzo is no structural geologist. He never did a decent day's field work. What does he know about faults?"

I was resorting to the tactics typical of a doctrinaire reactionary, combining ignorance with arrogance in a feeble attempt to win an argument. Van Andel was getting excited, but I was saved by the bell—Jerry Winterer came in and told us that I had to go give my talk.

The Vema fracture zone was first drilled during the Leg 4 cruise in 1969. Dick Bader and Robert Gerrard were co-chief scientists. About half of the sequence, 610 m thick, was penetrated at Site 26, but the sediment is still Upper Middle Pleistocene, or about 1 million years old. The thick sequence, at least the upper part, is relatively young, because of the very fast rate of turbidity-current sedimentation. The minerals in the sandy layers include hematitic quartz, rose quartz, beryl, and corundum, indicative of a derivation from the Precambrian shield of the Amazon region. The shipboard scientists thought that this part of "the Vema fracture did not originate until the Pleistocene" because it is situated near the axis of the Mid-Atlantic Ridge, and it is probably a transform fault. Van Andel was wrong in assuming a very old age for the sediments, but I was also wrong in dismissing the postulate of transform-faulting.

Site 26 was redrilled by JOIDES/DSDP Leg 39; Katharina Perch-Nielsen and Peter Supko were co-chief scientists. By then, the postulate of a transform-fault origin for the fracture zone was already a foregone conclusion. The objective was to sample the lower crust at the foot of the steep trench wall. The effort could be considered an "end run for Moho" because faulting had brought up close to the surface the lower crust and upper mantle. The first attempt to drill into the trench wall (Hole 353) was interrupted by a mechanical failure after penetrating some 400 m sediment. The other two attempts, made closer to the wall, terminated in a layer of basalt fragments of cobbles, which had fallen down from the steep slope.

Drilling in the Atlantic did not add much to our understanding of oceanic fracture zones.

A New Class of Faulting

Back in the 1960s, when we were still puzzling over oceanic fractures, one aspect was particularly troublesome: some oceanic fractures seem to find a continuation on the continents. The Mendocino fracture could be viewed as a continuation of the San Andreas, but, as I mentioned, the sense of movement was all wrong. Geologists, who could see faults and fracture zones with their own eyes on land were thus very suspicious of the findings of their oceanographer colleagues who mapped underwater features with fancy gadgets. We did not quite understand what they were doing, and we tended to take the attitude that what they found was irrelevant to what we were working on.

The concept of transform faults was an outrageous idea suggested by Tuzo Wilson in 1965, two years after he proposed the outlandish notion of "hot-spots." Wilson said that those oceanic fractures are not really strike-slip faults; they are "a new class of faults," to which he gave the name transform faults.

Geologists may not have believed in large displacements of strike-slip faults, but they at least knew what those faults are; the direction and distance of the offset along the fault can be measured by matching two equivalent points across the fault (Figure 12.3). Now Wilson wanted to tell us that there was another kind of fault. Furthermore, he had the guts to say that the way we determined fault displacement across fracture zones could be wrong. Yes, we should match magnetic lineation patterns, but the ground has not moved in the way we thought it did. The sense of movement along a transform fault is directly opposite to what our intuition tells us.

The key to this paradox is that an ordinary fault only displaces, but the new class of fault owes its origin to an increase in ocean crust surface area. Looking at the sketch in Figure 12.3, for example, a linear marker is displaced by an ordinary fault. The fault is a right-lateral (dextral) strike-slip fault if the block across the fault is moved to the right side, and left-lateral

12.3. Transform faulting and strike-slip faulting. Arrows indicate that transform faulting (above) causes motions of the seafloor different from those caused by strike-slip faulting (below).

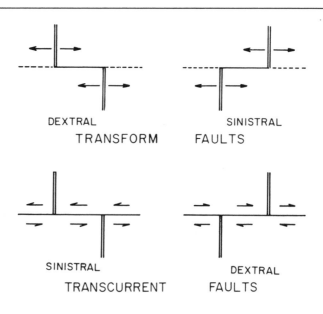

DEXTRAL   SINISTRAL

TRANSFORM   FAULTS

SINISTRAL   DEXTRAL

TRANSCURRENT   FAULTS

(sinistral) if the block moves to the left. The San Andreas Fault, for example, is right lateral. There is conservation of matter in strike-slip faulting; no new rocks have been added, or squeezed in on either side of the fault.

Wilson had been the first to accept the seafloor-spreading theory, advanced by Vine and Matthews in 1963, which pro-

12.4. Accretion of seafloor and transform faulting. Tuzo Wilson used this series of diagrams to demonstrate the transform faulting of the Atlantic seafloor. Note that $D'B' = D'C$, $DB = DC'$; the equal displacement is a manifestation of symmetrical seafloor-spreading south and north of the transform fault AA'.

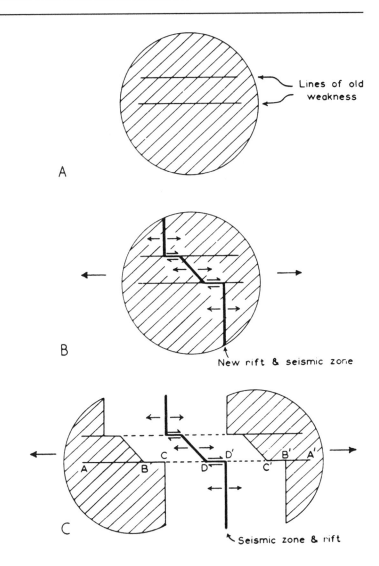

A

Lines of old weakness

B

New rift & seismic zone

C

Seismic zone & rift

posed the addition of new materials from the mantle to make ocean crust under the seafloor. Wilson visited Cambridge in 1965, talked to Vine, thought the problem over during a holiday, and came back with the idea that adding new materials caused a new kind of faulting.

Geologists were used to strike-slip faults. When they looked at the offset of the Mid-Atlantic Ridge across the Vema fracture zone of the Equatorial Atlantic, an intuitive conclusion was that the ocean floor to the north of the fractures moved westward relative to the floor to the south (Figure 12.2). Wilson said that it was not so! He made a diagram to portray the seafloor-spreading of the Atlantic and the nature of the transform faults (Figure 12.4). The initial crack that broke up Pangea did not go straight down the middle; there was a zigzag course, as manifested by the bulges of West Africa and South America. During the seafloor-spreading of the Atlantic, the movement of the continents was largely latitudinal. The axis of the Mid-Atlantic Ridge marks the position where the crack was. During the course of seafloor-spreading, materials added onto the mid-ocean ridge pushed the continents and the previously formed seafloor apart. The half on its right (Europe, Africa, and the eastern Atlantic) has been moving eastward; the half on its left (the Americas and the western Atlantic), westward. In other words, the eastern Atlantic seafloor north of the equatorial fracture zones (D' in Figure 12.4) has been moving eastward, not westward, relative to the sea floor south of the fault (D in Figure 12.4). The motion on a transform fault there is therefore directly opposite to what a geologist would intuitively assume on the basis of strike-slip faulting.

Wilson's idea was so unconventional that he convinced few and enraged many, especially those who had spent all their life mapping faults on land. When I first read Wilson's hypothesis in *Scientific American*, I was so furious that I could not force myself to finish reading the article.

I wrote the preceding paragraph ten years ago. Since then, I have had a chance to read Menard's account of Wilson's idea about transform faults. He wrote:

The immediate reaction to Wilson's paper generally ranged from positive to enthusiastic. It was elegantly organized and lucidly written. It gave a simple logical explanation for, or useful way of viewing, a host of well-known problems. Best of all there was no reason to oppose it. It fitted admirably into any hypothesis of global tectonics.

How could the judgments of land and marine geologists be so different?

If I had been more modest, I would have discovered that Wilson was not making idle speculations. He made precise predictions of displacements along oceanic fracture zones: if the Vema fracture zone, for example, is a strike-slip fault, the displacement of the Mid-Atlantic Ridge would be left lateral, and the ocean crust north of the fault should have been moving westward; if the fracture zone is a transform fault, the part of the ocean crust east of the ridge should have been moving eastward.

Displacement on a fault produces earthquakes, and the motion is registered by seismograms. Lynn Sykes of Lamont was a graduate student working on that very problem in 1962, but during the early 1960s Lamont was a bulwark of resistance to the theory of seafloor-spreading. Therefore, Sykes did not pay much attention to Wilson's new idea until May 1966, when he was shown a copy of that amazing *Eltanin* 19 profile of magnetic lineations (see chapter 4) and converted to the new creed. Dropping everything, Sykes processed available data from various stations of the Worldwide Standardized Seismograph Network, with the computer facilities of the NASA Goddard Space Flight Center. The displacements, as registered by 17 earthquakes from the Atlantic and Pacific, were worked out in a few weeks' time.

The answer was clear: the Vema fracture is not a left-lateral fault. The ocean east of the Mid-Atlantic Ridge moves everywhere eastward (Figure 12.2). In fact, every one of the 17 faults has moved exactly in the direction predicted by the Wilsonian theory.

I was in the audience when Sykes gave his talk at a Zürich meeting of the International Union of Geodesy and Geophysics in 1967. Sykes presented his results without fanfare, and he spoke in a monotone. I did not understand Tuzo Wilson, and I understood Sykes even less. His words of wisdom fell on deaf ears. Another person in the audience, Dan McKenzie, did understand, and was impressed. Later on, McKenzie was to tell us that Sykes's verification of the Wilsonian paradox was the revelation that made him a believer of the seafloor-spreading theory.

McKenzie's conversion was necessary before he could start thinking about plate-tectonics. Morgan also modestly claimed that his presentation was only "an extension of the transform-fault concept." We have thus witnessed a chain reaction, triggered by Hess's idea of seafloor-spreading, leading to the interpretation of magnetic anomalies by Vine and Matthews, to the innovation of transform faults by Wilson, to the confirmation of their predicted motions by Sykes, and to the formulation of the plate-tectonic theory by Morgan, and by McKenzie and Parker, which provided a theoretical basis for the postulated seafloor-spreading. The circle was complete!

Ninety-East Ridge

The Ninety-East Ridge was thought to be a transform fault. Unlike the Atlantic transform faults, the Ninety-East is not a trench but a ridge. The ridge was first recognized by S.R.B. Sewell in 1925 as a major topographic feature in the Indian Ocean. Heezen and Tharp called it "Ninety-East" because it runs along the 90° E longitudinal from 15° N to 31° S—almost 6,000 km! For most of its length, the ridge stands some 1,500 to 2,000 m above the adjacent seafloor.

Distinctive magnetic lineations of the Indian Ocean were discovered in the early sixties, and the anomalies on the Carlsberg Ridge, as I mentioned earlier (chapter 4), provided the inspiration for Vine's seafloor-spreading theory. Later Drum Matthews of Cambridge and Bob Fischer of Scripps continued to collect information with shipboard magnetometers. In less than a decade, the data had accumulated to such an extent that Dan McKenzie, working with John Sclater, could interpret the magnetic data to produce a scenario for the wandering of the fragmented Gondwana continents.

Glomar Challenger entered the Indian Ocean after leaving Darwin, Australia, on 13 January 1971, and traveled the whole ocean in a counterclockwise fashion, drilling at 64 sites during seven cruises (Legs 22–28) over all the ridges and basins from the Red Sea to the Antarctic. The program was completed on 23 February 1973, when *Glomar Challenger* arrived at Christchurch, New Zealand. The results were so satisfactory that no more drilling of the Indian Ocean was scheduled until the *JOIDES Resolution* came in 1987. Much of the work could be characterized as routine data-gathering. The first and the last of the cruises, however, brought back findings that made newspaper headlines—Leg 22's investigations of the Ninety-East Ridge and Leg 28's drilling in Antarctica.

John Sclater joined Chris von der Borch of Australia to lead Leg 22. Sclater was another of Teddy Bullard's "bright young men." He came to the United States after receiving his degree from Cambridge. I saw him often during my sabbatical year at La Jolla in 1972. Sclater later went to MIT and then to Texas. When I was named co-chief scientist for Leg 73, I called him up in hopes that he would join me as the other co-chief. Unfortunately, his wife was lying in a coma, and of course he could not leave her. She eventually passed away, and Sclater went down to Austin, Texas, to start a new life.

Back in the early 1970s, Sclater was a carefree young man, full of enthusiasm for all the good things in life. He had worked with Dan McKenzie on the seafloor anomalies of the Indian Ocean. A tour of duty on the *Challenger* gave him an opportunity to verify his own interpretation. After twenty-one deep-sea drilling legs, the dating of magnetic lineations had become routine and was no longer the exciting "numbers game in the South Atlantic" (see chapter 5). Some discrepancies from the predicted values were noted, and the new results helped improve the picture, but on the whole, Sclater found that the drilling results confirmed what he had known all along (Figure 12.5). About 105 million years ago, India, Australia, and Antarctica were part of the southern continent Gondwanaland. Then India and Africa took off. Seafloor-spreading along an east-west trending mid-ocean ridge created an ever-widening Indian Ocean. India was riding piggyback on the Indian Plate, like a container on a freight train, and moved northward until it met Eurasia in a collision that gave rise to the Himalayas. Meanwhile, Australia remained attached to Antarctica until 53 million years ago. The Ninety-East Ridge was situated on or near a transform fault! During the whole time that India was moving north, Australia stayed stationary on the other side of the fracture.

The Ninety-East Ridge was formed by uplift of the ocean crust (Figure 12.5). The ridge has a north-south grain south of 12° S and an en echelon nature north of this point. Cretaceous and Eocene oozes had been dredged from the sides of the ridge. Site 214 was chosen on an elevated region at the crest of the Ninety-East Ridge at 11°20′ N and 88°43′ E.

A surprise came when a lignite bed, almost a meter thick, was discovered on the top of the basalt basement at Site 214, now lying in 1,670 m water depth. Several holes were then drilled farther north on the ridge. They all showed the same pattern of subsidence. The elevations on the ridge all started out as volcanic islands. Lignite was deposited in ancient swamps, and coral reefs grew on the fringes of the islands. After the volcanism ceased and subsidence started, islands were turned into seamounts or guyots. Lignites and debris from coral reefs were ultimately buried under the nannofossil oozes of the Indian Ocean.

The Ninety-East Ridge is not exactly a fracture zone of transform faults; it is a ridge, a seamount chain. To the east of the ridge lies a series of long, linear north-south trending deeps and highs. The most prominent trough lies immediately east of the ridge (Figure 12.5), and this trough marks the remnant of an old transform fault. The ridge itself, Sclater and his associates believed, is a feature of volcanism, attached to the Indian Plate. The basement age increases to the north, and this is typical of

12.5. Ninety-East Ridge in the Indian Ocean. The ridge stands 1,500 to 2,000 m above the surrounding seafloor. Locations of DSDP Sites 211–218 are indicated.

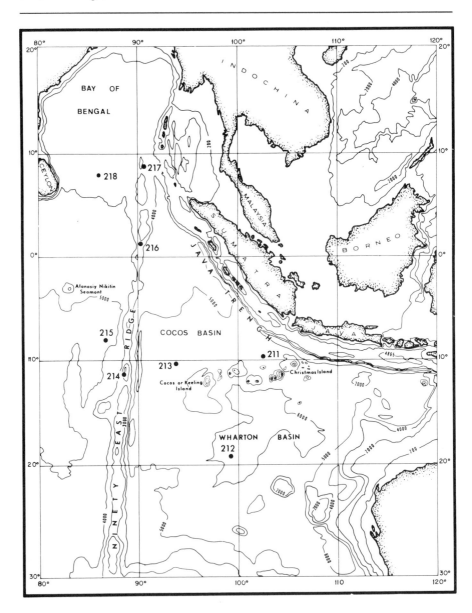

the northerly increase in age of the Indian Plate. The depth of basement on the ridge also increases to the north, similar to the direction of increasing depth on the Indian Plate. Furthermore, the movement of the volcanoes on the ridge has a kinematic history similar to that of the seafloor of the Indian Plate.

Invoking the hot-spot theory as a model, Leg 22 scientists related the origin of the Ninety-East Ridge to the migration of the Indian Plate along an old active transform fault over an isolated magma chamber. Lavas, derived from a hot-spot hidden in the deep mantle on the Indian side of the Ninety-East Ridge, were preferentially deposited near the old transform fault. After the Indian Plate moved north and arrived at the hot-spot 80 million years ago, the first volcanoes on the ridge rose above the seafloor. The chain of seamounts was produced by the successive northward motion of the plate. Drilling proved, as the theory had predicted, that the northernmost volcano (9° N) is the oldest; it has sunken to a depth of more than 3,000 m. The one at the southern end (31° S) is younger and thus stands considerably higher; it has gone only half of the way down to its final resting place.

Northward March of India

The Ninety-East Ridge traces the northward movement of the Indian Plate. The two extreme locations on the ridge are 4,500 km apart, and their age difference is 40 million years. The speed with which the plate moved over the hot-spot was thus more than 10 cm per year during this period of fast seafloor-spreading.

On the basis of geophysical and geological information, Sclater and his colleagues reconstructed the relative positions of the Indian, Antarctic, and Australian continents from the late Cretaceous until the present. In the early Cretaceous, India, Antarctica, and Australia formed the eastern part of the Gondwana land mass (Figure 12.6d). Around 100–105 million years ago, spreading centers became active, and India and Africa started to move away from Antarctica and Australia (Figure 12.6c). About 40 million years ago, Antarctica started to separate from Australia; India and Australia then became part of the same plate (Figure 12.6b).

Evidence of the northward march of the Indian Plate is provided by paleomagnetic studies of drill cores. The remnant magnetization of seafloor basalt, as discussed in the previous chapter, gives the latitude of the site at the time when the basalt first came out. Laboratory results indicate that the basalts from the Indian Ocean floor were erupted at a site far to the south of their present latitude. Basalt samples from equatorial Site 216, for example, have a remnant magnetization of 40° S latitude—in other words, the site has moved 4,500 km during the last 65 million years.

Further evidence of northward movement has been given by studies of fossil faunas and floras. Cold-water faunas and temperate floras are present in the oldest sediments of the boreholes now situated near the equator. This is understandable if we accept the theory of seafloor-spreading, which places Site 217 at a high southern latitude at the time when India was first separated from Antarctica. Organisms living in the earliest Indian Ocean belonged to the types that flourished in cold or temperate

12.6. Four stages in the fragmentation of Gondwanaland, reconstructed by Dan McKenzie and John Sclater on the basis of seafloor magnetic stripes: (a) present, (b) 35 million years ago, (c) 75 million years ago, (d) Jurassic Gondwanaland prior to the fragmentation.

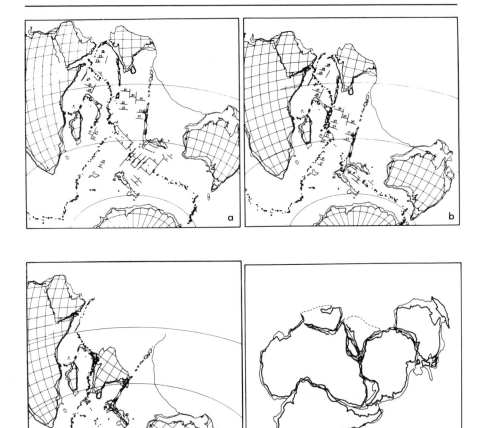

climates. The old crust was moved to the tropics during the northward march of the Indian Plate.

India's northward march, as determined by the results of the deep-sea drilling cruises, has been verified by paleomagnetic studies of rocks on land. Chris Klootwijk of Australia and his associates in India studied the remnant magnetism in rocks of India and Pakistan. Rocks older than 100 million years, formed when India still lay close to Antarctica, indeed showed a magnetic inclination that indicated a high southern latitude. Samples from successively younger formations indicated a systematic latitudinal change, recording the passage of India northward. Northern India started to cross the equator some 60 million years ago, preparing for the collision with Tibet.

Terminus of the Long March

What was north of India before the collision? The Tethys, of course. What happened to the Tethys?

I was able to get firsthand information on that when my wife, Christine, and I visited Tibet in 1981. We were told that the terrain around Lhasa was also once part of the great Gondwanaland, but it was separated from the mother continent as an island arc early in the Mesozoic. The Tethys was formed between this arc and Gondwanaland by seafloor-spreading during the Triassic and Jurassic. The Mesozoic sediments of the high Himalayas were laid down on the northern passive margin of the Indian subcontinent south of the Tethys ocean.

Eventually, the axis of the seafloor-spreading shifted to the south. India was split off from Antarctica to begin its long march. While the Indian Ocean expanded, the Cretaceous Tethys Ocean shrank in size as India moved steadily northward. Finally, during the Eocene, about 40 million years ago, India and Tibet collided, and the Tethys was eliminated (Figure 12.6c). The passive-margin sequence of India was peeled off from its basement to form folds and thrust faults, like the so-called Helvetic sequence in the Alps. The direction of subduction is, however, opposite: In the Alps, Africa or its vanguard, Italy, plowed over Europe; the subduction zone dipped to the south. In the Himalayas, India plunged under Tibet, and the subduction zone dipped to the north.

The northward march of India was slowed down by the resistance of Tibet. The terminus of India's long march, or the scar of the collision, is the Indus-Tsangpo Suture. Driving westward in the Tsangpo River Valley from Lhasa to Shigatze, we saw outcrops of green ophiolite or red chert on either side of the highway. They are the remnants of the mighty Tethys, and they are now found only as exotic blocks in a chaotic mélange.

The leading edge of the Indian plate has been subducted under Tibet. The melting of the ocean floor, swallowed in the bowels of the earth, gave rise to volcanism and igneous intrusion (chapter 6). The volcanic rocks are largely eroded, but the granites, or the root of volcanism, are widely distributed in central Tibet. The famous Potala of Lhasa stands on one such granite hill.

The key to understanding the Himalayas thus lay with the drilling of *Glomar Challenger*. The decline and fall of the Tethys was related to the birth and growth of the Indian Ocean.

Three Exploring New Territories

1973–1975

13 Antarctic Adventures

Maurice Ewing sent the D/V Challenger *to the South Pole, and Dennis Hayes had good-weather drilling in Ross Sea. The lucky success of the first Antarctic summer was followed by the unforeseen disasters of the second. The "invasion" of a new territory nevertheless broadened our horizons and furnished samples enabling a reconstruction of climatic changes during the last 60 million years.*

When Did the Ice Age Begin?

Cesare Emiliani's original purpose in proposing a long-core program (LOCO) with the use of a dynamic-positioning vessel was to obtain a record of the climatic changes during the last Ice Age (see chapter 2). Eventually LOCO was changed to JOIDES, and the aims of deep-sea drilling also changed. The first twenty-seven legs of the project were preoccupied with experiments to test the various predictions of the theories of seafloor-spreading and plate-tectonics. We did not quite forget Emiliani. I remember the last days of Leg 3, before our triumphant return to Rio de Janeiro from the successful drilling of the Mid-Atlantic Ridge. We had achieved our objectives, but orders were given to drill one more site over a submarine elevation called the Rio Grande Rise. The weather was bad, the equipment was malfunctioning, the inexperienced drilling crew was tired, and a frustrated operations manager wanted to get to Rio before the festival ended, but Art Maxwell persisted. Not only was it too early for the port call, but the JOIDES planning committee had given the mandate to obtain a long continuous core of Pleistocene sediments for Emiliani. Maxwell, a reserve officer in the U.S. Navy, was to carry out this command, rain or shine. Besides, Emiliani was a good friend; we could not disappoint him.

But disappoint him we did! Maxwell tried his best, but *Glomar Challenger* was not able to help Emiliani. He wanted a hundred meters of a continuous record, and he also expected no mechanical disturbance of the sediment core. This posed no problem when one used the piston-core device invented by Kullenberg (see chapter 2). However, the drill string coming from *Glomar Challenger* was not suited to taking undisturbed cores in soft sediments. The oozes deposited during the last Ice Age are very young sediments and have not yet been compacted. The rotary drill stirred up the oozes like a spoon in porridge. The first 50 or 100 meters of drill cores were always badly disturbed. Like a book with its pages torn out and rearranged randomly, a disturbed core, with older sediments mixed in with younger ones, is not a good document of history.

Maxwell got more than one hundred meters of Pleistocene sediments for Emiliani, but they were useless. Others, on other cruises, also tried to help him, but one does not make wood carvings with a butcher's knife, nor could we get good cores from soft oozes with *Glomar Challenger* until we had better equipment. Eventually the equipment was improved, and we did get cores for Emiliani (see chapter 20), but in the meantime he could not wait, and as he did not need a really long core, improvement of the Kullenberg-type piston corer was able to give him what he wanted, namely a record of the climatic changes during the last Ice Age.

Many of us geologists are interested in the climates on earth. While the *Challenger* was not able to provide a fine record of the more recent history, the drill string could penetrate deeper than any piston corer and could thus obtain earlier histories. Instead of worrying about the number of ice ages during the last million years, people began to ask questions about the beginning. There were glaciers on Gondwanaland 300 million years ago; then the climate got warmer. When did the last glaciation on earth begin? Why did it begin, and where?

On land, the earliest sign of the last glaciation was provided by the earliest moraine. Yet moraines seldom contain fossils that can provide clues as to their age, nor is the radiometric method (^{14}C) any use in dating older moraines. When Emiliani first attempted to correlate temperature changes of the oceans, there was a tendency to believe that the Ice Age began everywhere with Günz half a million years ago. Some of the less informed colleagues in archaeological studies were for years happily using Emiliani's 1961 time scale.

Some thought that the climate change causing the first glaciation was a globally synchronous phenomenon. They proposed, therefore, that the Pleistocene epoch should be defined not by fossils, as were the other geological epochs, but by criteria indicative of the first cooling. There was a suggestion that the earliest arrival of the mollusk species *Arctica islandica* in the Mediterranean be chosen as the time when the Pleistocene epoch began. Others preferred the microfossil *Hyalinea balth-ica*, another invader from the north to Calabria in Italy. Still others preferred to define the beginning of Pleistocene by studying pollen and spore assemblages in sediments, which give a record of vegetation changes in response to a cooling climate. Those proposals led to great confusion.

Even before *Glomar Challenger* steamed out on its first voyage in 1968, studies of ocean sediment samples obtained by long piston corers had corrected several misconceptions. The accurate dating of marine sediments proved that the marine creatures living in the cold waters of high latitudes did not all show up in the temperate regions at the same time. Nobody is foolhardy enough to say that summer begins when the first Nordic guests arrive on various Mediterranean beaches. Why then should we assume that the first arrival of the cold-water faunas in the south coincides with the beginning of the Pleistocene?

Studies of sediment samples from the oceans gave indications that glaciation may have started much earlier than was previously assumed, as longer and longer piston corers were developed. Several times, Emiliani had to revise his estimate of the timing of the initial glaciation on the continents. In 1961 he thought that the first glaciers came 0.3 million years ago. Three years later, the timing was revised to 0.42 million years before present. Then, in two papers published in 1966, Emiliani made revisions that pushed the timing further back, first to 0.6 and then 0.8 million years. Dave Ericson of Lamont and his associates thought that the first glaciers in North America could be dated at 1.5 million years of age. Others speculated on even older ages. But how much older? Long piston corers seldom obtain cores longer than 20 meters; there is a limit to their penetration into the seafloor. We needed *Glomar Challenger* to tell us the more ancient history of the earth.

Leg 12, led by Bill Berggren of Woods Hole and Tony Laughton of the National Institute of Oceanography of Great Britain, was scheduled to work out the history of seafloor-spreading in the North Atlantic. *Glomar Challenger* left Boston in July 1970 and drilled several holes near the Arctic Circle. At all but one site they found evidence indicative of glaciation in the Arctic. The most impressive was found in Hole 112, at 65° N in the Labrador Sea. There, the drill string penetrated 115 meters of clays, silts, and pebbly mudstones. The pebbles in the mudstones were not carried thither by ocean currents; they are ice-rafted pebbles, originally frozen in drifting icebergs. After the icebergs melted away, the pebbles dropped down to the mud bottom of the deep Atlantic. Berggren could date the sediments through his studies of microfossils, and found that the ice-rafted pebbles were embedded in Pliocene sediments, estimated to be about two and a half million years of age.

I came across Berggren when I went to Lisbon in the summer of 1970 to start the Leg 13 cruise. He and I were old friends, going back to the days when we both worked for Shell. I had just finished my doctorate and he was starting his studies, working half-time to support himself. He went to Sweden for graduate studies and eventually became one of the leaders in biostratigraphy; he produced a radiometrically dated Cenozoic time scale, defined on the basis of the first or last appearance of foraminifers or nannofossils. I remember that summer afternoon in the bar of the Florida Hotel at Lisbon. Berggren was very excited about their new discovery. Yes, many people had suspected that the Ice Age did not start at the same time everywhere, but he now had the hard evidence. The Arctic ice sheet had been built up and icebergs were floating south with the Labrador current two and a half million years ago, long before glaciers found their way to the Swiss meadows.

Drilling at the South Pole

While Berggren and I were chatting away about his newest finds, Maurice Ewing and his associate Dennis Hayes were drafting a proposal to use *Glomar Challenger* in the polar regions. Lamont scientists had acquired considerable experience exploring the Arctic and Antarctic waters with their research vessels *Vema* and *Conrad*. *Vema*, a lightly built schooner, converted from an old pleasure yacht, had ventured as far north as 80°30' N, to the area around Spitzbergen. More recently, Hayes had been gathering geophysical and geological data in the Southern Oceans with the naval research vessel *Eltanin*. Piston cores were taken, and seismic profiling and magnetic surveys were carried out. Numerous problems were identified, waiting to be solved by a drill vessel like *Glomar Challenger*.

In November 1970 Ewing and Hayes published a short appeal in *Geotimes* asking for public support of their request to drill in regions at high latitudes. They pointed out that it was entirely feasible for *Glomar Challenger* to operate for several months during the southern summers in the Antarctic.

The initial funding for the Deep Sea Drilling Project was 12.6 million dollars for 18 months of drilling in the Atlantic and the Pacific. A second phase of 30 months, costing 22.7 million dollars, was to extend the geographic coverage of the drilling to the Indian Ocean, before *Glomar Challenger* was to return, after Leg 25, to a home port in the United States. By 1970 the success of the project was impressive, and plans were being made for a second extension.

The proposal by Ewing and Hayes was timely. When Phase III of the drilling was granted for three years, with a funding of 33 million dollars, the Antarctic was the main target. The ship-track of _Glomar Challenger_ was rescheduled. The Indian Ocean program was prolonged, and Leg 25 was to end up on 22 August 1973 not in the United States but in Durban, South Africa. The _Challenger_ could then work her way through the Southern Indian Ocean to Freemantle, Australia, to begin her Antarctic program.

JOIDES Panel Structures and Their Functions

Denny Hayes of Lamont was a proponent of the Antarctic drilling, and he and Larry Frakes of Florida were the co–chief scientists of Leg 28. Denny Hayes personified the spirit of Doc Ewing and of Lamont. Like Doc, Hayes worked on marine seismic studies. Hayes was devoted to his science, as Doc was, and had exceptional intelligence, ability, and drive. Like Doc, he had a passionate belief in the worthwhileness of what he was doing. Unfortunately, Hayes did not have Doc's knack of persuasion, the ability to make people feel that "what he wanted was not only what he wanted but what the Good Lord would have done."

I got to know Hayes very well in the 1980s when we both served on the JOIDES planning committee. I would like to mention the following episode to counter the argument that JOIDES is an exclusive club, only for insiders. Denny Hayes, a representative of a founding institution of JOIDES, the member with the longest tenure on the planning committee, and a veteran of a most successful campaign in deep-sea drilling, did not succeed in getting approval for a drill proposal that, in my opinion, was one of the best proposals to investigate the passive continental margin ever submitted to JOIDES. Just the same, I felt that the decision—by a democratic majority—was fair and square. The episode only illustrates the irony of circumstances in a historical text.

I served on various JOIDES advisory panels from 1969 to 1988 and witnessed the evolution not only of their structures, but also of their functions. The funds for the first phase of the JOIDES drilling were given for specific proposals that had been reviewed by the National Science Foundation. The first three panels—Atlantic, Pacific, and Indian Ocean—were constituted for tactical planning and to organize geophysical surveys to pinpoint drill sites. When the Mediterranean Panel was first appointed in 1970 for the second phase of DSDP, a Mediterranean cruise had already been decided upon; the planning committee could judge, on the basis of informal discussions, that such a cruise would be very worthwhile. The mission of the panel was to generate ideas and propose drill sites, mainly on the basis of panel members' experience and knowledge. There was input from outsiders, of course, but it was mostly impractical and could be dismissed lightly. There thus seemed to be no need for peer review.

When I joined the JOIDES Ocean Paleoenvironment Panel during the early years of the International Phase of Ocean Drilling, we were also given the specific assignment of planning two cruises to the Central Pacific and three (later five) cruises to the South Atlantic. Most of the original proponents, who influenced the decision of the planning committee to send _Glomar Challenger_ to those regions, became panel members. The duty of the panels was then tactical planning. The panel did receive new proposals from the community at large. We had to discuss, for example, the merit of drilling "dipping reflectors" that had been identified on seismic reflection profiles. In that case, the panel took over the duty of a peer-review committee. The proposals of "outsiders" rarely received sympathetic treatment, however, because the panel members felt that those proposals were not designed to achieve the goals of drilling the South Atlantic as originally conceived by the planning committee.

While I was a representative of the European Science Foundation on the JOIDES planning committee, I attended meetings of the Mediterranean Working Group of the Atlantic Advisory Panel, and I noted a drastic change from the old days. The panel received no dearth of drill proposals, and the JOIDES panels were no longer primarily concerned with tactical planning, but rather they served more the function of a peer-review committee. Competing proposals from "insiders" and "outsiders" were judged on the basis of scientific merit, and panel members had to absent themselves when their proposals were being discussed.

I was appointed a member of the Tectonics Panel, 1986–1988. The tectonics problems fell into two groups, those on passive margins and those on active margins. The Atlantic margins are mainly passive; those of the Pacific, mainly active. The choice of drill proposals for the Atlantic was made during the early 1980s, and the Atlantic cruises were scheduled for 1985–1986. When I first attended a panel meeting in Ottawa in the autumn of 1986, the panel was already engaged in planning the Indian and Pacific Ocean drilling. Experts in passive-margin tectonics had been largely replaced by experts in active margins.

Denny Hayes had the bad luck to be interested in a passive

margin of the Pacific, namely the northern margin of the South China Sea. He pointed out the advantage of drilling such a young margin, which, unlike the Atlantic margins, is not buried under a thick cover of sediments. I was also enthusiastic, motivated in part by my interest in South China tectonics. Most other panel members, however, had been appointed because of their expertise in active margins. The passive margins had been drilled by the Atlantic cruises, they argued, and the shiptime in the Pacific would be better spent investigating problems of active-margin tectonics. The Hayes proposal was turned down by an overwhelming majority.

Denny Hayes was single-minded. He revised and resubmitted his proposal. It was again turned down. Again a new revision. Again a rejection. Perhaps Hayes had been too self-reliant. Several people had suggested to him that we should form a working group on South China Sea, even before he submitted his first proposal. The planning committee proposed an informal group to revise the proposal after the first rejection. There was in fact a need for such a working group. The JOIDES Ocean History Panel had considered South China Sea objectives a first priority; Chinese marine geologists were also interested. The working group was never constituted, partly because we could not find a suitable time to meet. When I visited China in February 1988, Chinese scientists asked me to send in a supplementary proposal to support Hayes's proposal to drill the South China Sea. The uncoordinated efforts were unfortunately weak and unconvincing. Finally, my friends on the Tectonic Panel lost patience; they had had to repeat for the "umpteenth" time that the active-margin problems had to receive top priority in the West Pacific drilling. When I resigned from the panel in October 1988, Hayes lost his last spokesman on the panel.

The passive margin of the South China Sea was not drilled.

An Ocean of Icebergs

Back in Hayes's more triumphant days, the young co–chief scientist on _Glomar Challenger_ set off on 20 December 1972 from Freemantle for the South Pole. The drill vessel was accompanied by two ice breakers, _North Wind_ and _Burton Island_ of the Coast Guard.

The _Challenger_ went as far as 77° S latitude in the vicinity of the Rose Ice Shelf. The drilling crew worked at times under very severe weather conditions, with icicles hanging down from the drill tower. Icebergs were the greatest nuisance. At Site 273, 74°33′ S, the _Burton Island_ had to push an invading iceberg away to prevent a collision with _Glomar Challenger_.

In view of the manifold difficulties, I was somewhat surprised to read the happy reports given by the co–chief scientists. Drilling at Site 270, at 77°07′ S, 176°46′ W, in water depth of 619 m, Hayes wrote the following resume:

The shallow water and relatively hard sediment . . . placed stringent requirements on both the positioning system and the coring operations. During the almost forty hours it took to bury the bottom-hole assembly with continuous coring, excursions of the ship off the beacon seldom exceeded 20 feet. The positioning requirements were maintained even in the face of 40-knot winds and a 7-foot swell on the evening of February 1. The drilling and coring operations proceeded with no problems except for a short delay caused by frozen air lines to the Bowen powersub controls . . .

Only someone who had been a co-chief, or an operations manager, on a deep-sea drilling cruise could appreciate the magnitude of the luck and the technical achievement. *Glomar Challenger* was a dynamically positioned vessel. She was not anchored on the bottom by a chain; her position was fixed by a beacon, which permitted a deviation measured in angular, not in linear, distance. Commonly, the vessel had to stay within one degree of the vertical axis defined by the beacon. One degree of angular distance is about 100 meters in 6,000 meter depth, but only 10 meters in 600 meter depth. To maneuver the vessel so that it stayed within a circle 10 or 20 meters across was difficult even with the best of equipment. Therefore, the *Challenger* could not drill in very shallow waters, especially when weather conditions were unfavorable.

I noticed that it was Captain Dill's turn on Leg 28. He was known to us veterans as the more permissive of the two captains who alternately took over the command on *Glomar Challenger*. During my three tours of duty as a co–chief scientist I had sailed with Captain Clarke; he was an excellent captain, but he seldom encouraged us to drill in waters shallower than 1,000 meters, especially if the weather forecast predicted 40-knot winds and 7-foot swells.

Hayes's luck seems to have held throughout the cruise. I found few complaints about the weather or about equipment in his reports. At Site 269 near the Antarctic Circle, a hole was drilled to a depth of almost 1,000 m below the sea bottom in about a day and a half. This is a good effort even for drilling under normal conditions, and was almost an unbelievable performance in the Antarctic.

The scientific achievements of Leg 28 were no less spectacular. The cruise accomplished its task of dating magnetic lineations, by now almost routine. The results permitted a reconstruction of the history of the Southern Oceans. New Zealand was first ripped off from Australia 80 million years ago, giving birth to the Tasman Sea. Then Australia itself began to separate from Antarctica, 53 million years before present, and was drifting northward at a rate of 5 cm per year.

Leg 28 drilling also afforded us the first real look into the climatic changes around the South Pole. In 1970 Stan Margolis and Jim Kennett had reported finding ice-rafted pebbles in sediments some 40 million years old. The boreholes of Leg 28 penetrated many layers of ice-rafted sediments. The oldest of those are late Oligocene to early Miocene, some 25 million years old. The icebergs seemed to have come from the Ross Sea or from Victoria Land, during a warmer period when previously formed glaciers were being melted away. On the strength of such clues, the scientists of Leg 28 speculated that the Antarctic glaciation may have started not later than the Oligocene, 25 or 30 million years ago. Subsequent drillings were to confirm this conclusion.

Another climatic event was revealed by the drilling of three boreholes in the region of the Ross Sea. Hayes and his colleagues found evidence that the continental shelf there, now submerged in seawater more than 600 meters deep, was once covered by a thick ice sheet. The advance of glaciers over the Ross Sea Shelf caused widespread erosion of sediments older than 5 million years of age. Soon the glaciers were melted, and the eroded shelf was submerged under the sea, permitting the deposition of the youngest marine sediments. The evidence of ice erosion on the Ross Sea Shelf signifies that the Antarctic ice sheet was much larger some five million years ago than it is now.

Birth of Paleoceanography

The euphoric operations narratives authored by Dennis Hayes may have reflected the unusually good luck that befell the first Antarctic drilling cruise. However, I have a suspicion that the overt optimism was motivated by the fact that Leg 28 was to be a test case for evaluating the feasibility of future drilling in high latitudes. Three seasons of Antarctic drilling had been planned, but much hinged on the performance of *Glomar Challenger* during this initial test. Hayes may have been thinking of his colleagues on the JOIDES planning committee when he underplayed the difficulties he encountered. In any case, the *Challenger* performed her feats better than anyone could have hoped. Leg 29 could now sail as planned.

Glomar Challenger left Lyttleton, New Zealand, on 2 March 1973, toward the end of the southern summer. Jim Kennett of Rhode Island and Bob Houtz of Lamont were co-chiefs. Katharina Perch-Nielsen, now with us at Zürich, was invited to participate as a nannofossil specialist. I ran into her in Tehran; she was on her way to join the cruise, carrying an Eskimo jacket under her arm. She was to need her parka, because the weather was severe down under, even though *Glomar Challenger* did not venture south of the 60-degree latitude during this leg.

At Site 276, southeast of New Zealand, the *Challenger* was hit by a gale, with wind velocities over 100 km per hour. Neither the vessel nor the beacon could keep its position. When the weather improved, *Glomar Challenger* found her beacon again, but the two had drifted 20 km apart during the storm. Fortunately, very little shiptime was lost during the cruise. Storms usually came while *Glomar Challenger* was under way from one location to the next. Perch-Nielsen organized a number of dancing parties, as the cruise was staffed with five female scientists and technicians; everyone seems to have enjoyed the rock-and-roll with the pitch-and-roll of the vessel during storms.

Jim Kennett was a New Zealander and a microfossil specialist. After a few years in Florida, Kennett joined the staff of the Department of Oceanography at the University of Rhode Island. Microfossils are useful not only in determining the age of ocean sediments, but also in giving indications of ancient ocean environments, because some groups lived only in the tropics, while others lived only in polar regions.

Cesare Emiliani, using the technique developed by Harold Urey, analyzed the oxygen-isotope composition of microfossils to determine ancient ocean temperatures. His samples were planktonic microfossils that lived in surface waters, and his results gave indications of temperature changes of the ocean surface (see chapter 2). There are also foraminifers, which lived only on the cold ocean bottom, but they constitute only a few percent of the bulk of the oozes. For the older mass spectrometers to give significant results, many individuals of the benthic species had to be picked out.

Nick Shackleton of Cambridge succeeded during the early 1970s in developing a new generation of mass spectrometer to analyze samples of very minute size. The time was now ripe for analyzing both the planktonic and benthic foraminifers of deep-sea drilling cores.

Kennett shared his prized possessions from Sites 277, 279, and 281 with Shackleton. First the foraminifers had to be picked out. Planktonic species were separated from one another. Individuals from benthic species were analyzed as a mixed assemblage, although later the species were also separately analyzed. The combination of micropaleontology and isotope geochemistry gave us the basic tools for paleoceanography.

Analyzing planktonic foraminifera, Shackleton and Kennett found that the surface waters at this high latitude (50 degrees south) were actually quite warm during the Paleocene epoch some 55 million years ago. After that, there were many short-lived fluctuations, but the trend was a relentless decrease, from

13.1. Change of ocean temperatures. The oxygen-isotope record of Tertiary ocean sediments indicates numerous stepwise decreases in the ocean temperature at high latitudes during the last 60 million years. Nick Shackleton and Jim Kennett produced this curve on the basis of their studies of the DSDP Leg 29 cores.

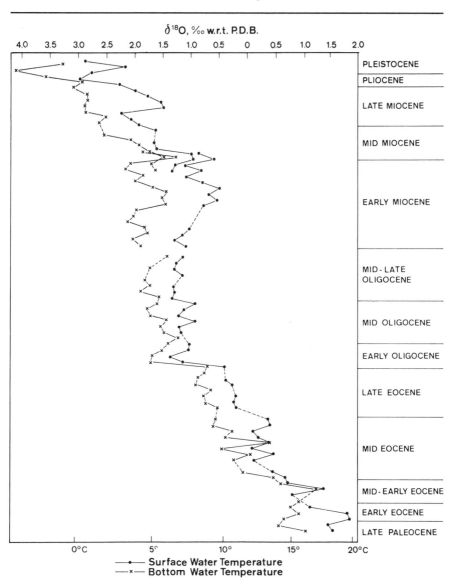

about 20 °C then to about 0 °C now. Three epochs of accelerated cooling were identified: early Oligocene, middle Miocene, and late Miocene, 35, 12, and 5 million years ago, respectively. The trend was reversed only once, early in the Miocene epoch, 20 million years ago. The oxygen isotopes of the benthic microfossils indicated a similar temperature decline of the ocean waters at depth, from 15 °C to 0 °C during the last 55 million years (Figure 13.1).

Shackleton's data on temperature changes of ancient oceans, combined with other geological data from the drill cores, permitted Jim Kennett to reconstruct the history of the climatic changes for the 65 million years of the Cenozoic era. In the beginning of the era, the global temperatures were on the average very warm. The Arctic continent had no ice cover at the time, but rather vegetation similar to that of cold, temperate regions of today. Local mountain glaciers may have come into existence near the South Pole during the Eocene period, 40 or 50 million years ago. The end of the Eocene brought a remarkable change. Tropical faunas and floras, which had until then flourished in middle latitudes, were replaced by new species that could survive a great cooling. Numerous geologists now believe that the Antarctic glaciers became firmly established at that time, 30–35 million years before present, and ice sheets extended down to the sea level. The second epoch of accelerated cooling during the middle Miocene added major ice sheets to the Antarctic. Icebergs drifting northward carried ice-rafted pebbles to the vicinity of New Zealand. Leg 29 drilling also confirmed the observations made by the preceding cruise that the Antarctic ice sheet expanded farther and became much thicker five million years ago. Many glaciers came and went on the Antarctic continent before the Arctic and the continents of Europe and North America were covered by ice sheets. We might say that the Ice Age started 35 million years ago!

"Disasters" at Sea

The success of deep-sea drilling during its first Antarctic season was followed by near disasters during the second. Leg 35 had bad luck at the very beginning; the cruise was delayed for ten days for repairs at Lima, Peru. *Glomar Challenger* finally left the port of call on 13 February 1974, already quite late to start a venture for a southern summer, and she had to dock again at Valparaiso, Chile, a few days later to take on additional personnel, equipment, and supplies. Creeping forward at a modest speed of seven and a half knots, *Glomar Challenger* did not get near the first drill site until 27 February. She also had to sail 5,000 nautical miles to end her voyage at Ushuaia on 30 March.

It was thus an unusually short leg, and the drilling program was further curtailed by bad weather and mechanical problems. In contrast to a normal deep-sea drilling expedition, when 25 to 35 days were available for operation on location, Leg 35 was able to realize a total of only 8 days drilling.

Chuck Hollister of Woods Hole and Campbell Craddock of Wisconsin led the cruise. Terry Edgar, the chief scientist of the Deep Sea Drilling Project, joined them to handle the routine of describing sediments. I could imagine their disappointment and frustration. Edgar, who usually had the authority to grant requests for extensions of cruises, was a model of selflessness; he cut his own cruise short so as not to upset the plans of the scientists on the next leg.

It was a short leg, but the initial report of the cruise was not a thin volume. With its 930 pages, it is about as thick as the other volumes in the series, almost reaching the limit imposed by the National Science Foundation. I was amused when a corollary of Parkinson's law was formulated by an editor of the project: "The number of written words in a deep-sea drilling report will expand to fill the maximum number of allowed pages of print."

The scientists of Leg 36 were to have experiences even more harrowing than those of their colleagues on the preceding leg. The choice of Ushuaia, Argentina, as the port of call almost led to a disaster. The difficulty in docking the *Challenger* did not appear in any of the reports by the Deep Sea Drilling Project. However, as Captain Clarke told me later, the small port was there for fishing boats; it was not built to handle large drilling vessels. Furthermore, the *Challenger* was greeted on her arrival by a gale with cross winds over 100 kilometers per hour. Captain Clarke had to use all the thrusters in full power to prevent the vessel from being grounded, or being wrecked by harbor installations. After the near miss, the captain filed a long report with Global Marine, Inc., asking for contractual specifications on the adequacy of port facilities.

The start date of Leg 36, coming after Leg 35, was delayed even more by unforeseen repairs. *Glomar Challenger* did not leave Ushuaia until 4 April, long past the high summer of the Southern Hemisphere. Furthermore, she was sent to drill holes in the Drake Passage, known to all mariners for its savage weather. Lady Luck did not smile this time. The operations manager's report complained:

The elements, in general, played havoc with operations during Leg 36. The *Challenger* was in three hurricane-force storms plus being subjected to hundreds of icebergs while on site as well as while under way. Due to the influence of the elements, six of the ten holes spudded during Leg 36 had to be abandoned before reaching their scientific objectives . . . An interesting statistic . . . is the total time lost directly or indirectly due to the elements during Leg 36. A very close estimate of this lost time will amount to a total of 13.88 days or 26.4% of the total leg time . . .

The elements, indeed, dealt harshly with drilling operations throughout the leg, but the thing that was finally to turn the lights out on Leg 36 was the *Challenger*'s dynamic-positioning system. When positioning over the final drill site on the Rio Grande Rise, the power supply to the vertical reference gyro failed, and all efforts to restore the power to the gyro were unsuccessful, thus ending operations for Leg 36.

Glomar Challenger experienced bad weather at the very start. Soon after she departed the sheltered waters of the Beagle Channel, 8 hours after leaving Ushuaia, the sea conditions began to build up. By midnight of 4 April, the vessel was rolling 20 to 25 degrees. As the *Challenger* approached the first drill site, 150 km southeast of Cape Horn, in the early hours of the fifth, the weather conditions improved a little. However, the winds were still strong and the currents fast, and after a consultation the leaders of the cruise decided to steam on past the site, in order to wait for better weather. Toward evening, the winds died down to 20 or 30 knots, and swells were reduced to 3 m. The *Challenger* returned, and the acoustic beacon for the positioning of the vessel was dropped onto the sea bottom, marking the location of the ill-fated Site 326.

Site 326 lies in 3,812 m of water in the northern part of the Drake Passage, 80 km off the South American continental margin southeast of Cape Horn. The main objective was to date the opening of the Drake Passage, when circum-Antarctic deep circulation started.

On the evening of 5 April, the crew began to lower the drill string toward the seabed. But the winds started to blow harder again as the bit was getting near the bottom. It was difficult for the vessel to remain in a stationary position. Nevertheless, having come this far, the leaders agreed that a first core, punched into the soft oozes of the seafloor, could be easily and safely taken. A half-barrel of gravels was obtained, proof of the very strong currents in the Drake Passage, which had removed all finer sedimentary particles from the bottom. Meanwhile, the northwest winds blew harder and harder and were reaching 50 knots, while swells exceeded 6 m at times. The vessel, heading into the wind as usual, had rolls up to 10 degrees, and pitches of

19 degrees! A strong current, estimated to be as much as 4 knots, was also pushing the vessel.

Shortly after the first core arrived on deck, at about six o'clock in the evening on the sixth, part of the overworked equipment for dynamic-positioning failed. Repairs were made, and the captain made a valiant effort to bring the vessel over the beacon again after a drift of a kilometer or so away from the site. During all that time, the drill string was hanging above the seafloor and was drifting with the vessel. Finally, the stresses induced by the pitch and roll of the vessel and by the drag of the drifting vessel were too much for one weak joint. Shortly before seven o'clock, a pipe with a five-inch outer diameter broke. The broken pipe was located near the drill ship. Almost the whole drill string, 4,000 m long, and the bottom-hole assembly were lost to Neptune!

The last sentence of the operations resume at this site reads: "The remaining pipe was recovered, and the ship sailed for more sheltered water at 2325 on April 6, 1974."

The emotionless tone of the report belies the anguish and disappointment of all, particularly the two co–chief scientists, Peter Barker of the University of Birmingham and Ian Dalziel of Lamont. It was such a traumatic experience that Dalziel never talked about it during the entire year of his sabbatical leave that he spent with us at Zürich. The consequences of the loss were to be felt by us all.

I was writing this chapter on *Glomar Challenger* while taking a break from my chores as co-chief of Leg 73, because the drilling operations had been suspended for hours. The suspension was necessary because of the new rule, put into effect after the disaster in the Drake Passage: "The maximum limit under which the drill pipe is to be worked is a vessel roll and/or pitch of 7 degrees. In no case are operations to continue where occasional vessel rolls and/or pitches exceed 9 degrees." During the drilling of the remaining sites of Leg 36, and during all subsequent cruises, this rule was diligently enforced. We lost many days of shiptime, but until Leg 73, at least, never a drill string again.

The object of drilling Site 326 was to determine the time when the Drake Passage between the Pacific and Atlantic Ocean first opened, permitting the circulation of ocean waters around the Antarctic. The circum-Antarctic current is of great oceanographic and climatic importance. This current circulates around the Antarctic continent and is the only current that mixes the waters of all oceans. Guessing when exactly the circumpolar circulation started was a favorite game, and estimates ranged from a ripe old age of more than 35 million years (Oligocene) to a very young age of 3.5 million years (Pliocene).

If Leg 36 had been successful in drilling through the sedimentary sequence in the Drake Passage itself, at Site 326, we might have known for sure. The unfortunate turn of events at high latitude in 1975 induced Global Marine, Inc., to specify in their new contract with the Deep Sea Drilling Project that _Glomar Challenger_ would not drill any holes south of 50° S latitude.

Neither the _Challenger_ nor the _JOIDES Resolution_ returned to drill the Drake Passage. JOIDES scientists found a safer approach to the question of when the passage was first opened to permit the deep circulation of the circum-Antarctic waters. Anyway, in the passage itself, the bottom current is, and has been, so strong that the sedimentary sequence there is not likely to be preserved to record such an event. One could instead find indirect evidence through a study of the sediment distribution outside of the passage. After the deep strait was opened, there should have been a northward advance of the polar front, and this advance should be recorded by the sediments of the South Atlantic.

The _Challenger_ returned to the Southwest Atlantic in 1980, during Leg 71, led by W. J. Ludwick and Valery Krashenini-kov. They had hoped to acquire a continuously cored Oligocene and Miocene sequence on the Falkland plateau to answer the question of the opening of the Drake Passage. Unfortunately, as Woody Wise, a shipboard paleontologist, summarized, the core coverage was not as complete as it might have been, the recovery was incomplete, and the preservation of the microfossils was poor. Nevertheless, the evidence suggested that the deep-water circulation through the Drake Passage was early Oligocene or earliest Miocene.

The final solution had to wait for the _JOIDES Resolution_, which visited the Southern Oceans during the southern summer of 1986/87. Peter Barker came back with Jim Kennett as co–chief scientists of Leg 113, but the drilling at Site 696 south of the Orkney Islands did not yield unequivocal results. Later, Paul Ciesielski of the University of Florida, Gainesville, and Ingve Kristoffersen of the University of Bergen, Norway, led Leg 114 and drilled the sub-Antarctic South Atlantic during March–May 1987. The history of a shifting polar front is recorded by the sediments in drill holes around the Falkland

Islands. "The advance of the polar front (in the earliest Miocene) appears to have been intimately related to the opening of the Drake Passage," the shipboard scientists reported in 1988. "Soon after [that] . . . a major increase occurred in the intensity of Circumpolar Deep Water," producing gaps in the sedimentary record.

Scientists now believe that the Drake Passage was opened to deep-water circulation some 25 million years ago.

The End of JOIDES/DSDP?

The postman always rings twice, so the saying goes. It is an American superstition that disasters come in pairs—or is that the Joseph effect of Mandelbrot's fractal geometry, denoting the common clustering of catastrophes. I was sitting in Terry Edgar's office one spring day in 1974, discussing with the project's chief scientists the planning of the forthcoming cruise to the Mediterranean. It must have been 8 April. Edgar had just come back from Argentina. After the disappointments of his Leg 35 experiences he had taken a holiday in South America, mountain climbing with Chuck Hollister. A telegram was waiting for him when he got back, informing him of the loss of the drill string in the Drake Passage. We were still chatting about the aftermath of that catastrophe when his secretary walked in and handed him another telegram. It was the usual morning report, he told me. It was a morning report all right, from _Glomar Challenger_, but not a usual one. Edgar's expression changed after he glanced at it, and he handed me the message:

GLOMAR CHALLENGER SEIZED BY ARGENTINE NAVY. END OF JOIDES/DSDP, BARKER/DALZIEL.

Discreetly I left Edgar's office, as he was asking his secretary to make a long-distance connection to the State Department.

Naturally it was not the end of JOIDES/DSDP. It was all a misunderstanding on the part of an overzealous gunboat lieutenant. After the misfortune in the Drake Passage, the captain of _Glomar Challenger_ had established radio contact with the Argentine Navy to gain permission to anchor in sheltered waters. The Argentines directed the drill vessel to the Bahia Aguirra, where no pilotage was required to enter the bay. The _Challenger_ was anchored on the evening of 7 April, and the crew started making up a replacement drill string. Meanwhile, a small vessel of the Argentine Coast Guard was cruising in the stormy sea when the lieutenant saw a drilling rig. He thought he had caught some private company drilling for oil on the sly. At midnight he and his troops, armed with submachine guns, stormed on board _Glomar Challenger_. They invited themselves to a hearty dinner, which was always available on our beloved vessel, while the officer told Captain Dill that the drill ship was to remain at anchor until she was given clearance to leave the area by the Argentine Coast Guard.

The _Challenger_ was detained for 35 hours, before she was released to get under way on the evening of 9 April.

The shipboard scientists recovered their wits. _Glomar Challenger_ went on to drill 10 holes at 6 sites, and made a major contribution to our geological knowledge of the sub-Antarctic Atlantic. The cruise finally ended at Rio de Janeiro on 22 May.

The cruise report of Leg 36 was dedicated to Maurice Ewing, and the dedication reads:

While _Glomar Challenger_ was drilling Site 330 on the eastern end of the Falkland (Malvinas) Plateau, word of the sudden death of Dr. Maurice Ewing was received on board. Founding Director of Lamont-Doherty Geological Observatory of Columbia University, "Doc" conducted much of his early work at sea in the southwestern Atlantic, and throughout his life he continued to study the region. In recognition of his outstanding contributions to earth science and in particular his roles in the study of the southwestern Atlantic Ocean Basin, and in the initiation of the Deep Sea Drilling Project, this volume is dedicated to him. The shipboard Scientific Party of DSDP Leg 36 proposes that the eastern prolongation of the Falkland (Malvinas) Plateau, on which Sites 327, 329, and 330 are located, be named the Maurice Ewing Bank.

The death of Ewing coincided with the end of an era. The breakthrough of the earth science revolution had been made. JOIDES was being internationalized. There was the mopping up, and there were new green pastures to conquer.

The misfortunes of Legs 35 and 36 did not terminate JOIDES/DSDP, but they may have led to the cancellation of plans for a third Antarctic drilling season. After a successful campaign during Leg 38 in the Norwegian Sea, _Glomar Challenger_ raced from the North Pole to the South Pole for a third round of Antarctic drilling. However, Global Marine, Inc., which leased the drill vessel to the Deep Sea Drilling Project, would not risk sending their vessel to the Antarctic without the protection of an ice breaker; someone had to tow the icebergs away if they came barging in when _Glomar Challenger_ was on station with thousands of meters of drill pipes under her belly.

To make matters worse, the preparations for the third drilling season were inadequate. The persuasive "Doc" was no longer there to help, and the proponents did not have enough data to convince the members of the JOIDES safety panel that the proposed drilling would not carry a risk of marine pollution. I saw Dennis Hayes in the autumn of 1974 at Lamont, when I was trying to get my Mediterranean proposals approved by the same safety panel. He was coming out, after a harrowing session with the gentlemen of the panel, as I was called in. He had been making a last-ditch effort to rescue his program of Antarctic exploration. He had even managed, as I recall, to find an ice breaker from the Union of South Africa to be a chaperon of *Glomar Challenger*. But, no go. Without the approval of the safety panel, the Antarctic legs had to be canceled. The drill vessel had to race back north for her Mediterranean appointment.

Glomar Challenger never drilled in waters south of 50 degrees parallel again; no insurance company would take the risk. The successes and failures of the Antarctic drilling program made us realize her limitations. The third Antarctic season of Ewing and Hayes had to wait for more than a dozen years, when JOIDES returned with *Resolution*.

14 Mid-Cretaceous Anoxia

The planning of the logistics of the deep-sea drilling was a full-time job, and not all legs were given the most interesting scientific assignments. Much transit time was necessary to bring the drill vessel from one place to another so that weather was optimal for drilling. During one such transit leg up the West African Coast, Bill Ryan, in his frustrated attempts to sample a salt formation, was made aware of the significance of black shales. Was the Cretaceous ocean bottom stagnant?

The Other Side of the Coin Called Engineer

To dispel any notions that *Glomar Challenger* always managed to get to the right place at the right time, I have just told the horror stories of Legs 35 and 36. In fact, the preparation and scheduling of the deep-sea drilling cruises is a very difficult task.

I was sitting in the ship's office of *Glomar Challenger*, writing this book, while drilling was suspended because of some mistakes in planning by engineers on the beach.

We had been on board the drill vessel for almost six weeks, after leaving Sao Paulo, Brazil, early in April. Drilling a transect of holes near the 30° S latitude, we enjoyed good weather. As the start of the southern winter approached, we became nervous about what the weather would bring. We had had to shut down once already, at the start of drilling Site 524, waiting for a small storm to blow past us. After the coring operations began, under marginal weather conditions, the captain gave us the honor of visiting our laboratory every evening with the forecast of an oncoming storm and the warning of possible suspension of operations. For two nights in a row, I went to bed worrying, but was greeted with sunshine when I got up. The technicians were laughing at my "Chinaman's luck." The first storm was due at 0600 local time, 24 May 1980. It did not come, having made a right-angle turn to the east at a position 60 nautical miles south of us. The second storm was predicted for dawn the next day. It came closer, but it also turned to the east, after having come within 45 nautical miles. We were lucky, all right. Looking at the weather maps we could see a whole string of storm centers down there at about 40 degrees south latitude, while we were enjoying sunny weather under the protection of high-pressure zones. However, we were where we were because we had looked at those weather maps of the South Atlantic when we did the planning. For optimum scientific results, we would have preferred to drill our transect down south, but we had to compromise with the dictates of the weather.

When I took my sabbatical leave at Scripps in 1972, I went quite often to the Deep Sea Drilling Project headquarters to visit Terry Edgar. I had learned to appreciate the enormous difficulty of his task, and could understand why he resigned after the end of the DSDP Phase III drilling. The JOIDES planning committee gave the project's chief scientist a set of guidelines, and he had to come up with a schedule that took into consideration not only science, but also logistics, not to mention "politics." Each cruise was planned as tightly as possible, and each unforeseen delay might endanger the scientific objectives of a whole cruise. Often Edgar would be bombarded by telegrams from *Glomar Challenger*—co–chief scientists asking for an extension of the cruise. Usually the request was justified. The drill vessel had had to sail a thousand nautical miles to drill a one-thousand-meter hole. Now they were down to 950 meters. Perhaps they could reach their goal if they could drill just 30 or 50 meters deeper, but time was running out. Could they have two more days? Yes, Edgar was a reasonable person. We could not scuttle an experiment just two days before the completion of the experiment, after so much had been invested. So two days were granted. So they drilled down 50 meters deeper in two days, but then came another telegram asking for one more day to drill 20 meters deeper. What should Edgar do?

Then there was the logistics to consider. The repairs prior to Leg 35 were all unforeseen. Yes, the malfunction of the thrusters of the vessel was not exactly expected to happen just then. However, much work in port was done on the so-called "Antarctic modifications." Some valves had to be installed to control the temperature of the seawater used in cooling the engines. A special type of gas injection had to be installed on air lines to prevent their freezing. Four new heaters had to be installed on the rig floor. All exposed air and water lines had to be covered with insulation material. An additional generator had to be brought on board for emergency situations. And so on. The Antarctic modifications were not unexpected before Leg 35, of course, but they were not anticipated until *Glomar Challenger* had gathered experience during the first Antarctic season. The scheduling of those cruises, however, had to be done long before the first Antarctic cruise, Leg 28, even started.

Often a scientist's bitterest competitors for shiptime were the engineers of the project. It is true that we could not have done any drilling at all if the engineers had not come up with all their wonderful inventions. However, I had seen enough time wasted that I always fought against the engineers' attempts to "steal" shiptime. During Leg 3, the engineers came up with the idea that they could raise the acoustic beacons used in dynamic-positioning from the ocean floor so that they could be used again. It was a reasonable attempt to minimize waste. They spent two or three hours at almost every site trying out their experiments. Finally they succeeded; they could raise the beacons from the abyss for possible reuse. However, only then did they start thinking about the economics—the cost of the shiptime needed for the recovery process was far greater than the value of a used beacon. They had the satisfaction of a successful feat, but beacons were never reused during all those years of deep-sea drilling. Another time, during Leg 13, the engineers wanted to try out a new type of sampling device to take cores from the side wall of a borehole. I was an inexperienced co-chief then, and did not even know what they were doing. Again, hours and hours were taken away from the time allotted for scientific drilling. The engineers had their successful tests, but this sampling device was never used, because no chief scientists were told of its availability.

After seeing all that wasted motion, I became known as a very tough person to deal with, if anyone wanted to try a new engineering experiment on "my cruise." A few weeks before I was due to join Leg 73, I received a phone call from the manager of the Deep Sea Drilling Project. His engineers wanted shiptime to test the up-and-down motion of the drill string on the sea bottom. I told him that I received my orders from the JOIDES planning committee, not from the Deep Sea Drilling Project. The manager recognized the lines of command, but he pleaded on the phone for three quarters of an hour, San Diego to Zürich. The engineers wanted only six hours. It was not worthwhile to bother the gentlemen on the planning committee, he said, for a "puny" six hours. I was adamant; not a single hour of shiptime on *Glomar Challenger* is puny for me. On a two-million-dollar cruise, the time available for drilling may be only 500 hours, and thus the cost of six hours is not puny. The project manager finally asked the planning committee to give me the order to make available the puny six hours, with a promise that they would not take one minute longer, regardless of the outcome of their experiments.

I was almost sure that the six hours would be wasted, but even I did not expect that they would have no way to keep their promise. Before we got to this site (524), they had one complete bust that ate up more than two hours. The reason was simple: the engineer forgot that the ocean-bottom temperature here was not the same balmy air temperature of San Diego, California, and the battery in the gadget did not work in the near-freezing temperature of the ocean bottom. Now they wanted to have their second trial here. They dropped the engineering package down in a steel-cylinder, but they could not pull it up. Applying tension to the rope caused the rope to break. They tried to fish the package out with various tools. No success. With the package blocking the opening of the drill pipe, no coring was possible. The whole string of pipes, more than 5,000 meters long, had to be raised on deck again. Yes, they now knew what happened. The engineers on the "beach" had not communicated with each other. One person had designed a housing for an engineering package with an outer diameter identical to the inner diameter of the drill bit designed by another. After the package was pumped down, the steel cylinder fitted perfectly into the drill bit. It was so tightly clamped that not even the most skillful fisherman could fish it out. Raising up a drill string closed watertight at its lower end also had its effects on the roughnecks— they were subjected to a cold shower each time they unscrewed a segment of the pipe. With more than 150 joints, they had to take more than 150 cold showers under the open air of the southern winter, with winds blowing at 20 or 30 knots. The engineers wasted thirty hours of our shiptime, not the puny six hours promised by our dear project manager.

Perhaps I was too harsh with the engineers, but it is an undeniable fact that a poorly designed engineering program may have been mainly responsible for the failure of Legs 35 and 36. Although the operating conditions of the first Antarctic season were good, the engineers thought that a piece of equipment could be installed on *Glomar Challenger* to neutralize the devastating effects of weather. They discovered that there was something called a "heave compensator" (heave being the up-and-down motion of a vessel). It was thought that the drill string could operate in bad weather if the heave of the drill vessel could be largely taken up by the "compensator." The engineers did not think that the installation of the compensator on the *Challenger* would take very long, and Edgar was advised to schedule a port call of eight days at Honolulu before the start of Leg 33. When the scientists showed up in Hawaii, they were all housed for "the night before the cruise" in luxury hotels on Waikiki beach. But they did not leave the next day, nor the day after. The sailing had to be postponed because "a little more time" was needed for the installation work. Kerry Kelts, one of my assistants at Zürich, did enjoy the first few days of his holiday in Hawaii. However, when the departure was delayed again and again, the scientists became uneasy. Finally, after three weeks of port call, the scientific party issued an ultimatum to the project representative—they would all go home if the vessel did not leave Honolulu within 48 hours. The ultimatum worked; the installation of that monstrous instrument was completed in 23 days, and the vessel departed on 10 October. But it was too late. The South Pacific leg ended in Tahiti just before Christmas, when *Glomar Challenger* had to be in Valparaiso, Chile, to get ready for the second season of Antarctic drilling. Further delay was unavoidable when the drill vessel had to sail to Lima, Peru, near the equator, where a dry dock had to be found in which her thrusters could be repaired. Perhaps one of the drilling cruises should have been canceled—Leg 34, for example. However, the ship schedule of *Glomar Challenger* represented a delicate compromise after the long and intense infighting among the various special-interest groups on the JOIDES planning committee, and few dared to "upset the applecart." Besides, one could hardly schedule a transit of 4,500 nautical miles without drilling a few holes *en route*. Consequently, the start of the Antarctic venture was pushed back to the waning days of the southern summer, leading to the difficulties during Leg 35 and the catastrophe in the Drake Passage.

To add insult to injury, the heave compensator was of no use to us. To put that piece of equipment into operation would cost a great deal of shiptime. The investment might be worthwhile, if it worked, but after a few initial trials, it was admitted even by the engineers that the heave compensator did not function well. After Leg 36 the rumor began to circulate among future chief scientists that one should avoid the "monster." As I was writing in the ship's office, it was still out there on *Glomar Challenger*, hidden under a canvas, a mysterious figure to newcomers, but avoided like the plague by veteran chief scientists. We had all learned to respect the weather god. When storms came, we closed down, because the "monster" would be of no help.

(*Author's postscript:* After serving in 1991 on the Performance Evaluation Committee of the JOIDES Ocean Drilling Program, I have acquired a more balanced view of the various activities. I think my critique is unfair to DSDP engineers; the continued success of the *Challenger* at sea in fact owed much to engineering innovations. Incidentally, the "monster" described in the preceding paragraphs turned out to be a "pillar of strength" for ODP; the heave compensator contributed significantly to the successful drilling and coring operations of the *JOIDES Resolution* under adverse weather conditions. I decided, however, to leave my critique in the manuscript to illustrate the emotions felt by participants during the "hour of battle.")

A Co-Chief on a Transit Cruise

The 44 expeditions during the first three phases of the Deep Sea Drilling Project could be grouped into two broad categories: some were to perform critical experiments to prove or disprove specific ideas; others were to collect new data from specific regions. The "idea" cruises, thematically oriented, were the spectacular ones—the legs that set out to test the hypothesis of seafloor-spreading, to evaluate the plate-tectonic theory, to determine the origin of back-arc basins, to look for "hot-spots," to investigate a transform fault, or to unravel the climatic history of the earth. To carry out those experiments, *Glomar Challenger* had to be in the right place at the right time of the year. She should be in the Arctic during July and August, and in the Antarctic during December and January. She should avoid the hurricane seasons in the Pacific, and the mistrals in the Mediterranean. The constraints of weather necessitated seasonal migrations of the vessel. On rare occasions, the *Challenger* had to steam in transit, with only one scientist and a few technicians aboard to take care of routine geophysical measurements under way. Bill Riedel was "baby-sitting" after Leg 4 to see that the *Challenger* sailed through the Panama Canal from the Atlantic to the Pacific. Peter Finckh, my assistant in Zürich, got an assignment when the drill vessel had to double back across the Mediterranean from Istanbul to the Azores. More often, a drilling leg was planned during transit. The transit cruises had to do the less glamorous work of extending the geographical coverage of data-collecting. Their planning was severely limited by many constraints. Often, one hole had to be drilled here, and another there, and the cruise could not be devoted to a central theme. Leg 4 was one example, scheduled to bring *Glomar Challenger*, after her triumphant success in the South Atlantic, back to the northern waters of the Pacific. Leg 30 was another, to bridge the gap between the Antarctic adventure of Leg 29 and the exploration of the Philippine Sea by Leg 31. Leg 39 was yet another, a race from the Arctic to the Antarctic; the *Challenger* sailed from Amsterdam after a successful campaign in the Norwegian Sea during the northern summer and got to Cape

Town in time to prepare for an Antarctic cruise during the southern summer. Leg 39 was the longest DSDP cruise, both in duration (82 days, including time in port) and in distance traveled (almost 10,000 nautical miles). Yet, after the *Challenger* arrived at Cape Town, the news came out that the third Antarctic campaign was canceled. So Legs 40 and 41 had to be scheduled to take *Glomar Challenger* back up again for drilling in the Mediterranean.

Leg 40 was led by my Zürich colleague Hans Bolli and by my shipmate during Leg 13, Bill Ryan of Lamont.

I first met Ryan in 1969. I had come to the United States to attend a meeting at which we would discuss the scientific findings of Leg 3; such a post-cruise meeting was indispensable for the final preparation of the cruise report. I called my shipmate Tsuni Saito, who lived near Lamont. Saito told me to take a bus to the Washington Bridge terminal, where he would pick me up. While Saito was driving me across the bridge, he told me that he had also invited Bill Ryan over. "He is brilliant, and will be Doc's successor," Saito said.

I had never met Ryan, but we had been corresponding. Brackett Hersey, Ryan's former professor at Woods Hole, was then the chairman of the Mediterranean Panel. Hersey was much too busy with his administrative duties for the U.S. Navy, and he had given Ryan the "proxy" to, in effect, chair the panel. After Saito's remarks, I had an image of a man of ambition, arrogance, and possibly conceit. I could not have been more surprised, therefore, when I stepped in Saito's apartment and saw a young graduate student there to greet me; he was introduced to me as Bill Ryan. He was soft-spoken and unobtrusive.

Bill Ryan was a protégé of Bruce Heezen's. Ryan had acquired Heezen's intensity in science, but not his zeal, probably because they came from different family backgrounds. Ryan and his wife came from families that voted Republican. Family friends had been members of President Eisenhower's and President Ford's personal staff. Ryan went to a private school, Williams College, before he showed up at Woods Hole for a degree in marine geophysics. He was very good at electronics, and he went with Brackett Hersey on several Mediterranean cruises. The research results were eventually written up in his dissertation at Lamont.

Years later, I asked Ryan why there had been rumors that he was to be the director of Lamont; Ryan had given me no sign of any ambition for administrative positions. "Oh, that was funny. I went to see Doc after I came to Lamont. He asked me what I wanted to do. I said that I would like to do what he was doing, meaning marine geophysics. They made a joke of it, because Doc was by then better known for what he was doing as the director of Lamont."

In 1969, Ryan, aged 29, had not yet submitted his Ph.D. thesis. When he and I were named co–chief scientists for Leg 13, the sarcastic joke circulated in Europe that a student and an amateur had gotten the job of leading the important expedition to the Mediterranean Sea.

Ryan and I became very good friends during the cruise, although we had different backgrounds—he was a marine geophysicist, and I was a geologist. He did all the navigation involved in getting the _Challenger_ on site. I was more familiar with the carbonate and evaporite rocks, which are better known to geologists on land. We shared, however, the same scientific philosophy—that science predicts.

After the success of Leg 13 drilling, when both of us agreed on an unorthodox interpretation of the Mediterranean Salinity Crisis, there was the general feeling that the second cruise to the Mediterranean had to be led by "impartial" scientists. One of the co–chief scientists for the next cruise had to be an "outsider." In other words, the Ryan-Hsü team had to be broken up.

I always appreciated Bill Ryan's generosity. He was, at the time, as interested in the Mediterranean as I was, if not more. Nevertheless, he gave indications that he would find interests beyond the Mediterranean. He also saw to it that not all "leg-thirteeners" would be banned from the next cruise: I was to be a co–chief scientist for Leg 42, and Maria Cita was the senior shipboard paleontologist. Ryan himself took an assignment as a co–chief of a transit cruise, Leg 40.

Inspiration by "Nemesis"

During the spring of 1975, there were rumors that the participants of the Law of the Sea Conference of the United Nations were about to come up with an agreement to restrict the freedom of scientific exploration. A 200-mile exclusive economic zone offshore was about to be accepted, and drilling within this zone had to be approved by the onshore nation. Ryan wanted to take advantage of this last opportunity to drill a few holes close to the coast, where he could investigate the geological history of the African continental margins.

Africa started to separate from South America 140 million years ago, giving birth to the South Atlantic in the process. In the beginning the ocean was a narrow seaway, like today's Red Sea. Evaporation of the seawater in restricted basins produced salts (Figure 14.1). Through drilling on African and South American coasts, salt deposits had been discovered and geophysical surveys offshore indicated their presence under the continental margins. Ryan hoped to drill through the sedimentary cover to reach this salt bed, which lies more than a thousand meters below the ocean floor. But he could not reach the salt bed. He was getting very close on his last trial, and received, in fact, a four-day extension of his leg in a last effort to reach the objective. Unfortunately, the drill bit was worn out,

14.1. Mid-Cretaceous Angola Basin. The Atlantic Ocean was a narrow gulf during the middle Cretaceous, about 110 million years ago. The restriction from time to time caused oxygen deficiency in benthic environment.

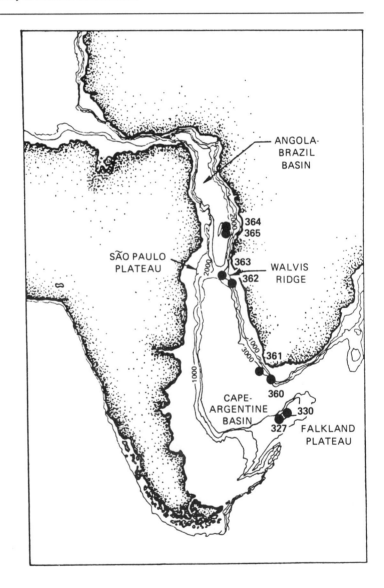

and time was also running out. He did not make it, and the *Challenger* never made it either, because the site now lies within the economic zone of Angola. The government of Angola would not have given us permission to drill there.

Ryan wanted to penetrate the salt sequence because he wanted to know if the Angola Basin was ever completely desiccated—like the Mediterranean Sea—or whether a constant supply of seawater had been available to immerse shallow shelves and tidal flats where the salts precipitated. Not having sampled salt, he failed to resolve the question. Drilling for oil in areas offshore Angola did eventually penetrate salt beds. The mid-Cretaceous Angola Basin was a narrow desiccated basin that once hosted a giant lake (Figure 14.1); salts precipitated when the climate was arid.

The *Challenger* left Cape Town on 17 December 1974 and returned to Abidjan, Ivory Coast, on 15 February 1975. Even though no salt samples were taken, Leg 40 was far from being a failure. Ryan did not reach salt because he was mired in black shale. With all that delay, he had time to think, and he recognized what is now known as the Mid-Cretaceous event.

The Cretaceous is the name of the geological period that ranges from 145 to 65 million years before present. The period is named after the chalk of England, the white cliffs of Dover. Thanks to zonation by fossils, 12 stages have been identified, namely Berriasian, Valanginian, Hauterivian, Barremian, Aptian, Albian, Cenomanian, Turonian, Coniacian, Santonian, Campanian, and Maastrichtian.

In the southern Alps, a Lower Cretaceous formation is called Maiolica, because it is made up of a dense white rock resembling the maiolica porcelain. Thinking of chalk and Maiolica, we tend to associate the color white with the Cretaceous. However, a few black layers have been found in the white Maioloca of Italy. They contain more muddy materials and are thus black marls or black shales. The same kind of Cretaceous black marl had already been encountered in piston cores and during Leg 1 drilling. They were encountered again during the Atlantic drilling, Legs 11 and 14. The sediments are black because of the presence of organic matter, which is derived either from dead marine plankton or from terrigenous detritus. The occurrences were noted in the cruise reports, but there was not much discussion of their significance. Cretaceous black shales were later found in the Caribbean Sea during Leg 15, in the Indian Ocean during Legs 25, 26, and 27, and finally in the Pacific during Leg 32. All of these black sediments seemed to have about the same age. An earlier group was deposited during the Aptian, 110 to 115 million years ago, and a later one near the Cenomanian/ Turonian boundary, some 90 million years ago. In fact, Ryan and I had encountered the Aptian black shale while drilling the Gorrange Bank during Leg 13. Like those before and after us, we dutifully recorded the occurrence, but we did not think more about it.

While Ryan was sitting on the rig floor waiting for the breakthrough to the salt, his nemesis was the Cretaceous black shale. At Site 361 in the Cape Basin, the hole encountered the Aptian black shale at about 1,000 m subbottom. It was 9 January almost the halfway mark for the cruise. Ryan had to drill through the shale formation to reach the salt. Penetration was slow, because coring had to be continuous. It was monotonous work, and the tool-pusher got so bored that he let the co-chief, Ryan, take over the rig operation.

After a little more than 100 m penetration of the black shale, even Ryan's patience wore thin. The operation was rescheduled: 10 m drilling and 9 m coring. The end came at noon on the twelfth, and that was it; the drill bit was worn out at 1,314 m subbottom depth. There was no reentry yet, to change the drill bit, and Ryan could not get to the salt of the Cape Basin.

At Site 364 in the Angola Basin, Ryan penetrated the Cenomanian/Turonian black shale and got into the Aptian black shale at about 1,000 m subbottom. Again the black shale was to prevent him from reaching the salt of the Angola Basin.

The black shales were, furthermore, threatening his operations. Black sediments commonly contain hydrocarbon gases. The JOIDES safety panel had made a ruling that "drilling should be stopped immediately whenever hydrocarbons are encountered in concentrations markedly above normal base-level and under circumstances suggesting the possibility of substantial petroleum accumulations." The panel members realized the difficulty of spelling out exactly what constituted a dangerous level, and left the decision "in large part to the common sense and experience of those in charge of the operations."

To carry out the mandate of the safety panel, Ryan and the operations manager would rush to the laboratory after each core was taken. Chemical analyses were made to determine the composition of the gases in the black shale. One quantitative measure was the ethane/methane ratio. Some previous legs had adopted a cutoff value of 1/500; others, 1/300. When the ratio of a black-shale sample exceeded 1/300, there was a long consultation involving the shipboard geochemists, the operations manager, the drilling supervisor, the ship's master, and the co–

chief scientists. The consensus was to proceed cautiously with continuous coring, checking the hydrocarbon content of sediments until the ratio again became less than 1/500. The caution meant, of course, delay and patience.

While Ryan was sitting in the drill shack, night after night, musing over his hard luck, it occurred to him that those black shales looked familiar. He had done his dissertation research on the Recent sediments of the Mediterranean Sea, and he had encountered such black layers in the eastern Mediterranean.

The black ooze of the Mediterranean was called sapropel by Eric Olausson, because of its similarity to the black mud of Norwegian fjords. The name is derived from the Greek word *sapros*, which means rotten; the mud is rotten because of the presence of organic debris in the sediment. The organic-carbon content of sapropels exceeds 2% of the bulk, and is there because the seawater on the bottom did not have enough oxygen to oxidize the dead bodies of plants and animals. There was not enough oxygen, according to one theory, because not enough oxygenated water could be brought over by bottom-currents, as the sea bottom was stagnant there (see chapter 19). An alternative theory invoked "supply-side economics" to explain the origin of the oxygen deficiency; this theory suggested that the production of marine plankton, and/or the sedimentation of terrestrial plant debris, was so rapid that the oxygen in bottom waters was quickly used up in oxidizing the excessive influx of organic carbon.

Surface waters of the oceans descend when they are heavier than the bottom waters. Ocean waters are heavier when they are more salty, or colder, or both. In the Mediterranean today, the new seawater comes through the Strait of Gibraltar from the Atlantic with a salinity of about 36.5‰. The surface water flows east along the African Coast, and becomes more and more saline because of evaporation. When it gets to the Levantine Coast near Israel, the salinity exceeds 39‰. This water becomes so dense that it descends to an intermediate depth of about 400 m when it flows back westward, south of Greece. At the Strait of Oranto, the salty water from the Levantine meets the cold water coming out of the Adriatic, which is cooled by the winter winds from the Alps. The refrigeration of the salty water makes it colder and still denser, so that it sinks down to the abyss as the bottom-current of the eastern Mediterranean Sea.

This is the origin of the bottom-circulation of the Mediterranean today. The Mediterranean sapropel layers were intercalated in normal ocean oozes of the last 5 million years, and the last was laid down 7,000 or 8,000 years ago, shortly after the end of the last Ice Age. Why should there have been stagnation then?

The situation then was quite different. The post-glacial influx from the Black Sea to the eastern Mediterranean Sea was extraordinary. A mixing of this brackish water with the salt water of the Levantine made a surface layer that was far less salty than the Levantine water of today. The less salty water could not become denser than the bottom water, even after its refrigeration by the cold current coming out of the Adriatic. Consequently, Eric Olausson of the Swedish Deep Sea Expedition, the first to study the Mediterranean sapropels, concluded that the bottom of the eastern Mediterranean was at times stagnant. With no bottom-currents to bring in new oxygenated waters, the little oxygen that had been in the water was soon used up. The carbonaceous material, which usually is eliminated by oxidation, could thus be preserved in the sediments, making them black.

Now Ryan was looking at the black shale, the "sapropel of the Atlantic." Had the Atlantic bottom also been stagnant once? The ocean floors are very well ventilated today, as C. Wüst first found out with his many current measurements during the expeditions of the German research vessel *Meteor*, 1925–27. Currents with a speed of 20 cm per second are not uncommon on the bottom of the Atlantic. The running waters of the bottom originate in the polar region. Denser waters spill out of the Norwegian Sea from the Arctic and the Weddell Sea from the Antarctic, and descend down to the depth to ventilate the Atlantic. Those deep water masses then make their eastward turn across the Indian Ocean to the South Pacific before they finally come back to the surface again in the North Pacific. The Norwegian Sea and the Weddell Sea could thus be compared to a person's two lungs, and the bottom-circulation to the blood circulation: lungs take in oxygen, and the circulation ensures the supply of oxygen. Deprived of oxygen intake because of pneumonia or lung cancer, a person could die. Did the Atlantic once suffer from pneumonia?

Mid-

Cretaceous

Anoxia

Freshwater has its greatest density when the temperature is 4°. Salt water is, however, densest at its freezing point, which is less than 0 °C. The polar waters sink today not only because they are very cold, but also because they are more salty. Investigations in the Weddell Sea region have shown that sea ice begins to form where the water from the shelf cools below freezing. Since the ice does not take up any salt, the water that has not been frozen becomes more salty. It is this colder and saltier water that descends to form the Antarctic Bottom Water (see chapter 19). The situation is similar in the north, where the North Atlantic Deep Water originates.

Studies of the ocean temperatures suggest that during the Cretaceous the polar regions were so warm, they were ice free. Obviously the respiration of the Cretaceous oceans had to take place via another mechanism. Many scientists believe that the present type of bottom-circulation of the oceans began only during the relatively recent Oligocene epoch, some 35 million years ago. The driving mechanism of Cretaceous bottom-circulation could be similar to that of the Mediterranean today, namely, a combination of evaporation and refrigeration. When equatorial salty waters met cold polar waters, they sank, driving the bottom circulation.

When there were no polar ice caps, production of bottom-currents thus depended upon chance encounters of salty waters with cold waters. When geographic constraints prevented such encounters, an ocean could become stagnant. With this line of thinking, Ryan and his colleagues proposed that the Atlantic Ocean had suffered repeated stagnation during the middle part of the Cretaceous period. The Cretaceous black shales of the Angola and Cape Basins were deposited during those Mid-Cretaceous Events, when the bottom waters of those narrow seaways were poorly oxygenated or entirely devoid of oxygen (Figure 14.1).

Studies by my students Helmut Weissert and Judy McKenzie of rocks in the Alps that were sapropels of the Cretaceous Tethys Ocean gave support to this postulate. They found that several Tethyan basins became stagnant during times of more humid climate, when reduced evaporation minimized the production of salt water. The circulation became weak or came to a halt when the sources of bottom-currents were shut off.

The discovery of the black shales is of much more than academic interest, because the Cretaceous black shales have a rich organic-carbon content. Thirty years ago, when I was a young research scientist with Shell, my senior colleague Ted Philippi asserted that all the rich oil fields of Venezuela came from a Cretaceous black sediment—the La Luna formation. He maintained that this was proved by the match between the chemistry of Venezuelan oil and that of the organic materials in the La Luna. Many of us thought he was crazy. However, his view was eventually to prevail. Indeed, black shales of a mid-Cretaceous age were the source beds for not only Venezuelan oil. When I was in Trinidad last year, my friends in the oil industry took me to an outcrop of Cretaceous black marl not far from the port of San Fernando; I was told that this, the La Luna, was the mother bed for the oil of Trinidad. Cretaceous black shales have also delivered all the oil to the Golden Lane of Mexico. We could go on down the line, speculating on the Cretaceous origins of the oil of Nigeria, or of the oil of the Gulf Coast of the United States. There seems almost a consensus now among experts that most of the oil in countries on both sides of the Atlantic had its genesis in the periodic stagnation of the Atlantic during the Cretaceous.

Return to Angola Basin

The recognition of the importance of the Mid-Cretaceous Events was a chance discovery during a transit voyage; it soon became the hottest subject of study by the scientists of the JOIDES community, and by geologists outside of our community. Eugene Seibold of the University of Kiel and Yves Lancelot of the University of Paris, co–chief scientists of Leg 41, followed up, and drilled at several sites in more northern basins off the African coast. They found the same Cretaceous black shales. Later Atlantic cruises drilling on or near continental margins, such as Legs 43, 44, 47, 50, 51, 52, and 53, all devoted part of their efforts to investigating Cretaceous black shales.

The stagnation of the Atlantic during the Cretaceous is not all that improbable, because the early Atlantic was a narrow ocean, not much larger than the present-day Mediterranean. The finding of the black shales in the Pacific presented another problem, and even Ryan did not go so far as to postulate the dying of the whole Pacific. In fact, comparing sediments of the same age, Mike Arthur of Princeton, among others, noted that the Cretaceous black shales of the Pacific were found on submarine elevations, while white chalks were deposited at greater depths, where the bottom water was well oxygenated. The black sediments of the Pacific were apparently deposited in an oxygen-minimum zone, where the ocean water has the smallest oxygen supply and where current circulation is at its weakest. In the oceans today, a zone of minimum oxygenation lies at a depth of 1,000 to 2,000 meters.

Ryan agreed that this hypothesis could best explain the Pacific sapropels, but he still argued for a total stagnation of the Atlantic bottom. Ryan's holes were drilled on the continental slope, which could have lain within an expanded oxygen-minimum zone. To test the various ideas, Leg 75 was scheduled in 1980 to drill the Angola Basin again, where Ryan first called our attention to those wondrous sediments. A borehole drilled into the middle of the basin should settle the question of whether the black shale could accumulate only in the oxygen-minimum zone.

Bill Hay and Jean-Claude Sibuet were the co–chief scientists for Leg 75. Hay was a micropaleontologist and did his postdoctoral studies in Switzerland in the early 1960s investigating nannofossil stratigraphy of the Alpine Flysch formations. I used to admire his linguistic talent: I could not make any headway with my Swiss German during my post-doc tenure in Zürich, while he was speaking the Basel dialect fluently. Hay left the University of Illinois and became dean of the School of Oceanography at the University of Miami. He was a most active member of the JOIDES community—he served on various panels, was chairman of the JOIDES planning committee, president of JOI, Inc., and so on. During the International Phase of Ocean Drilling, he was the best friend of students of ocean paleoenvironments, and he was often the messenger who brought us the bad news that our proposed cruises were postponed or canceled.

Sibuet was a geophysicist on the staff of the Centre Océanologique de Bretagne. I did not know him very well, and saw him only occasionally at panel meetings. One small incident left a deep impression on me. The JOIDES tectonics panel had its 1987 fall meeting near Samedan, Switzerland. I invited my colleagues on the panel and their spouses to my vacation house for a drink before dinner. We all had such a good time that nobody wanted to go down to the village for supper. Seeing that we could not continue on empty stomachs, I made spaghetti for some 30 guests. When the evening was over, I suddenly realized that I did not have to clean up—Jean-Claude and Dave Howell had quietly done the KP duty. They were equally thoughtful in panel deliberations.

The *Challenger* sailed from Walvis Bay, South Africa, on 27 July and returned to Recife, Brazil, on 6 September 1980. The results of their investigations were reported by Walter Dean and the co–chief scientists. The Cretaceous sequence consists mainly of red and green claystones at Site 530 in the Angola Basin. Intercalated are 262 beds of black shale that contain up to 19‰ of organic carbon. There were clearly two time intervals of maximum carbon-accumulation, the Aptian/Albian and the Cenomanian/Turonian boundaries. The organic carbon has been derived mainly from marine plankton, although some terrestrial components are present. There were many factors that contributed to the oxygen deficiency, or anoxia, in the Cretaceous bottom waters of the Angola Basin. The Cretaceous climate was warm and evaporation was excessive, and the oxygen solubility in seawater was reduced because of its high temperature and salinity. There was an abundant supply of terrestrial plant debris and unusually high organic productivity. Finally, there may have been a density stratification in a relatively small silled basin; it was difficult to move the dense saline water up once it got to the bottom. The combination of those factors led to the circumstances in which "bottom waters in the deep Angola Basin became anoxic or nearly anoxic at least 262 times during the middle Cretaceous," even though Cretaceous bottom water in the basin had enough oxygen to support a bottom fauna most of the time.

The Leg 75 cruise report did not pronounce the last judgment. Black shales continued to be encountered in the Atlantic, Pacific, and Indian Ocean drill holes. Enough data have been obtained indicating that black sediments in open oceans are restricted to oxygen-minimum zones, but silled basins may have sufficiently restricted circulation that they became entirely anoxic. Meanwhile, proponents of the two schools of thought—stagnation and oversupply—continue their arguments. The complexity of the problem must be comparable to those in economics, and we may never be sure if a tax cut or spending reduction is the recipe for balancing a national budget. (*Editor's note:* This was written during Reagan's presidency.)

15 When the Mediterranean Dried Up

The clues that the Mediterranean had once dried up consisted of a bucket of pea gravel, and the first verification was provided by the Challenger *cores, which looked like a miniature "Pillar of Atlantis." A second cruise to the Mediterranean served merely to verify the seemingly preposterous idea.*

A Bucket of Pea Gravel

In 1969 an American research vessel, *Chain* from Woods Hole, entered the Strait of Gibraltar with the newly developed continuous seismic profiler to explore the Mediterranean. The new tool made possible a discovery: it was found that the deep Mediterranean is underlain by salt domes (Figure 3.1), the same array of pillar-like structures under the seafloor that Ewing had found in the Gulf of Mexico (see chapter 3). The scientists on board also discovered a hard formation a few hundred meters below the seabed; the top of the formation was an excellent acoustic reflector, sending echoes of sound signals back to the ship (Figure 15.1). Brackett Hersey, the leader of the expedi-

15.1. M-reflector. This composite diagram shows the parallelism of the bottom and the subbottom (M) reflectors. This parallelism indicates that the sedimentary layer under the reflector was deposited on a seafloor that had a relief very similar to that of the present Mediterranean. The vertical scale indicates two-way travel time of vertical waves in seconds.

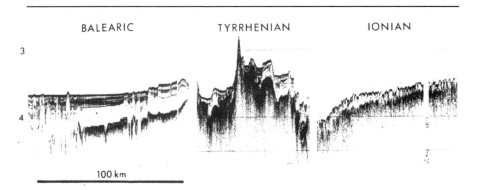

BALEARIC   TYRRHENIAN   IONIAN

3

4

100 km

tion, was puzzled. Ocean sediments are commonly hard oozes. They may become hardened into rocks after they are buried a kilometer or two, but we do not expect to find a hard formation at such a shallow depth beneath the seafloor. Hersey named this mysterious layer of unknown origin the M-layer, and its top the M-reflector, M standing for Mediterranean.

During the next decade, American and French scientists continued the seismic profiling surveys of the Mediterranean. Wherever they sailed, they could identify on their record the ubiquitous M-reflector. Furthermore, the geometry of this reflecting surface closely simulated the bottom topography of this inland sea (Figure 15.1); the sediments above covered the reflector like a thick blanket of snow draping over a mountainous plateau.

Bill Ryan went as a student on the Mediterranean cruise on which the M-reflector was first discovered. He had puzzled over it ever since, and he was finishing a thesis on this subject when he came to join me as co–chief scientist on Leg 13, in 1970. One of the main objectives of the drilling campaign was to investigate the M-reflector. Why was it so hard? When was the sediment deposited? And under what special conditions?

It was the early hours of 24 August, and both Bill Ryan and I were discouraged. More than 10 days had gone by, but we were as much in the dark as ever. This second Mediterranean site was positioned over the flank of a submarine volcano, but the drill string got stuck in sand and gravel before we could reach the basement.

Not having anything better to do, Ryan passed the time away mechanically washing and sorting pea gravels from the bucket of sands brought in the night before, while I sat on a high stool watching. As the morning wore on and the collection grew, I became more and more amazed by what I saw. A dark pile consisted of sedimentary grains of ocean basalt. A light pile consisted of white, hardened ocean ooze, and some glistening crystals of gypsum—a mineral precipitated out of evaporated seawater. Fossil shells of very small sizes were also found.

Gravel is not common in the deep sea, but coarse debris could be carried thither by turbidity currents (see chapter 6). Normally gravel or coarse sand consists of quartz, feldspar, fragments of granite, rhyolite, gneisses, schists, quartzites, sandstones, shales, reef limestones, and so on, namely, detritus from nearby land. This gravel was extraordinary. It consisted of four components: basalt, gypsum, lithified ocean ooze, and fossil shells of a dwarf fauna. That gravel alone led me to postulate

(1) isolation and desiccation of the Mediterranean Sea, (2) exposure and erosion of the seamount, (3) lithification of ocean oozes under the sun, (4) survival of a dwarf fauna in a briny sea, and (5) precipitation of gypsum.

The unusual gravel was also the clue that told us that the M-layer was a young evaporite formation—the residue of seawater when the Mediterranean dried up.

"You are too quick to jump to conclusions," Ryan chided me. "There is no need for a theory on the origin of an evaporite, when we have not yet found the evaporite formation!"

The "Pillar of Atlantis"

On the morning of 29 August, the _Challenger_ was positioned south of the Balearic Islands in the western Mediterranean Sea (Figure 15.2). We were drilling Hole 122 at a site in about 3,000 meters deep water. Ryan and I stayed up again until the wee hours of the morning, when the drill pipe apparently hit a hard layer. The drilling rate was reduced from several meters per minute to a meter per hour. Impatient with the slow progress, we went to bed just before dawn. We were not to rest long. Soon we were awakened by John Fiske, a marine technician, who came to report: "We found the Pillar of Atlantis!"

Quickly we dressed and rushed to the ship's laboratory for drill cores. Lying on the long worktable was a beautiful core that indeed resembled a miniature marble column (Plate X). I took one look and pronounced: "It is a chicken-wire anhydrite. It says what I said: the Mediterranean dried up then."

My shipboard colleagues were trained mainly as marine geologists. They were experts in analyzing ocean sediments, but most of them did not have to work with sedimentary rocks on land that were deposited in ancient shallow seas.

"What is a chicken-wire anhydrite?" Ryan asked.

"Oh, it is a calcium sulfate precipitated by groundwater under a sabkha." I managed only to confuse my friends a little bit more.

"What is a sabkha?"

"Sabkha is an Arabic word for a salt marsh, but they also call their coastal sand flats sabkhas."

"Why do you call it a chicken-wire anhydrite? How do you know that it was deposited under a sabkha?"

"Oh, it is obvious from the way it looks!"

My friends were not very happy with my chauvinistic approach. Not having encountered this type of rock before, they all thought I was pulling their leg. So I decided to pile it on: "Look at this, the stromatolite. The water must have been so shallow then that you could wade across the brine pool."

15.2. Location of deep-sea drilling sites in the Mediterranean Sea.

"Stromatolite?"

"Yes, stromatolite. See the wavy bedding, the dark laminae? They are traces of algal mats, which trapped lime muds on a tidal flat. The blue-green algae are cyanobacteria; they commonly require sunlight for photosynthesis and cannot grow in the deep ocean bottom, where it is eternally dark. We found many such algal mats in the intertidal zones seawards of the sabkhas of Abu Dhabi Island in the Persian Gulf."

Ryan remained skeptical, but he began to perceive that I was not joking: "Could you really tell just like that what those rocks are?"

"Why not, Bill? I am a sedimentologist, after all."

Sedimentologists are students of sediments; they describe and analyze sediments and sedimentary rocks. They cut off a chunk of carbonate rock and grind the chip into a transparent thin slice and examine it under a microscope. They crush a shale, pulverize it, and let the powder be bombarded by x rays to determine its mineral composition. They pound on a sandstone and shake it until the sand grains become loose enough to run through a series of sieves to analyze the sizes and sorts of disintegrated sand grains. They dissolve an evaporite (a chemically precipitated rock) and process it through a mass spectrometer to determine the isotope ratios of various chemical elements. Their purpose is to learn more about the origin of a sediment. Is it a beach deposit, or a lime mud laid down on a tidal flat, or an oceanic ooze?

In some instances, one does not have to go through complicated procedures, nor does one need sophisticated instruments. One can immediately tell the genesis of a rock and its environment of deposition by the way it looks. This particular approach, comparative sedimentology, was developed shortly after the Second World War, and the financial backing of this type of research by the oil industry contributed considerably to our success. Teams were sent out to study Recent sediments in various environments—river sediments on coastal plains, deltaic sediments at the mouths of major streams, marine sediments on open shelves, oceanic sediments on abyssal plains, and so on. Distinguishing features were recognized and described as "sedimentary structures," and those structures served to recharacterize suites of sediments deposited at various places. When a core of an ancient sedimentary formation is obtained from a borehole or an oil well, one could compare its sedimentary structures with a known standard, pretty much the same way that an expert identifies a purported Rembrandt by comparing its composition, its coloring, its shading, its brush strokes with the known Rembrandts. Sometimes, the comparison is purely empirical. In other instances, there are good reasons why a sediment should look the way it does.

Take our "Pillar of Atlantis," for example. This type of sediment has been found only on arid coastal flats. Prior to the cruise, my associates and I at the Swiss Federal Institute of Technology had studied the sabkha of the Arabian Gulf, supported by a research grant from the American Petroleum Institute. We dug many trenches on the sabkhas of Abu Dhabi and found "chicken-wire anhydrite" only in those places where the groundwater table was sufficiently close to the surface (Plate XI). There the saline groundwater could be heated to temperatures exceeding 30 °C or so. Where the groundwater table was deeper and the water cooler, gypsum, or hydrated calcium sulfate ($CaSO_4 \cdot 2H_2O$), would be precipitated out, in place of anhydrite. This finding is in accordance with chemical studies in the laboratory, which revealed that the transition temperature for calcium sulfate precipitated from saline groundwaters should be about 30 °C, or almost 90 degrees Fahrenheit. We thus have good reason to believe that anhydrite is not likely to be found in any environments other than the hot and arid sabkhas, because surface temperatures and groundwater chemistry elsewhere rarely permit anhydrite precipitation. We are almost certain that anhydrite could not be settled out of a deep sea. Even the Dead Sea is too deep a body of water to be heated hot enough to precipitate anhydrite; on the bottom of that salt lake only gypsum crystals have been found.

Ryan's next question was why the sabkha anhydrite looks like "chicken wire." We do not have a good answer, because we know too little about the mechanisms of crystal growth in nature. We do know that anhydrite under the sabkhas was precipitated by groundwater as concretions in arid soil. Fine-grained anhydrite would accrete and grow together as nodules underground, replacing preexisting carbonate sediments. Those nodules may be as much as several centimeters long. As the replacement proceeded toward completion, anhydrite nodules would coalesce to form a layer in which only wisps of preexisting carbonates could be discerned. Those dark wisps of carbonates in a white background of anhydrite resemble the wire mesh used by farmers to make chicken-wire fences (Plate XII). For this reason, the petroleum geologists who first encountered such anhydrite in their study of borehole cores used the descriptive expression "chicken-wire anhydrite." We really do not know why anhydrite should grow in this particular form, no more than an art expert understands why Fra Angelico chose to paint the way he did. We have to rely on the repeated observations by sedimentologists during the last few decades that this variety of anhydrite is typical of precipitation in Recent and ancient sabkha environments. Until we find evidence to the contrary, we feel complacent enough in considering "chicken-wire anhydrite" a signature of sabkhas.

Stromatolite is another sedimentary structure. It was not known for certain if it was a fossil or an inorganic structure of chemical precipitation until the 1930s, when a British sedimentologist, Maurice Black, waded across the tidal flats of the Bahama Islands. On the flat shores, covered intermittently by high tides, Black found a dense growth of blue-green algae, now called cyanobacteria, covering the ground like a thin mat (Plate XIII). After a severe storm the mat would be buried under a thin cover of mud, but the algal growth would return and a new mat would be constructed. This alternation ultimately results in the laminated sediment called stromatolite, which literally means "layered stone" (Plate XII). Since the very existence of algae depends on photosynthesis, the presence of a stromatolite structure is considered evidence of deposition in very shallow waters, commonly less than 10 m deep. In fact, repeated observations have confirmed that algal mats are a characteristic feature of intertidal environments. In the coastal areas between low and high tides, and the intertidal zone of Abu Dhabi, we found the current crop of lush growth in algal mats. We also found old algal mats, formed a few thousand years ago, buried under the windblown sand of the coastal sabkhas. Transpiration of groundwater led to the precipitation of gypsum or anhydrite in these fossilized intertidal sediments. When I was called to admire the "Pillar of Atlantis," I saw in this rock the same phenomenon as in Abu Dhabi, a stromatolite being partially replaced by nodular anhydrite. What better indication could there be that those sediments were formed on the tidal flat of a desiccated Mediterranean!

If the Mediterranean stromatolites are indeed fossil algal mats, we should find traces of algal filaments in those "flat stones." None of us on board ship were algae specialists. Later we sent specimens to Bob Parks of Liverpool, who had spent several years in Abu Dhabi studying the algal mats there, and indeed he was able to find fossil algae. Now all but diehard skeptics believe that the Mediterranean was a shallow evaporating pan at the time of stromatolite deposition. However, on that August day in 1970, Bill Ryan was still far from being convinced. When he came out of his reflective mood, he added:

"I cannot really buy your story, Ken. The M-layer was found everywhere in the Mediterranean draping over basement like a blanket. The geophysical evidence says to me that the M-layer was deposited when the deep basins of the Mediterranean Sea had already been created. Your chicken-wire anhydrite and stromatolite should have been deposited at depths where we now find them, thousands of meters below sea level. How could they be shallow-water or subaerial deposits?"

"They could be if all the waters of the Mediterranean were removed by evaporation! They could be if the bottom of the Mediterranean was a salt desert! By the way, haven't you found all those salt domes with your seismic profilings?"

At this point our dialogue was interrupted by Maria Cita. She had come up from the paleontology laboratory to give us a report on the age of the "Pillar of Atlantis" and had listened patiently to our debate. Now she exclaimed with excitement:

"Yes, why not? I was just about to tell you two that we have drilled into the Messinian. The Messinian was a time of salt deposition, as every Italian geologist knows. We have salt mines in Sicily, in Calabria, and in Tuscany."

"What is Messinian?" Ryan asked.

"Messinian refers to a rock formation near the city of Messina in Sicily. About a century ago, Professor Mayer-Eymar, one of Ken's illustrious predecessors at Zürich, studied the fossils collected from a marl between some gypsum layers in this formation, which incidentally, is called *solfifera sicilienne*. He concluded that those fossils were characteristic creatures of an age just before the end of the Miocene epoch. So he proposed a Messinian stage to designate this particular time interval. It was a relatively short stage, lasting perhaps less than one million years. Our latest estimate would place the interval from 6 to 5 million years ago.

"Salt-bearing and gypsum-bearing formations like the *solfifera* are common in other Mediterranean countries, like Spain, Algeria, Tunisia, Greece, Turkey, Cyprus, Israel, and so on. Many of those have also been dated as Messinian in age."

"Are you two telling me that five or six million years ago the Mediterranean Sea was a salt desert 3,000 meters below sea level?" Ryan asked.

"Why not?"

This was the question foremost in our minds during the remainder of the cruise, but our doubts were removed when we returned to Lisbon in October. The drill cores from the Mediterranean provided a resoundingly positive answer to our inquiry. We even found the salt that was deposited on a salt flat at the last site of our cruise (Plate XIV; see also Figure 15.3).

15.3. A salt flat in the abyss. Deep-sea drilling at Site 134 hit the edge of a salt flat in 3,000-m water at the foot of the continental slope west of Sardinia. The salt pan was fringed by alluvial fan sediments penetrated in Hole 133. *A, B, C, D,* and *E* indicate the positions of 5 other holes east of Hole 134.

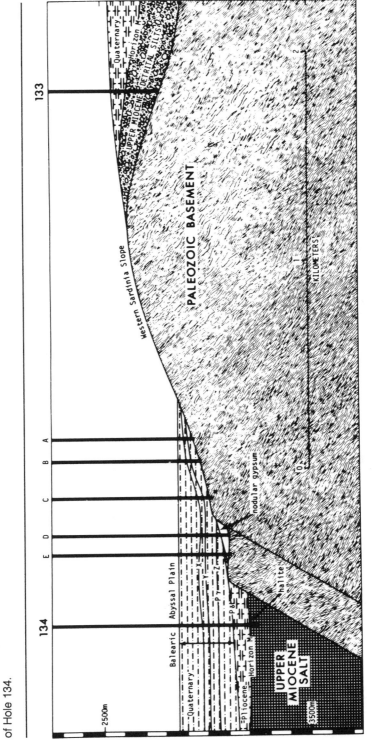

When the Mediterranean Dried Up

After Ryan, Cita, and I returned from Lisbon and started to study the geology on land, we realized that there had been many clues that could have prepared us for that discovery. As Cita told us, Upper Miocene salt deposits are very common in the Mediterranean countries. Blinded by the traditional theory, proposed in 1877 by C. Ochsenius, a German mining engineer, that salts could be precipitated only in coastal lagoons, geologists had naively assumed that those Miocene deposits on land were formed locally. No one had ventured to speculate that the scattered salt beds are the uplifted fringes of a giant salt formation, which remains largely buried under the Mediterranean.

The idea that the Mediterranean may have dried up should have occurred to the oceanographic investigators who found submarine canyons under the Mediterranean Sea. Some researchers thought that those canyons were once river gorges on land, and they now lay submerged because a whole segment of continent had sunk down to the abyss. Others thought that those valleys were the handiwork of submarine currents. Only one young French scientist ventured a guess that those submarine valleys were cut out on the margins of desiccated Mediterranean basins, by rivers cascading down into the central salt flats. Naturally nobody paid any attention to such a fanciful idea.

Paleontologists had long noticed something odd about the biological evolution of the Mediterranean. The terms Miocene and Pliocene, as I mentioned previously, were given by Charles Lyell to designate the fossil assemblages he collected in 1820 during his wedding trip to Italy. The Pliocene faunas were very similar to the present-day faunas of the Mediterranean, but the Miocene faunas were not. Unfortunately, Lyell, being the high priest of the doctrine preaching against catastrophic changes, overlooked the big difference. A firm believer in the Darwinian principle of survival of the fittest in the struggle for life, Lyell was unable to see that all had to perish when the Mediterranean dried up. He could not imagine that the Pliocene faunas of the Mediterranean were immigrants from the Atlantic, after the Strait of Gibraltar was opened by an invasion of marine waters from the Atlantic. During the early decades of this century, several paleontologists did note that the last Miocene fossils were strange creatures that did not live in seawater of normal salinity, but none were bold enough to postulate that the change in salinity was to lead to total desiccation.

The geographic distribution of the present-day faunas and floras of the Mediterranean countries also seemed very puzzling. Endemic species of land plants and freshwater faunas around this inland sea seemed to have dispersed from a common origin in the Mediterranean Sea. Not being experts in geology, biologists tended to rely upon their classic hypothesis of a sunken continent to explain this odd distribution.

Return to the Mediter- ranean

After the press release announcing the surprising discovery of Leg 13, the reactions of the scientific community were varied. The discovery of a huge salt deposit was an unusual finding. The interpretation that a deep sea had dried up was almost too incredible to be taken seriously. Leg 13 did bring back drill cores and scientific data, but evidently not enough to convince even our shipmates. Only Maria Cita joined the co–chief scientists in authoring an article postulating the desiccation of a deep Mediterranean Sea. Several of our co-workers on Leg 13, as well as other scientists, preferred other explanations. Vlad Nesteroff and many of his French colleagues accepted the evidence that the stromatolite, gypsum, and salt were deposited on sabkhas and lagoons, but they thought that the Mediterranean had a shallow bottom then, like the Baltic Sea today. Only after the evaporite deposition did the shallow bottom suddenly collapse and sink to the abyss like the proverbial Atlantis. Other scientists, mainly American, recognized the physical impossibility of changing the earth's crust from continental to oceanic during the few tens or hundreds of thousands of years represented by the contact signifying the Pliocene deep-marine inundation of the late Miocene desert. They accepted our postulate that the deep Mediterranean Sea was already there when the evaporites were precipitated, but maintained that they were precipitated in a deep brine pool. What was the basis for their preference of the alternatives? I knew of no scientific basis. All arguments had been expressions of a doctrinaire view: to postulate that the Mediterranean dried up is a violation of Lyell's uniformitarianism, which was the basic foundation of classical geology.

I spent a sabbatical leave at La Jolla in 1972, when Phase III of DSDP was being planned. The excitement of the Leg 13 findings was still fresh in everyone's memory, and controversies raged. My friends in the JOIDES organization were fascinated by the discovery, and a second drilling cruise to the Mediterranean was high on everyone's list. It was scheduled as Leg 39, and the *Challenger* was to sail from Brest after returning from the Arctic in the summer of 1974.

The atmosphere changed drastically, however, a year later, when the "fight" for shiptime was getting to the stage of "cut-throat competition." Our competitors found a weak spot in the plans for the Mediterranean; they tried to convince the decision-making bodies of JOIDES that drilling into the Mediterranean might encounter an oil or gas accumulation, causing pollution of the sea. A tentative decision to cancel the cruise was made.

The JOIDES planning committee met in Zürich in 1973. Bill Hay of Miami, who was then the chairman of the committee, invited me to present our case. My arguments may have been persuasive. Also, the Black Sea, being "virgin ground," was still an alluring target for all, and we all know one has to sail through the Mediterranean to enter the Bosporus. Anyway, I was happy to learn from Hay after the meeting that the Mediterranean cruise was again "on," now scheduled as Leg 42A. It would start after the _Challenger_ returned from the Antarctic in the spring of 1975.

The on again, off again plans for a third Antarctic drilling season were to create havoc with the scheduling of _Glomar Challenger_. The tight schedule was to bring about a cancellation of the Mediterranean cruise again. Fortunately for us, the Antarctic cruise was dropped even as the _Challenger_ was on her way to Cape Town. The Atlantic Advisory Panel scheduled two drilling legs (Legs 40 and 41) on the West African margin, and Leg 42A became the second Mediterranean drilling expedition.

The Hsü-Ryan-Cita model of deep-basin desiccation made some precise predictions. The Mediterranean was a deep sea long before it dried up. We should, therefore, obtain normal marine deep-sea sediments under the salt formation. The gradual desiccation also should have led to a concentric distribution of saline minerals: the least soluble salts of carbonates and sulfates should be in the periphery. The zonation of saline minerals had been indicated by the Leg 13 drilling data. We now planned to drill into the "bull's-eye" to sample the more soluble salts of potash and magnesium.

The shipboard scientific staff gathered at Malaga on 8 April 1975. Lucien Montadert of the French Institute of Petroleum (I.F.P.) shared the co–chief scientist duties with me. Maria Cita of Milan, Ramil Wright of Beloit College, Wisconsin, Carla Müller of the Bundesanstalt für Bodenforschung, Hannover, and Germaine Bizon of I.F.P. were the paleontologists. Daniel Bernoulli of Basel, Bob Garrison of the University of California, Santa Cruz, Robert Kidd of Scripps, Fred Mélières of the

University of Paris, and Frank Fabricius of the Technical University, Munich, were the sedimentologists. Albert Erickson of the University of Georgia came on board as a geophysicist. We were delayed by some unforeseen repair work, but _Glomar Challenger_ finally sailed on 14 April.

In contrast to Leg 13, the operations proceeded smoothly, with few surprises, few difficulties, and little excitement. Tension, however, was always in the air, especially when Cita and I were impatient with shipmates who were reluctant to accept the "obvious." Many conclusions were obvious to us, who had had the benefit of a previous experience, even if I had not been called "Instant Hsü." In the end, the predictions were verified, and the evidence was persuasive enough to convince all but the most hardened skeptic.

The first test was to verify the supposition that the pre-Messinian Mediterranean Sea was a normal marine inland sea like it is now. To do this, we had to sample sediment older than the evaporite. As we had planned, we drilled through the evaporite formation at Site 372 south of Menorca, and again for a second time at Site 375 west of Cyprus. The thick deposits under the salt beds are deep-sea sediments—oozes and muds laid down on both the western and eastern Mediterranean Sea were normal marine. The Mediterranean basins were much older than the Messinian Event; the western basins are at least 20 million years of age, whereas the eastern basins, a relic of the Tethys, could be as much as 100 million years older. The Mediterranean basins are definitely older than the age of desiccation.

The concentric pattern of the saline-mineral distribution was also verified. Several core samples of potash and magnesium salts were obtained, as predicted, at Site 374 in the center of the Ionian Basin east of Sicily. The water is now more than 4,000 meters deep there, but the abyssal plain was a subaerial salt flat some 5 million years ago.

If we had any real surprises during the cruise, it was when our drillers pushed the drill string into a cavern below the ocean floor. One day while drilling Site 376, I was having lunch in the mess hall. I saw Jim Ruddle, the drilling superintendent, rush in. He found the operations manager, Mike Pennock, and the two engaged in a serious consultation. Ruddle's face was pale with excitement, and Pennock's expression also changed when Ruddle whispered into his ear. I moved over, and asked what was up. Pennock told me that we might have drilled into a big "gas pocket."

"Why?" I asked.

"We have lost circulation!" Pennock answered.

Lost circulation means that water pumped down through the drill string of a hole is not coming back. Usually, lost circulation happens when one drills into a formation containing much natural gas; water pumped down the drill pipe displaces the gas in the pore space of the rock formation and thus does not come back up the drill pipe to the seafloor, or to the rig floor. The same phenomenon would happen, however, if one drilled into a system of water-filled caverns; water pumped down would displace the water in the caverns, and would not return either. Caverns are common on land, where the movement of groundwater can cause dissolution of limestones and make caves. However, water in ocean sediments is commonly static, and there is no water circulation under the ocean floor to carve out caverns. Therefore, experienced drillers would conclude that they have drilled into a "gas pocket" if the lost circulation happens in a hole spudded on sea-bottom.

The Mediterranean has an unusual history. When the inland sea was a desert, groundwater made its way from the humid north and was able to cause extensive dissolution of salts to produce caverns. In fact, the topography of the Mediterranean Ridge was pockmarked with solution holes like the karst country of Yugoslavia, where limestone caverns abound. Ryan and I, therefore, predicted that there might be caverns under the Mediterranean Sea. This prediction was made in a *Scientific American* article written for lay readers, and Pennock was able to find a copy in the ship's library. He was relieved after he became convinced that we had indeed drilled into a cavern under the deep sea. All other indications on the gauges of the rig floor led him to the same conclusion.

The Mediterranean-desiccation theory found more adherents among the shipboard scientific staff of Leg 42A; 10 out of 12 were willing to sign up as coauthors of an article advocating the drying up of a deep marine basin. The most enthusiastic support came, however, from unexpected quarters.

One day in 1978, Gilbert Bocquet of the Botany Institute at our university rushed into my office, crying "Eureka!" He had been doing research on the origin of the endemic Alpine floras on the coasts of Mediterranean islands—Corsica, Crete, Cyprus. How did the floras get there?

The desiccation of the Mediterranean is the answer! The Messinian vegetational zones were lowered when the Mediterranean abyssal plain was a salt marsh. The present continental rises and continental slopes of the Mediterranean were Messinian savannahs, steppes, temperate or subtropical forests.

Messinian Alpine floras flourished in coniferous forests that were established on the rims of the desert, rising some 3,000 m above the salt marsh (Figure 15.4).

Then came the deluge at the beginning of the Pliocene. The Messinian salt marshes, steppes, and subtropical forests were destroyed. The Alpine floras survived on coastal plains of Mediterranean islands. Isolated from their relatives on the continent, the island floras evolved independently during the last 5 million years and became endemic.

Bocquet now had the solution to one of the enigmas of the century in botany.

Lago Mare   Although we did not have many scientific surprises, we did clear up one important aspect of the history of the salinity crisis. During Leg 13, we obtained a few cores of laminated sediments from Site 124 in the western Mediterranean (Plate XII). We found no marine microfossils, but a flora of freshwater diatoms.

15.4. Floral zonation in desiccated Mediterranean. The Mediterranean vegetation zones were much lower in elevation compared with global sea level at the time when the Mediterranean abyssal plain was a salt marsh. This diagram was reconstructed by Gilbert Bocquet on the basis of paleobotanical evidence.

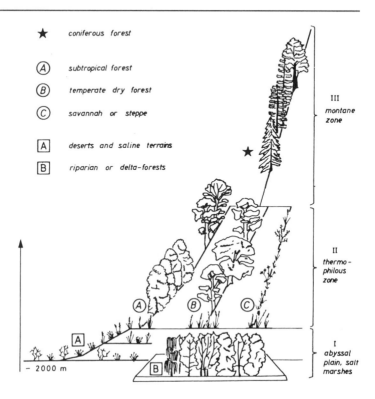

Diatoms are small photosynthetic one-celled plants, and their skeletons are made of silica (SiO_2). Being light, their dead bodies could be transported very long distances in the air. A transported flora of diatoms has been found, for example, in the sediments way out in the middle of the Atlantic Ocean. However, the diatoms in our samples do not belong to the kind that could be transported very far; those diatoms are a special kind that commonly lived on the bottom of shallow lakes.

The discovery of lake sediments suggested that the desiccated Mediterranean bottoms were not always salinas or salt marshes. At times, an excessive influx of freshwater converted desert basins into freshwater lakes.

During Leg 42A, we were able to establish the fact that the Mediterranean basins were indeed host to a number of lakes at the very end of the Miocene. In addition to the diatom floras, bottom-dwelling freshwater faunas were found in the lake sediments. These faunas resemble those living today in the Caspian Sea. The Great Lakes, or Lago Mare, did not survive very long; they may have existed for about a hundred thousand years or so. The lakes were eliminated when the gate at Gibraltar was "crashed" 5 million years ago, when the desiccated Mediterranean was inundated by the flood from the Atlantic.

Where did the freshwater, which changed the giant desert into beautiful mountain lakes, come from? This will be the story of the next chapter.

Was the Mediterranean a Desert?

In the May 1988 issue of *Geotimes*, a publication of the American Geological Institute and subscribed to by all professional American geologists, was a comment by Robert S. Dietz and Mitchell Woodhouse entitled "Mediterranean theory may be all wet." The article opens with the following statement:

> We have been disturbed lately by the repeated media promotion accorded the deep-dry-basin model developed by Kenneth J. Hsü (ETH Zurich, Switzerland) and William B.F. Ryan (Lamont-Doherty Observatory). Their model supposedly explains the "giant salt" . . . beneath the Mediterranean Basin. It has been featured in *National Geographic*, . . . "The First Eden" television series with David Attenborough, and in a full episode of "The Ring of Truth" . . . with Philip Morrison. . . . The Hsü-Ryan model has achieved textbook stature . . . All these accolades are deserving if true. But we, among others, find the evidence flimsy.

I was surprised by the comment, and I was surprised that the first published critique of our theory should have come from Robert Dietz.

Dietz is no doctrinaire reactionary. He published a theory of seafloor-spreading a year before Hess's article came out. He was an iconoclast when he challenged the establishment theory of the volcanic origin of what we now know as impact craters. These, and his many other contributions to geology, earned him the Penrose Medal of the Geological Society of America, the highest award in North American geology.

Geology became a science after the earth-science revolution of the 1960s. Was geology science prior to the revolution?

"There is science, which is physics, and there is stamp collecting, which is . . ."

This statement has been attributed to a famous physicist, but it would have long been forgotten if it were merely an expression of arrogance. Implicit in the statement is a commentary on the difference in scientific methodology between mathematical and natural sciences in the nineteenth century.

Philatelists collect stamps, and naturalists of Lord Kelvin's time collected observations, using the Baconian philosophy of induction. The pitfall of the approach is obvious. If a stamp collector is a specialist in American stamps, he could conclude, on the basis of his studies of stamps issued by the United States government, that only deceased persons are ever portrayed on postage stamps. His conclusion would be supported by billions of American samples, and by many issues of Chinese stamps. But the first British stamp in his collection would prove him wrong. Criticism of inductive reasoning was charmingly formulated by Karl Popper in the 1930s: you may have seen hundreds or thousands of white swans, but your conclusion that all swans are white collapses when you see your first black swan.

Observations are, of course, necessary in science. They are, however, only the clues that give ideas; they do not prove. The proof lies in disproving the alternative. "When you have eliminated the impossible, whatever remains, however improbable, must be the truth."

Dietz and Woodhouse claimed that our "evidence for a bone-dry Mediterranean consists of such things as alleged desiccation cracks, . . . stromatolite layering, and great submarine canyons." They thought that Ryan and I had used the same inductive approach as they themselves do. Not so. We started from the very outset with a Popperian approach.

It is true that I started to formulate the deep dry basin model on the basis of very flimsy evidence: I was inspired by the gravel at Site 122. Ryan protested then. Like Dietz and Woodhouse twenty years later, Ryan considered it "most unlikely" that the Mediterranean Basin dried up. There could be another explanation for the origin of gravel, or for the occurrence of gypsum, he argued. Again like Dietz and Woodhouse, Ryan then preferred a theory of evaporite precipitation in deep water, or a giant Dead Sea model, for the genesis of the Mediterranean evaporites.

"What seems likely or unlikely is personal judgment, not science," I replied caustically. "Besides, we still have six weeks to go. We can make predictions now and see which set of predictions is shown to be false. If your postulate of deep-water deposition is correct, we should see very little facies changes; the sediments settled down on the ocean bottom should be about the same everywhere. If, however, I am right that the basin was desiccated, we shall see all kinds of shallow marine, coastal, or even terrestrial sediments in the next few drill holes."

Ryan agreed to this test. Having studied Mediterranean cores for a decade, he knew that conditions for deep-sea sedimentation are very uniform. He could correlate laminated sediments in piston cores of the eastern Mediterranean with those of the west, thousands of kilometers away.

We drilled almost a dozen sites after the experiment was formulated, and we penetrated a different sequence of evaporite at every site. At the base of the continental slope, we found gravels and silts laid down on desert alluvial fans. On the margins of abyssal plains we found sediments typical of deposition on coastal marshes and lagoons. Finally, at Site 134, the salts were found on the edge of a shallow salina in the deepest of desiccated depressions, where the last brines were evaporated dry (Figure 15.3).

Ryan was not convinced by any single item observed, neither stromatolites, nor mud cracks or alleged mud cracks, nor submarine canyons. Stromalites could theoretically be formed by some kinds of cyanobacteria that needed no light, only chemical energy, for their carbon fixation. Mud cracks may not be desiccation cracks; they could be formed subaqueously. Submarine canyons are not necessarily drowned river valleys; they are commonly eroded by turbidity currents in marine realms. A greater collection of random observations would not have convinced Ryan; he would have an answer for each. Yet Ryan was convinced at the end of Leg 13 because the predictions of the deep dry basin model were verified, while the predictions of his deep briny pool model were disproven.

After the two drilling cruises, geologists working on land were to obtain a detailed scenario of the Mediterranean Salinity Crisis. The work by my students Daniel Müller in Spain and Judy McKenzie in Italy during the last decade has been particularly instructive. They have found evidence of an expansion of the Antarctic ice cap some 5.5 million years ago. While water was being locked up in ice, the worldwide sea level was lowered 50 m or more. Finally the level dropped below the height of an isthmus linking Africa and Spain. The isolation was at first complete; desert soil formed on the floor of the desiccated deep basins. Eventually, Atlantic seawater cut a pathway through southern Spain and debouched into the Mediterranean. The evaporative residue of this influx formed the thick Mediterranean salts.

There was an intra-Messinian inundation. Some 5.2 million years ago, the retreat of the Antarctic glaciers caused a rising of the sea level and marine inundation of the desiccating basins. Coral reefs grew not only in southern Spain but also in Cyprus, while marine oozes were deposited on the deep-sea floor. The episode was, however, very brief, and the climate changed again, causing a readvancement of Antarctic glaciers. The Mediterranean was again isolated. There was a Lago Mare in the eastern Mediterranean, and numerous lakes in the west.

The final inundation took place at the beginning of the Pliocene, about 4.8 million years ago. The gate at Gibraltar was "crashed" as a consequence of faulting. The strait was eroded down to 400 m subsea. There was a brief warming trend, which could be either the cause or the consequence of the inundation. The Mediterranean was never again isolated, even though the sea level dropped down to 130 m subsea during the last Ice Age.

I was a member of the JOIDES planning committee at the start of the new Ocean Drilling Project in the early 1980s; a French colleague and I lobbied for two legs of Mediterranean drilling. Unfortunately, the history of the Mediterranean Salinity Crisis had become so well established by then that we could not convince our JOIDES colleagues of the need for another major campaign in the Mediterranean. We had our answers, and we had no more predictions that had to be verified or disproven by working with the new drilling vessel. My position was weak, and my proposals were rejected.

JOIDES Resolution did go through the Strait of Gibraltar, and drilled in the Tyrrhenian Sea from December 1985 to February 1986. Kim Kastens and Jean Mascle were co–chief scientists, and my colleague Judy McKenzie was on board. The goal of Leg 107 was to study the genesis of a back-arc basin. There was, of course, still some mopping up to do, and they drilled into a Lago Mare in the Tyrrhenian Basin, but there was no longer much excitement when the Mediterranean evaporite was being penetrated.

16 The Black Sea Was Not Always Black

Geochemists were disappointed, because they did not obtain samples of hydrocarbon-rich sediments. Biologists were delighted, because they now had answers to their century-old puzzles.

To decipher the sedimentary history of this inland sea may seem relevant only to persons preoccupied with local problems, but the evolution of the Black Sea may in fact serve as a model for understanding some of the most prolific petroliferous basins of the world.

Daughters of Paratethys

Many caves on the Yugoslavian Adriatic Coast hide a great number of "living fossils." They are the last descendants of an ancient fauna that once inhabited large parts of eastern Europe. After a prolonged isolation, these species became inbred or endemic. The wealth of such endemic underground inhabitants was a puzzle to zoologists, who referred to them as the "enigma of the Adriatic corner."

In 1891 two Viennese zoologists, Steindacher and Sturany, visited Lake Ohrid, a little-known mountain lake on the border between Yugoslavia and Albania. They collected fish, mollusks, and other specimens from this freshwater body, and described many new species. The Lake Ohrid faunas resembled the underground "enigma of the Adriatic corner," and were thus also portrayed as "living fossils." The discovery at Lake Ohrid proved that the preservation of ancient relics is not limited to subsurface terrains.

The occurrence of endemic faunas in Lake Ohrid is particularly surprising because lakes are commonly short-lived. Many lakes of Europe and North America are "young," having been carved out by glaciers during the last Ice Age, and they are destined to be silted up in the not too distant future. Their inhabitants moved in during the last 10,000 years of the "wholly recent" Holocene geological epoch. Such young races have had little time to become inbred. The endemic faunas of Lake Ohrid and the caves of the karst region imply a much older age for these geomorphic features.

Still more perplexing was the fact, noted by the Yugoslavian biologist Stankovic, that relic faunas were found in isolated spots in southern France, Spain, North Africa, and other Mediterranean countries. Like those in Lake Ohrid, these animal groups are not consanguineous to their neighbors, but rather have relatives in the distant Caspian Sea. Furthermore, the "living fossils," the relic faunas, and their Caspian cousins all seem to have a common ancestry in the fossil faunas of a brackish sea called Paratethys.

Early in the Miocene epoch, some 15 to 20 million years ago, continental collision led to the fusion of Eurasia and Africa and caused the rise of a chain of mountains—the Alps, the Dinarides, the Hellenides, and the Taurides. The old Tethys was divided into two inland seas, the Mediterranean and the Paratethys (Figure 16.1). For some time, connections between the two existed north and south of the rising Alps, and the Mediterranean was still linked to the Indian Ocean through a shallow strait in the Middle East. Eventually, the Paratethys was completely separated from the Mediterranean, but it may have had access to the Pacific through a northeastern passage across the Arctic.

Starting out as a brackish sea, like the Baltic of today, the Paratethys was named *lac mer* by a French geologist, Maurice Gignoux, in 1920 to emphasize its intermediate nature between a lake and an open ocean. Stretching from the Vienna Basin to regions beyond the Aral Sea, the Paratethys severed its link to the Mediterranean during the middle Miocene, 15 million years ago. After that, the Paratethys lost much of its water and disintegrated; all that remains now is the Caspian Sea and the Black Sea.

In 1973 I chaired a JOIDES advisory panel that evaluated proposals to drill the Black Sea. It was high time that we drilled a few holes to clarify the relationship between the Mediterranean and the Paratethys.

Another reason to drill the Black Sea was to study the origin of petroleum. We know that the Black Sea is linked through the Bosporus Strait to the Sea of Marmara and thence via the Dardanelles to the Mediterranean. A two-layer current system is present in the Bosporus; the Black Sea water of about 17.5‰ salinity flows south at the surface, while the Mediterranean water of about 38.5‰ salinity flows north along the bottom. The incoming Mediterranean water mixes with the brackish water of the Black Sea; the mixture, being heavier, sinks, becomes stagnant, and finally becomes anoxic. Today the water mass of the Black Sea 200 m below the surface is rich in toxic H_2S, so the deep bottom can support no life other than anaerobic bacteria. Organic materials laid down on such anaerobic bottoms tend to be preserved. Eventually they are buried and

16.1. Mediterranean and Paratethys. Fifteen million years ago, continental collision caused the rise of a chain of mountains in Europe, and the old Tethys Ocean was split into two seaways, the Mediterranean (M) and the Paratethys (P), as shown in the upper figure. When ocean water could no longer enter the Mediterranean, the inland seas were changed into numerous inland lakes.

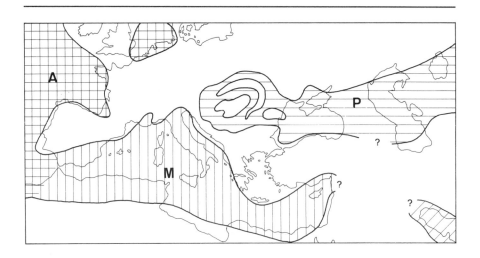

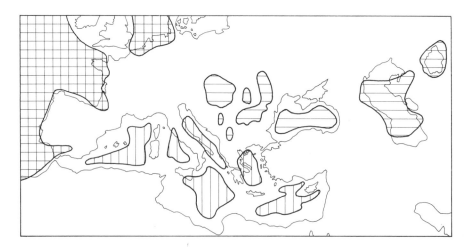

converted into hydrocarbons through chemical changes at elevated temperatures and pressures. Thus it seemed to us that the Black Sea would be a natural laboratory in which to study the process of generating crude oil.

Still another consideration that inspired the Black Sea drilling was the chance to obtain a record of the climatic history during the last Ice Age. Studies of continental glaciation in central Europe have, as I have discussed, revealed four glacial stages, Günz, Mindel, Riss, and Würm. Studies of ocean sediments have found, however, many more alternations of cold and warm climates (see chapter 2). The question was raised, Were the glacial advances on land taking place at the same time the ocean water was being cooled? The Black Sea is surrounded on all sides by continent, and may provide us with a link between the climatic records on land and those under the sea.

Soviet-

American

Cooperation

Phase III of the JOIDES Deep Sea Drilling Project was coming to an end during the mid-1970s, and the program was to be internationalized. German and Soviet institutions were the first to join when the International Phase began in November 1975.

David Ross of Woods Hole and I were appointed co–chief scientists in 1974. Ross was a veteran of the Woods Hole cruises to the Black Sea, and he was very knowledgeable. I was the chairman of the planning panel, and my lack of expertise was partially compensated for by my enthusiasm. Since our first priority was to examine the organic geochemistry of the Black Sea sediment, we were advised during panel discussions by Egon Degens, John Hunt, and Frank Mannheim of Woods Hole. They later also joined the cruise.

For shipboard paleontology, we preferred a "general practitioner" instead of several specialists, because we anticipated that there might not be much of a normal marine sedimentary sequence in the Black Sea. Steve Percival of the Mobil Oil Company, Dallas, had been a shipmate of mine on Leg 3. I was most impressed by his professional competence and his warm and selfless personality, and I secured his appointment for Leg 42B. Expecting that pollen and spores might be the only microfossils in the nonmarine sediments of the Black Sea, I also persuaded Alfred Traverse to join us; he was a distinguished palynologist, having been chairman of the Paleobotanical Section of the American Botanical Society, president of the American Palynological Association, president of the International Palynological Association, and so on. Albert Erickson from Leg 42A stayed on as the heat-flow specialist. Peter Supko was the Scripps representative, and was to be the editor of the cruise report. Peter Stoffers and Egis Trimonis were sedimentologists, representing a German and a Soviet institution about to become JOIDES members. Finally, Muhittin Senalp came on board as an observer from Turkey, because the *Challenger* had to sail through Turkish territorial waters. Senalp turned out to be a great help, acting as an interpreter during the many emergency telephone calls between the ship and the shore. He had to help us call the City of Istanbul before we returned; the traffic on the intercontinental bridge was to be stopped so that the *Glomar Challenger*, with her high tower, could squeeze under the structure.

Shortly before the cruise was to start, I got a phone call from Terry Edgar. He had a problem, he told me. Ross and I had been appointed the two co–chief scientists for the Black Sea cruise before JOIDES secured the Soviet's agreement to join the Deep Sea Drilling Program. Now the Soviets wanted a dowry. To endorse the Leg 42B drilling of the Black Sea, they wanted Yuri Neprochnov, of the Soviet Academy of Science's Institute of Oceanology in Moscow, to be a co–chief scientist of the expedition. The bottom line was plain. They could not ask Ross to step down, because the leg could almost be considered a Woods Hole undertaking. Edgar did not have to finish his introduction, and I tendered my resignation as a co–chief scientist. Like Jerry Winterer on Leg 17, I served as a sedimentologist on Leg 42B.

Glomar Challenger entered the Bosporus on 19 May 1975. The vessel was not docked, but was anchored in the Sea of Marmara. We had a change of the ship's scientific staff. I moved out of the cabin for co-chiefs and roomed with Alfred Traverse.

Traverse was a dear personal friend, dating back to the days when we both worked for Shell. One afternoon in the early 1960s, he came to my office and told me that he was leaving. It was quitting time; I was surprised he had come to tell me that.

"No, Ken. I am leaving Shell. I am entering a seminary, a theology seminary at Austin. I shall study to be a priest. My congregation gave me a scholarship. Betty and the kids are supportive."

"But why?"

"I got the call."

"How are you going to manage it?" I asked, thinking that they had to leave their air-conditioned "mansion" at Spring Branch to move, with all four children, into a run-down shack, densely populated by cockroaches and poorly isolated from the heat of Texas. He would also have to give up the top salary of an industry scientist and live on the meager scholarship of a first-year student.

"The Lord will provide."

When Traverse came to the *Challenger*, he had long since finished his seminary studies—with highest honors, naturally. He had gone to see a bishop, expecting an appointment to a parish. He was told, however, that he would serve God more if he resumed his career as a scientist.

After a five-year interruption, it was difficult to make connections again. Fortunately, a friend from his Harvard days invited him to join the teaching staff of the Pennsylvania State University. He has been there ever since.

I greatly treasured the chance to renew my friendship with Traverse. I spent most of my time in his paleo lab downstairs, ironing out the little wrinkles of misunderstanding he initially had with his technician, Patricia, and learning from him the history of the Ice Age.

Steppes and

Forests

Came and

Went

Three sites were drilled in the Black Sea during a short voyage of about three weeks. Of those, Site 380, located in about 2,000-m-deep water at the base of the western continental slope, had the deepest penetration, 1,073 m, and provided the most complete geological record.

From the very outset, the drill cores told us that marine waters did not always find their way across the Bosporus. We had had some hints prior to the cruise that the Black Sea might not always have been a brackish marine body. During a 1969 cruise there by the Woods Hole R/V *Chain*, sediment cores about 10 m long were obtained. Studying these materials, David Ross and Egon Degens found that the Black Sea was a freshwater lake during a 12,000-year span in the last Ice Age. The changing environment of the Black Sea was related to the rise and fall of the sea level. Ross and Degens emphasized that the Bosporus is a very shallow strait with a sill lying less than 35 meters below sea level. During cold stages of the Ice Age, much water from the oceans was locked up in continental glaciers, so that the sea level was more than 100 m below its present level. In those times the Bosporus was not a marine strait, but a meandering river draining a freshwater lake (the Black Sea of the Ice Age) into the Mediterranean Sea.

Ross and Degens did not know, and did not expect, that the lake was already there millions of years ago. They assumed that the nonmarine sediment in their piston cores was from a rare happening, being limited to those glacial stages in which global sea level dropped beneath the sill depth of the Bosporus. Under normal circumstances, the Black Sea deposits of the past should be similar to those of today, and there should have been a great deal of black sediment to be sampled for geochemical studies.

The drill cores of *Glomar Challenger* were to present us with an unpleasant surprise: freshwater conditions turned to be more the rule than the exception. Not only John Hunt was disappointed; we all realized that we had gotten into a very difficult situation because we could find almost none of the marine fossils that are useful for determining the age of oceanic sediments. We were thus mostly in the dark concerning the chronology of the sequence we penetrated.

The Black Sea sediments do contain fossils. Some floated like planktonic diatoms, dinoflagellates (cysts of a kind of one-celled algae), and pollen and spores of land plants. Others were remains of bottom dwellers; they include ostracodes, mollusks, small crustaceans, and some unusual foraminifera that could live in a brackish sea. Practically all of the fossils belong to long-ranging species and are of little use in dating sediments, but they tell us a great deal about the changing environments of the Black Sea.

Plants, for example, do not evolve very fast, but their growth is sensitive to climate. Plant fossils such as pollen and spores are good indicators of ancient climates. Traverse identified plant fossils from dozens of genera that originally grew in a wide range of habitats. Some were pollen shrubs, belonging mainly to the steppe vegetation _Artemisia_ and _Chenopodiaceae_. Others were tree pollen from pines, oaks, beeches, and so on. The pollen was either carried by rivers or flown through the air before finally settling on the bottom of quiet bodies of water. Normally pollen would be oxidized on the ocean bottom, and thus fossil pollen are found only in black shales, like the ones encountered by Bill Ryan during Leg 40. On a lake bottom, however, the deep water would be poorly oxygenated or entirely devoid of oxygen, and fossil pollen could be preserved. My foresight in having my friend Al Traverse along was now recognized by all. We might get some idea of the age of the sediments after all, if we could correlate them with the advance and retreat of glaciers.

Traverse invented a "steppe index" as a quantitative measure of the floral composition. The index is the percentage of steppe pollen in a total assemblage. A value of 100 suggests that the land surrounding the Black Sea was almost completely covered by steppe vegetation, and a value of zero paints a landscape of pine and mixed forests.

The cause of the change is not difficult to find. During an interglacial warm climate, the surrounding land was largely forested. With the advance of continental glaciers, forests disappeared, to be replaced by steppe vegetation. Thus the steppe index records not only the vegetational but also the climatic history of a region. Individual values indicate short-lived climatic oscillations. Using five-point averages to smooth out the curve, Traverse recognized three major stages of glaciation, which he named Alpha, Beta, and Gamma (Figure 16.2).

One important observation made by Traverse is that steppe pollen are rare or absent in sediment cores taken from greater depths. At Site 380, there is an increase in the amount of pollen from elm trees and a corresponding decrease in the amount of shrub pollen in the so-called pre-Alpha sediments below 650 m

16.2. Climatic record of Black Sea sediments. Using various event markers, a history of the Black Sea sedimentation was reconstructed. Units of sediments, alternately rich in tree pollen and in steppe pollen, indicate alternating interglacial and glacial climates. Numbers 1, 3, 5, and so on indicate successively older interglacial stages that correlate with those discovered by Nick Shackleton in analyzing ocean cores of the Quaternary period.

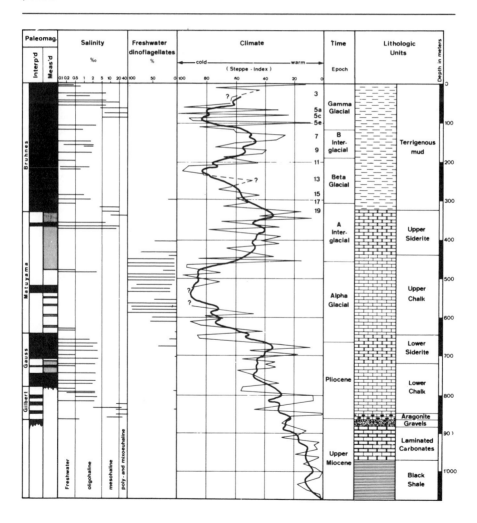

subsea. Further down, below 890 m subbottom, practically no more shrub pollen are present. Near the bottom of this hole, below 988 m subbottom, the pollen assemblage was derived from semitropical flora, which flourished in Europe long before the Ice Age. A similar trend is found in Hole 381.

Not only were we geologists disappointed, not able to decipher the sedimentary history of the Black Sea; John Hunt, our organic geochemist, also came up empty-handed. Except for the near-surface layer, we recovered no sediments particularly rich in organic carbon. Apparently the toxic bottom-water of the Black Sea is a recent phenomenon, related to the intrusion of marine waters through the Bosporus. In former times, the Black Sea bottom was often aerated, especially when a freshwater lake occupied the basin. The environment then was enlivened by bottom-dwelling creatures like ostracodes and mollusks. Others left no skeletons, but their feeding tracks and trails have been fossilized.

There were times of stagnation, especially during the preglacial age, shown by the alternate sedimentation of microorganisms (such as diatoms) and suspended clay particles, which produced a rhythmic repetition of light and dark laminae. This laminated structure was not disturbed, because no benthic animals existed. Even under those conditions, however, not much organic material was deposited. Hunt had come with all kinds of gadgets to freeze-dry the cores, but he seldom needed the equipment.

When we returned to Istanbul in June 1975, we carried with us more than 2,000 meters of Black Sea muds, but we were more than a little puzzled. We perceived that the Black Sea was different in the past than it is now. It was a deep freshwater lake before the Bosporus became a marine strait. We also knew that the lake floor was a site for the accumulation of chemical precipitates for a considerable period of time. Finally, it appeared that we had a long geological record, extending back to the time before the Ice Age, but we had no key to read this history. We had to wait for years before the message of the sediment cores was decoded.

Event Stratigraphy

Geologists date rocks on the basis of events. Paleontological time-markers were events of biological evolution. Shipboard paleontologists used the expression FAD, signifying the First Appearance Datum (the date on which a species first appeared), and LAD, signifying the Last Appearance Datum. The moment when a species first came into existence was an event, and the moment when the species became globally extinct was also an event. Biostratigraphy is, in a broad sense, event stratigraphy.

The term event stratigraphy, sensu stricto, has been introduced during the last decade to designate a methodology of dating sediments on the basis of well-known events other than those of biological evolution. Sudden global changes of climate, for example, are events. Using this event-stratigraphy approach, Traverse's pollen data give a chronological correlation between the Black Sea climatic events and those elsewhere in the world.

During the early seventies, long piston cores were obtained from the Pacific and Atlantic, providing excellent material for investigating climatic changes during the last few million years. They were the "long cores" Emiliani had wanted (see chapter 2). Nick Shackleton of Cambridge studied the temperature changes by analyzing oxygen isotopes, and Neil Opdyke determined the ages by measuring the remnant magnetism of the sediments. They extended the climatic record started by Emiliani back to more than 2 million years. Emiliani was right. There were many more oscillations of the ancient climate during the Ice Age than had been suggested by Penck. For the last 0.7 million years alone, there were 19 changes: 10 warmer epochs alternating with 9 glacial stages (Figure 16.3).

My student Federico Giovanoli worked on the magnetics of the Black Sea cores. Detecting field reversals is not easy in lake sediments, because many of the iron-containing minerals in a lake are not chemically stable: the magnetic orientation they exhibit today might not reflect the direction of the earth's field when the minerals were first deposited, but that at some later time of chemical transformation. Nevertheless, Giovanoli and I were able to recognize in the magnetostratigraphy of the Black Sea cores the transition from the Matuyama epoch of reversed

16.3. Glacial and interglacial stages during the last Ice Age. Using isotope geochemistry to determine paleotemperatures and magnetostratigraphy to date sediments, Shackleton and Opdyke presented this curve in 1976 to show the rapid alternation of glacial and interglacial stages. The ages of the magnetostratigraphic epochs Bruhnes and Jaramillo (Jar.) are expressed in thousands of years.

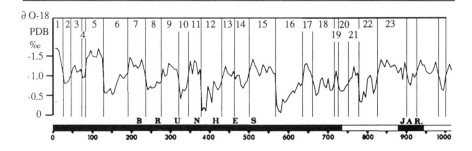

polarity to the Brunhes epoch of normal polarity; that was 0.7 million years ago (Figure 16.2).

I saw Cesare Emiliani in 1977 when I visited Miami and discussed with him our Black Sea results. I thought that Traverse's Alpha, Beta, and Gamma glacial stages might correspond to three of Penck's famous stages of continental glaciation; there were perhaps not so many epochs of climatic changes on the continent as Shackleton's marine record suggested. Emiliani disagreed; he told us that we had misled ourselves by taking five-point averages.

When I came back, I took out Traverse's manuscript. Giovanoli's studies of magnetic anomalies gave us a date for the start of the Beta glaciation. There were indeed more fluctuations of climate during the last 0.7 million years if we did not take five-point averages. In fact, we recognized the 19 glacial and interglacial stages that had been identified based on studies of ocean sediments (compare Figures 16.2 and 16.3).

What was the age of the Alpha, then (the beginning of the Black Sea glaciation)?

Giovanoli's work in 1979 suggested that the Alpha started at about the same time as the polarity reversal between the Gauss epoch, of normal polarity, and the Matuyama, some 2.4 million years ago. Would the records elsewhere lead us to the same conclusion? As it turned out, we did not have to wait very long for an answer.

Helmut Weissert, a former student of mine at Zürich, went with me on the Leg 73 cruise to the South Atlantic in 1980 and studied the oxygen-isotope composition of Upper Cenozoic deep-sea drilling cores. He found a major cooling event, signifying expansion of continental glaciation about 2.5 million years before present. His prediction was verified by the 1981 drilling in the North Atlantic. "Oxygen-isotope analysis of the sequence together with nannofossil and magnetostratigraphic time control indicate that the first major glacial event (in Northern Hemisphere), as represented by coincident ice volume and ice-rafting horizons, occurred at 2.37 m.y. ago." That was the conclusion by Herman Zimmerman, Nick Shackleton, Jan Backman, Dennis Kent, and others of the Leg 81 scientific staff.

The conclusion that the continental glaciation of the Northern Hemisphere started about 2.4 million years ago is further supported by studies of continental deposits in northwestern China. Liu Dongsheng, a specialist on Quaternary geology, was in the audience when I gave a talk at Peking University in 1977 urging a study of loess and soil horizons. A Lamont scientist, W. Kukla, had studied Central European loesses and found that they were the windblown deposit of the glacial stages; the intercalated red soil horizons were, in contrast, the product of interglacial weathering. Liu initiated a program to study the loess plateau of northwestern China, and indeed found cycles of glacial and interglacial climate correlating with those on the marine record. The beginning of the glaciation could thus be dated on the basis of the first deposition of the loess. Working at our Geophysics Institute in Zürich, Liu and Heller found that the beginning of the Northern Hemisphere glaciation took place at the same time as the polarity reversal between the Gauss and Matuyama magnetostratigraphic epochs, or 2.4 million years before present.

Using the approach of event stratigraphy, we were thus able to conclude that the beginning of the continental glaciation, or the Alpha glacial stage, as manifested by the first widespread occurrence of shrub vegetation in the lands around the Black Sea, took place during the late Pliocene, 2.4 or 2.5 million years before present.

When Did the Danube Arrive?

The sedimentary record of the Black Sea is distinguished by an abrupt transition between two very different types of sediments. The sediments deposited during the last 0.7 million years are mainly terrigenous silts and muds, the fine detritus carried down by rivers emptying into the Black Sea. Prior to that time, the material deposited on the bottom of the Black Sea was mainly chemical precipitates.

Pure chemical precipitates can accumulate in a lake only if clays, silts, and other products of weathering in the surrounding landscape, collectively called *clastics*, are excluded. At Lake Zürich, the clastics are trapped in other lakes upstream. The Black Sea, on the other hand, now receives a heavy burden of clastics, much of it contributed by the Danube. In earlier times, when the sediments were mainly chemical precipitates, some barrier must have prevented the Danube clastics from reaching the Black Sea Basin.

The nature of that barrier became apparent to me in 1976, when Dan Jipa, a Romanian geologist, guided me on an excursion to examine a thick section of clastics in the foothills of the eastern Carpathians. The material is now exposed at the surface, but it was deposited in an ancient lake of Pliocene and Pleistocene age. During a subsequent tour of the Danube delta I was told that boreholes there reached bedrock under a thin cover of Pleistocene sediments. Apparently the Danube River once passed through an eastern Carpathian lake, where most of the suspended solids had time to settle out. Some time during the Pleistocene, the lake became filled with silt, and the Danube was forced to change course. Deprived of the "settling tank" upstream, the Black Sea was flooded with clastics, and the period of chemical sedimentation came to an end 0.7 million years ago.

The chemical sediments of the Black Sea are carbonate minerals—calcite ($CaCO_3$), aragonite ($CaCO_3$), dolomite ($CaMg(CO_3)_2$), and siderite ($FeCO_3$). Freshwater lakes, especially in temperate limestone terrains, are commonly saturated with carbonates. Chalk made up almost entirely of calcite is now deposited every summer on the bottom of Lake Zürich. Photosynthesis of floating one-celled plants in a lake uses up dissolved carbon dioxide, and thus decreases the amount of calcium and carbonate ions that can be dissolved in water. Calcite is, therefore, precipitated as lacustrine chalk. A lake in the Black Sea Basin during the Ice Age would have had an environment similar to that of Lake Zürich today: glaciers were some distance away, and annual biological activity was sufficient to alter the solubility of calcite and to cause its precipitation.

The formation of carbonates other than calcite has to be explained by changes in the chemistry of the lake water. The iron-carbonate siderite, for example, was deposited mainly during warm, humid interludes, when meandering rivers carried much dissolved iron into the basin.

Bob Garrels, a geochemist well known for his work on Precambrian iron ores, was visiting Zürich shortly after I came back from the Black Sea cruise. He was very fascinated by our report of the laminated iron carbonate and silica ooze.

"This is an actualistic model for Precambrian iron formations," Garrels said enthusiastically. "We could learn much about the primeval ocean by studying the chemistry of the Black Sea."

Precambrian "banded iron ores" also consist of alternating laminae of silica and an iron mineral. The ocean had a different chemistry 2.5 billion years before present. The ocean water had little or no oxygen, so two-valence iron was precipitated, probably as iron carbonate, like the siderite of the Black Sea. Diatoms, of course, did not exist at that remote age, but there might have been some kind of silica-secreting bacteria.

Iron carbonate precipitated before the glaciers came. To have precipitated iron instead of calcium carbonate requires that the Black Sea was impoverished in calcium. Calcium ions are commonly much more abundant in river waters than dissolved iron ions, and thus lake carbonate precipitates are commonly chalk.

Iron instead of calcium will precipitate only when the lake water has a calcium-to-iron ratio of less than 20. Only in regions of lateritic weathering, such as the southeastern United States, could streams have an unusually high iron concentration. The siderite precipitation indicates a warm and humid Pliocene climate in the Black Sea region, verifying the conclusion by Traverse, the palynologist.

When the Black Sea Was Drained

The studies of fossil pollen and remnant magnetism yielded a geologic record for the Black Sea through the last few million years. A glimpse into the more remote past was obtained by Musat Gheorghian of the Romanian Academy of Sciences, who found in the oldest samples from the Black Sea a fossil fauna typical of the Paratethys. The fossils of several inbred species of bottom-dwelling foraminifera native to a brackish sea, and their evolutionary development, suggested that the oldest sediments recovered from Site 380 were between 8 and 10 million years old. The paleobotanical evidence supported this conclusion; in the same oldest layers Traverse had found abundant pollen characteristic of plant life from a warm upland habitat; the plants in question disappeared in Europe some six or eight million years ago, when the climate began to cool appreciably. More recent work by Giovanoli on the remnant magnetism of the oldest sediments confirmed that the oldest sediments are indeed 10 million years old. We can now safely conclude that our record of the Black Sea should extend through the late Miocene interval, some five to six million years ago, when the Mediterranean dried up. What happened to the Black Sea when its neighbor was reduced to a desert?

Before the 1975 expedition of the *Challenger*, a French oceanographic cruise in the Black Sea had identified a strong acoustic reflecting layer 1,000 meters or more below the seafloor. Deeply buried structures resembling salt domes were also found. It was the discovery of such a reflecting layer, of course, that had provided the first clue to the desiccation of the Mediterranean. During our campaign, the *Challenger* did drill through a reflecting layer 865 meters below the seafloor at Site 380, but the layer is not salt, nor other evaporating residues from seawater; the sediment consists of cemented gravel.

The Black Sea is now 2,000 m deep, and the deep basin has been there for a very long time, hosting either a deep lake or a deep sea. The usual sediments are the very fine-grained mud or the chemical precipitates. To encounter gravel on the deep basin floor was thus a surprise.

Submarine slumping could have transported sands and gravels from coastal regions to a deep abyssal plain, but the Black Sea gravel showed none of the features characteristic of that type of deposition. The mystery deepened when the shore-based research was carried out. Peter Stoffers of the University of Heidelberg described a dolomite between gravels. The dolomite was stromatolitic, owing its genesis to the growth of blue-green algae in shallow waters (see chapter 15). Stoffers also found oolite similar to that found near the shores of the Great Salt Lake. Furthermore, Hans Schrader of the University of Kiel identified diatom species from a very shallow habitat in sediments just below the gravel. The Black Sea has always been a deep basin, but the evidence implies that the basin held very little water when the gravel, stromatolite, and oolite were being precipitated, and when the shallow-water diatom covered the bottom. Could those sediments be an indication that the Black Sea was also desiccated during the late Miocene?

We had few fossils to date the event. We also had little reason to assume that the Black Sea could have been desiccated. Unlike the Mediterranean, the Black Sea does not have, and may never have had, a deficit of its hydrographic budget: rivers flowing into the Black Sea carry enough water to compensate for its evaporative losses. Besides, the Black Sea was one of the deeper Paratethys basins. If it was desiccated, the others must have also gone dry. Do we have any evidence of that?

In the late autumn of 1976, while I was traveling in eastern Europe as an exchange scholar, a coherent model began to emerge. Sediments from regions of the Paratethys include many layers of volcanic ash that can be dated radiometrically (see chapter 4). It was found that the salt beds found in the Paratethys basins are mostly more than 15 million years of age, much older than the salts of the Mediterranean. In late Miocene times, no salt was deposited in what are now the Balkan countries, so it does not appear that this western part of the Paratethys was then evaporated dry. However, R. Jiricek of the Czechoslovak Academy of Sciences was able to show that the Paratethys did indeed undergo a crisis when the Mediterranean dried up. This great brackish sea disappeared in the late Miocene, leaving only scattered freshwater lakes at the bottom of the Paratethys basins in the Balkans. Marls and muds were deposited, and freshwater faunas lived in those lakes. Lake Ohrid is apparently a descendant of one of those lakes, and its living faunas can trace their lineage to the Paratethys faunas of those days. We know that the local lake-basins were never evaporated dry, because we have found no salts there. Much of the water of the Paratethys disappeared; it seems to have gone down a drain.

Jiricek's report provided the clue we needed to reconstruct the closely entangled histories of the Mediterranean and Black Seas. In the late Miocene, after the Mediterranean was evaporated dry, there was a salt plain at the bottom of each Mediterranean depression, like enormous Death Valleys. Rejuvenated streams cut deep canyons into the continental slope, which had been laid bare. (Some of those canyons can still be seen today by deep-sea submarines cruising down the continental slope, which is once again submerged.) The erosion of headwaters lengthened the rivers draining into the Mediterranean, with the effect that the watershed was pushed steadily northward. Finally the divide was breached, and the Paratethys *lac mer* was emptied, by way of deep ravines that ran through Hungary and Yugoslavia to the Adriatic. The "death valleys" of the Mediterranean were flooded by fresh waters, creating a series of great lakes, including the Lago Mare, on the floor of the previously desiccated basins; the typical Paratethys fauna came with the flood.

The Lago Mare postulate was warmly embraced by zoologists concerned with the dispersal of freshwater organisms in circum-Mediterranean countries. As I was writing these pages, I received a copy of a new publication from an Italian colleague, Pier Giorgio Bianco, on the dispersal of Euro-Mediterranean fishes. In northern Spain, southern France, Italy, Yugoslavia, Albania, and Turkey, the freshwater fishes could be recognized as descendants of a Messinian fauna that originated in the Danube drainage basin (Figure 16.4). How the fishes had crossed the divide to inhabit rivers draining into the Mediterranean had long been an unsolved puzzle in biogeography. Bianco now had the answer: the Danubian fishes came with the Caspian waters, which spilled into the Lago Mare during the last millennia of the Mediterranean desiccation. They survived the Pliocene marine inundation and found refuge in the rivers of the Mediterranean drainage.

Other zoologists, specializing in snakes, freshwater mollusks, freshwater birds, and terrestrial animals, had all communicated to me their excitement in finally understanding the peri-Mediterranean dispersal of those terrestrial faunas.

With the disappearance of the Paratethys, the drainage system of Europe underwent a major reorganization. Whereas the Black Sea had received much of the runoff from the humid regions of central Europe, that water was now pirated by the Mediterranean drainage basin. The Mediterranean's gain was the Black Sea's loss. Deprived of its due share of the river influx, the Black Sea no longer had enough water to compensate

16.4. Invaders from Eastern Europe. This map shows the fish-faunas provinces of the Mediterranean countries. The faunas of Central Iberia (4), Ebro-Cantabric (5), southern France (6), Albania (10), and the Aegean (12) have a predominance of the so-called Danubian species. They are the descendants of invaders from the Paratethys of eastern Europe, and their ancestors lived in the late Miocene Lago Mare.

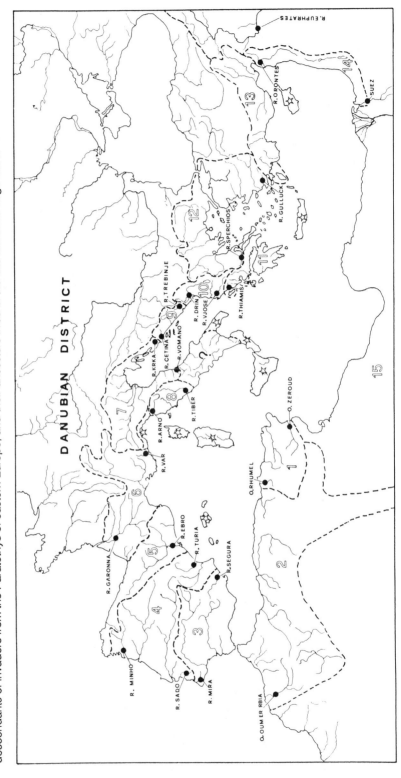

for evaporative loss. The great freshwater lake was gradually evaporated down, until the 2,000-meter-deep basin barely held enough water to cover a shallow bottom. The Black Sea became a great salt lake, at least three times as salty as normal seawater. We are not certain if the basin was ever completely dry, because we were not able to reach the salt deposit in the central depression of the Black Sea. The evidence indicated that a shallow briny lake must have existed for some time. Gravels were deposited on alluvial fans by flash floods coming out of the dry canyons that cut into the exposed continental slope of the Black Sea. Oolites and stromatolites were formed on lake shores. Finally, the brines of the salt lake were buried with the sediments; they are now found in the pore space between the pebbles and sand grains.

In late 1976, Schrader discovered a late Miocene diatom flora in sediments a little below the gravel. We now had paleontological confirmation of the event stratigraphy: the Black Sea gravel was deposited during the period when the Danube water was pirated by the desiccated eastern Mediterranean, causing a deficit of the Black Sea budget.

Birth of the Modern Black Sea

The Mediterranean Lago Mare was short-lived; it was inundated when seawater entered the Mediterranean at the beginning of the Pliocene, 5 million years ago. An event of this marine flooding was also recorded by the Black Sea cores: a layer of marine mud overlies the gravel bed, testifying that seawater had returned. The Paratethys basins of the Balkans were meanwhile silted up; what was once the floor of that ancient sea is now farmland.

There were mountain-building movements, and the eastern European drainage was reorganized. The Black Sea's tenuous connection to the Mediterranean was soon broken, but receiving more than enough water to balance evaporative loss, the basin now hosted a deep lake. With an excess input of freshwater, the Black Sea lost its salinity and became a freshwater lake; this is when the epoch of chemical sedimentation began.

The strange tales of the Mediterranean and Black Seas provided the key to the longstanding zoogeographical puzzle. The draining of the Paratethys carried animal species typical of that sea into the Mediterranean basin, and their descendants lived in a late Miocene Lago Mare that stretched from the Levant to Spain. When the Strait of Gibraltar was cut open, a marine fauna invaded, and most of the immigrant species were exterminated. Only scattered survivors found refuge in isolated

Relic

Back-Arc

Basins and

Oil Fields

freshwater habitats. The "enigma of the Adriatic corner," the "living fossils" of Lake Ohrid, and the relic faunas in coastal regions of circum-Mediterranean lands all bear a resemblance to the faunas of the Caspian Sea, because the Caspian is all that is left of the great brackish *lac mer*. After the recent inundation of the Black Sea by the Mediterranean, the Caspian is the last home of the descendants of the Paratethys.

Why should we be interested in the Black Sea? We went to the Black Sea because source beds for petroleum deposits are being laid down there today. The modern Black Sea could be considered a stage in the life history of one type of petroliferous basin. How did those basins originate? Where do those basins produce oil? These are questions of general interest.

The Black Sea is a large, deep basin situated between the Russian Platform to the north and the Pontic Mountains of Turkey to the south. How did this deep hole originate?

We did not expect to get a direct answer by deep-sea drilling; the sediment sequence is 12 or 15 km thick, and we could not expect to sample the oldest sediment of the Black Sea. We had to study the land geology. I had my first clue when I was a National Academy exchange scholar in Bulgaria. Vassil Vuchev and Ivan Nachev took me on an excursion. Most of the sedimentary strata we saw were deposited on land or in shallow seas, but I was surprised to see deep marine Cretaceous rocks in the Srednogorie region near the Black Sea coast. The sequence consists of submarine volcanic flows and tuffs, oceanic pelagic sediments, and deep-water turbidity-current deposits. The sequence is very similar to that at the JOIDES Philippine Sea Sites 53 and 54, which had been penetrated during Leg 6. The latter was laid down in a back-arc basin (see chapter 9). Was the Srednogorie once a back-arc basin?

The east-west trending Srednogorie structures plunge under the Black Sea. I soon came to the idea that the Srednogorie was only a part, the western extremity, of a Cretaceous back-arc basin. Here, the deep-sea bottom was heaved up to became part of the Carpathian Mountains. Yonder, the Black Sea continued to subside and is now buried by more than 10 km of sediment. The frontal arc of the Cretaceous basin was the volcanic arc of North Anatolia. The arc collided with a fragment of Gondwanaland that is now southern Turkey; the Taurus Mountains mark the suturing of the colliding plates. Trapped behind the newly risen mountains, the Black Sea is a relic of its former self, surrounded on all sides by continent.

What is going to happen now that the Danube and Volga are bringing much sediment to the Black Sea?

Sooner or later, the Black Sea is going to be silted up, changed into a large desert basin inside a continent. This fate is inevitable.

Are there any silted-up back-arc basins today?

I did not think of asking such a question until I was giving a talk on the origin of sedimentary basins to a group at a provincial geological survey in Qinghai. Qaidam Basin in Qinghai is a very large desert basin that now stands 2,800 m above sea level. Oil has been found in this basin, which is surrounded on all sides by high mountains. Wells drilling to 7,000 m depth have penetrated only late Cenozoic continental sediments. How could such a desert come into existence? Did God punch a hole into the Tibetan Plateau to make Qaidam?

Going down the list, I was forced to the conclusion that Qaidam must owe its origin to seafloor-spreading behind an island arc; it was a "Black Sea," but it is now silted up. In fact, the other two great desert basins of northwestern China, the Junggar and Tarim, must have had a similar origin (Figure 16.5). I then reviewed the geology and found that the back-arc-basin hypothesis could explain many of the facts that had been puzzling before. The final test was an aeromagnetic survey: if those basins were formed by seafloor-spreading, there should be the "magnetic signature." When I visited Tarim last summer, my Chinese host told me the verdict. There are well-defined magnetic lineations; the crust under the Taklimakan Desert of Tarim is largely oceanic.

The genesis of large basins in continental interiors has more than academic interest. One example is the West Siberian Basin in the Soviet Union. The oil-bearing formations are continental deposits, and the source of the oil was always a mystery. After my work on the Tarim Basin, I was convinced that the Siberian Basin must have been a relic back-arc basin, a silted-up Black Sea. No other postulate could explain the very thick sedimentary sequence and the extraordinary petroleum occurrences in this basin. I was right. Sergei Aplonov, a Soviet geophysicist, wrote me in 1989 that magnetic lineations are present under West Siberia. The basin was formed by seafloor-spreading behind an island arc; the West Siberian Basin was a "Black Sea" in Triassic times! In fact, Aplonov added, almost all major basins of hydrocarbon occurrences in the Soviet Union are relic back-arc basins; those include, in addition to the West Siberian, the South Caspian in the Caucasus, the Pre-Caspian in Kazakhstan, and the Barents-North Kara and South Kara in Siberia. We have always known about these deep holes in the interior of continents, in which many kilometers of sediments were laid down. Only recently, through our understanding of the genesis of the Black Sea, are we beginning to understand their significance.

My prediction in 1986 that the Tarim Basin should be one of the most petroliferous areas in the world has encouraged Chinese authorities to expand their exploration activities there. I estimated the potential reserves of Tarim at 50 billion tons of

16.5. Relic back-arc basins. Back-arc depressions of the Philippine Sea (upper figure) and the relic back-arc depressions of Tarim (lower figure) are shown together for comparison. The Kyushu-Palau Ridge and the Central Tarim Ridge are relic island arcs. The floor of the Tarim basin north and south of the relic arc is underlain by ocean floor, which is now buried under sediments more than 15 km thick.

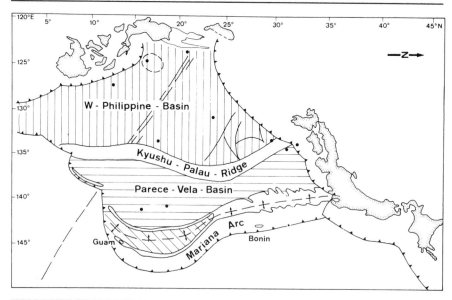

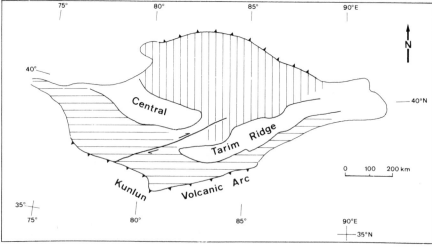

hydrocarbons. I was told, during my last visit in 1990, that some 18 billion tons may have been found; the wildcat success rate there during the last two years was a fantastic 80%. Little did I expect, when I first sailed with *Challenger* through the Bosporus, that the Black Sea drilling was to yield dividends in the exploration of the Taklimakan Desert.

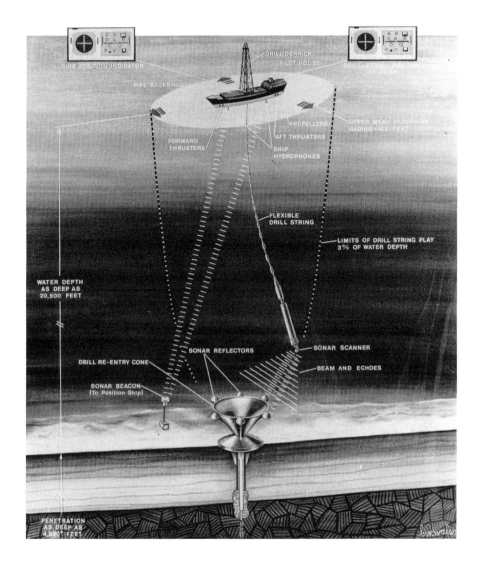

I. Dynamic-positioning and reentry system of *Glomar Challenger*. *Glomar Challenger* on station can stay within a radius equal to 3% of the water depth (for example, 60 m in 2,000 m depth) thanks to the dynamic-positioning system. The thrusters maneuver the vessel back to a position fixed by signals sent from a sonar beacon on the sea bottom. When a drill bit wears out, the flexible drill string is taken out of the hole and the new drill bit is put on the end of the string, which is lowered back into the same hole with the help of the reentry system.

a

b

c

d

II. Deep-sea fossils. The bulk of the deep-sea oozes consists of nannofossils, which are one-celled plants that floated near the surface of the oceans. Two common types are shown (a, b); they have a diameter of about 10 microns. Also present in the oozes are skeletons of foraminifera, one-celled animals that either swam in near-surface waters or were bottom-dwellers on the seafloor. Foraminifera are larger; the two shown (c, d) are about 400 microns in diameter.

III. *Glomar Challenger* **at sea.** The derrick is 45 meters tall and a temporary landmark in every harbor.

IV. Pipe rack on *Glomar Challenger*. Seven thousand meters of drill pipes are stored on Glomar Challenger.

V. Bowen unit on *Glomar Challenger*. The Bowen unit is a weight-lifting device, capable of lifting a string of drill pipes seven thousand meters long up from the abyss. Here the roughnecks are connecting the Bowen unit to the derrick tower.

VI. Taking a wire-line core on _Glomar Challenger._ The roughnecks have already un-screwed the drill string near its upper end, and secured the core barrel to a clamp on the rig floor. Then they had to unscrew the "overshot" (a hook-like device) at the end of the sandline, which had been sent down to fish out the core barrel from the bottom of the hole. After this was done, the core barrel could be taken out of the drill string, as shown here.

VII. Radiolarian ooze. Radiolarians are one-celled swimming animals with skeletons of silica (SiO$_2$) and are commonly found in tropical waters. The individuals in this radiolarian ooze are a fraction of a millimeter in size. When the ooze is lithified, it becomes a hard rock called radiolarite, which has been found in the Alps and other mountain ranges.

VIII. Nannofossil ooze, the most common deep-sea sediment. As shown by this photograph, the sediment consists almost exclusively of nannofossils, several hundredths or thousandths of a millimeter in size.

a

b

IX. Two examples of mélange: (a) near Piedras Blancas Lighthouse, San Simeon, California; (b) on San Simeon Coast, California. The rocks called the Franciscan are not stratified like normal rocks. The originally horizontal sedimentary layers have been crumpled, broken, sheared, and mixed up like the debris under a bulldozer. We call such rocks *mélanges*, and they are typically found in the Benioff zone where an ocean plate plunges under a continent.

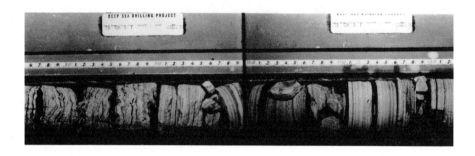

X. "Pillar of Atlantis." This sediment core containing evaporite residues of seawater gave us the first indication that the Mediterranean Sea was a dry desert five million years ago. The core was taken from Site 124, south of the Balearic Islands, and was playfully referred to as the "Pillar of Atlantis" by technicians on Leg 13.

XI. Stromatolite and anhydrite. Dark and light laminations are stromatolites. White irregular layers consist of anhydrite precipitated by groundwater. This picture from the Recent coastal mud flat of Abu Dhabi illustrates the depositional environment of the Mediterranean evaporites.

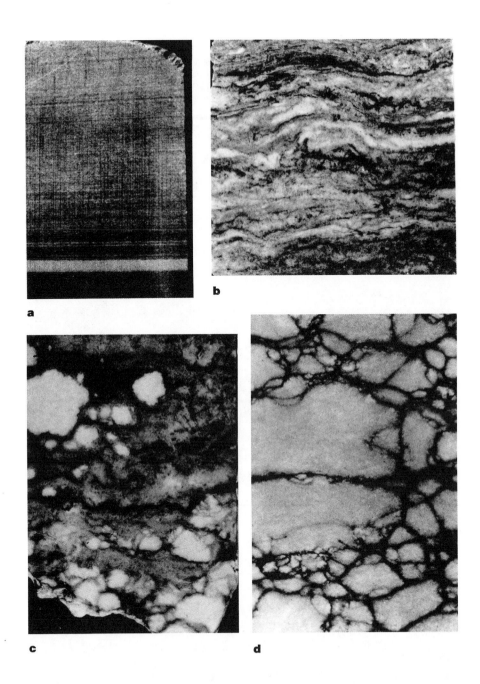

XII. Evaporites in the Mediterranean: (a) laminated sediment deposited in a brackish-water lake after the Mediterranean Sea was desiccated; (b) stromatolite, alternating layers of blue-green algae (dark) and carbonate sediments (light) deposited on the shore of a saline lake; (c) nodular anhydrite, precipitated by groundwater flowing through shores around a saline lake; and (d) "chicken-wire" anhydrite, formed during a more advanced stage of anhydrite replacement.

XIII. Algal mat. On coastal mud flats inundated daily by tides, one finds algal mat covering up mud-cracked ground like this. Digging a trench into the mat, one can see alternating layers of dark (algae) and light (carbonate) sediments like the stromatolites shown in Plate XII(b). This picture was taken in the Bahamas.

On following page:

XIV. Salt under the abyss. This core from Hole 132 west of Sardinia was taken from 400 m below 3000 m seafloor. The salt was deposited in a salt pond of the desiccated Mediterranean. The dark vertical line cutting through the lowest sediment is a mud crack, formed when the pond dried up.

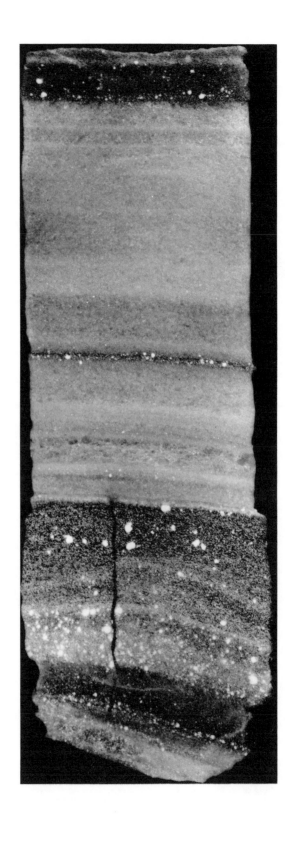

17 Getting Stuck in Ocean Crust

Thanks to the persistence of the planners and the ingenuity of the engineers, a hole was deepened to more than 1,000 m into ocean basement during four legs of IPOD drilling. The purpose was not so much to verify a model of ocean crust proposed by petrologists studying ophiolites, but to find differences in the chemistry of ocean crust, spatially and temporally. The significance of the results has eluded all but the specialists; perhaps the ocean crust was drilled for the same reason Everest was climbed, "because it is there!" Investigations of oceanic hot springs, however, took us to new territory.

JOIDES Goes International

My friends on the *Challenger* told me not to work on this manuscript today, because I was in a bad mood. I was still suffering emotional anguish brought on by lost shiptime caused by human errors. The word anguish is perhaps a most appropriate expression. When the weather is bad, operations have to stop; nobody rages against the weather god, and one can accept the fate. The anguish is, however, unavoidable when the operations are stopped by human errors.

Shiptime, measured in terms of the rental cost of *Glomar Challenger*, was worth 30,000 dollars a day. However, time had to be allotted for port calls, for getting from one place to another, for putting the drill string together; such allocations are indispensable, like the interest payments in a federal budget. Only the time allotted for science was expendable. Viewed from such a perspective, the cost of lost drilling time was much more expensive.

Time lost because of weather is simply part of the price one has to pay. Time lost because of engineering tests is annoying, but trial and error in experiments is to be expected, and is not all that costly. Time lost because of wasteful scientific planning, however, is inexcusable. Sometimes I had the uneasy feeling that we scientists were the greatest offenders in causing "loss" of shiptime.

Perhaps I was too closely involved with the project and did not have an objective view. However, the records were there. Successful legs not only produced headlines in the news media, they also acquired samples that were in great demand, and they led to many scientific publications that were to be frequently cited. Unsuccessful legs were soon forgotten. Some legs were very well planned scientifically, but did not achieve their objectives because of scheduling constraints, such as Legs 35 and 36 during the second Antarctic drilling season (see chapter 13). Others were poorly planned or simply poor science.

JOIDES started out as an undertaking by a few idealists to experiment with great scientific ideas. With the early successes, and the increasing cost of financing deep-sea drilling, a bureaucratic expansion became unavoidable. Politics crept visibly into science when IPOD, the International Program of Ocean Drilling, started in 1975.

Sometimes I wonder if my naiveté may have contributed to, or even accelerated, the internationalization of the Deep Sea Drilling Project. Back in the late 1960s, the Deep Sea Drilling was an American project. When I came to Zürich I was appalled at the lack of communication between earth scientists in Europe. Cooperation among international institutions was at a minimum, and contacts were rare. I remember to this day my embarrassment when I made a person-to-person telephone call to the director of an oceanographic institute in Europe and was told that he could not speak to me because he had been dead for four years. After I was appointed a member of the JOIDES Mediterranean Advisory Panel, I organized an informal group called "European Friends of JOIDES." We met regularly to plan for Leg 13, and the first Mediterranean cruise was the first dominantly European expedition of the Deep Sea Drilling Project.

After Leg 13 ended at Lisbon in October 1970, I was asked to give a press conference in Paris, arranged by CNEXO, the French National Center for the Exploitation of Oceans. On the airplane to Paris, I met Dan Hunt, an administrator of the U.S. National Science Foundation overseeing the Deep Sea Drilling Project. Hunt was bemused that Bill Ryan alone among the 10 scientists on board the ship was an American affiliated with an American institution; all the rest of us were residents of Europe. We all appreciated the generosity of the United States, of course. However, the time seemed to be ripe that the European institutions also carry some of the financial burden of this worthy cause. Hunt told me then that discussions were going on about possible internationalization of JOIDES.

Bill Nierenberg, then director of Scripps, headed the DSDP. He also came to Paris for the press conference. After I gave my report on the findings of Leg 13, Nierenberg made an official invitation for foreign participation in the Deep Sea Drilling Project. In fact, he started negotiations with the French the very next day, and went from there to Moscow for more exploratory talks. A year later, when President Nixon visited Moscow for a summit meeting with Brezhnev on détente, Soviet participation in ocean drilling was one item of scientific cooperation that could be agreed upon. It was a bargain for the Soviet Union. For a membership fee of a million dollars a year, they had an opportunity not only to learn of the latest advances in marine geology, but also to acquire knowledge of the newest technology in deep-sea drilling. A bilateral agreement between the two nations was signed, and Soviet representatives started to sit in on various JOIDES committees in 1973, even before a Soviet institution formally became a member of JOIDES.

The Bundesanstalt für Geowissenschaften und Rohstoffe of Hannover, representing West Germany, was the second foreigner to become a JOIDES member, in 1974. A year later, as we were waiting in Malaga, Spain, for repairs on *Glomar Challenger* to be completed prior to the start of Leg 42A, Mel Peterson, the project manager, came down from Paris; he told us that they were ironing out the last details of French participation. Just to throw in a "sweetener," a third French scientist had been appointed at the last moment to sail with the cruise. I don't remember now if the French signed first on the dotted line, or if the British or the Japanese did, but they all became officially or unofficially affiliated with JOIDES before the end of 1975.

Funding for the first phase of IPOD, from November 1975 to October 1979, was budgeted at 67.5 million dollars. Since the U.S. government was to contribute the lion's share, they deserved more votes on the JOIDES committee. Four American institutions were invited to join the original 5, so as to give JOIDES a U.S. majority of 9, with a minority of 5 foreigners. The new U.S. JOIDES institutions were Texas A & M, Oregon State, Rhode Island, and Hawaii. Later the University of Texas, Austin, became a tenth U.S. member of JOIDES.

Switzerland is a rich—but small and landlocked—country. With a population of six million and no coastline, ocean research cannot demand the same attention it does in some of the large maritime countries. Many Swiss scientists participated in the first three phases of DSDP, as individuals invited by the

National Science Foundation. As a matter of fact, there were more Swiss on the first 44 legs of DSDP than nationals of any other non-U.S. country. They were able to participate because of the generosity of the Americans, at a time when the selection of members of advisory panels and of shipboard scientists was based almost exclusively upon merit and technical competence. With the internationalization of ocean drilling, membership on the planning panels, and berths on *Glomar Challenger*, were distributed according to contractual agreements. Each country with a JOIDES institution could send one of its residents to sit on each of the panels and each of the committees. Each country with a JOIDES institution could send one of its residents to join each cruise of *Glomar Challenger*. To protect American interests, it was also specified contractually that no less than 50% of the shipboard staff had to be residents of the United States, including foreign nationals affiliated with an American institution. Unfortunately, American expatriates like me, who had to pay U.S. taxes to support the Deep Sea Drilling Project, were considered foreigners by the National Science Foundation. On Leg 73, my assistant, Judy McKenzie, a native-born American from Pittsburgh, and I were considered Swiss, but my Swiss student, Helmut Weissert, who was coming back after doing a post-doc at the University of Southern California, was considered an American.

The introduction of an international program of ocean drilling had many consequences, but the most obvious one was that there was no longer much room for scientists from countries that were financially not able to buy a share of the "stock" in JOIDES. A few were still invited to the IPOD cruises, but only rarely could the chief scientist of the project come up with a combination that permitted us to be invited as American guests.

I did make an attempt to see some kind of official affiliation of Switzerland with the JOIDES organization, perhaps with a reduced membership fee and correspondingly reduced privileges. Peter Fricker, general secretary of our National-fonds, negotiated with the U.S. National Science Foundation and came up with an agreement in 1975. I received a letter from Warren Wooster, representing the JOIDES executive committee, inviting ETH Zürich to become an associated member of JOIDES. Later on, Manik Talwani, the chairman of the committee, had to tell me that the decision had been reversed because of a protest by one or two foreign members; the pie was too small to leave crumbs for the underprivileged. My American friends were, however, generous enough to invite me to Orangeburg, New York, for the first and only joint meeting of the IPOD scientific panels in 1975; I went as a nonvoting observer from Switzerland.

I was not an impartial observer, or even a fair one. In my more emotional moments I had the feeling that bilateral contracts gave each foreign JOIDES member the privilege of wasting a million dollars' worth of shiptime per year, and that the representatives of more than one country seemed determined to carry out their duties in defense of this privilege. Those angry words are, of course, not true.

The panels had been reorganized. Instead of geographical panels (Pacific, Atlantic, Indian Ocean, Antarctic, Mediterranean), four site-selecting panels were constituted according to themes or specialization—Ocean Crust Panel (OCP), Active Ocean Margin Panel (AMP), Passive Ocean Margin Panel (PMP), and Ocean Paleoenvironments Panel (OPP). The rationale behind the reorganization was sound enough: there should be new faces and new ideas. However, it did not always work out that way. Instead of cooperation, there was to be fierce competition for shiptime between the different groups.

Ideally, the IPOD cruises were not to be assigned to planning panels, but would be scheduled for specific scientific objectives that cut across the interests of several panels. As it turned out, the cruises had to be "labeled" before much planning could be done. When shiptime was thus allotted to "thematic" panels, a consequence was that each panel tended to concentrate on its own field of specialization. Naturally, a person interested in the chemical composition of the ocean crust might want to drill directly into the basement while taking a minimum of sediment cores. Conversely, a project designed to study ancient ocean-currents of the South Atlantic would not wish to invest much time in drilling into the basalt.

To minimize "tunnel vision," the JOIDES planning committee had to make up a set of rules. All holes must be drilled at least 100 m into the basalt (or at least to the depth at which the drill bit wears out). All sedimentary sequences must be continuously cored. Measurements of the physical properties of the sediment in a borehole (by the well-logging method) must be made at every site. Only rarely were exemptions from those regulations approved by the planning committee. However, exemptions may have been few, but circumventions were many. During Leg 73, for example, people noted that the drill bit "wore out" very fast when we were drilling into basalt.

With all the political constraints, all the petty jealousies, it is remarkable that IPOD was a success. It was definitely the most outstanding example of international cooperation in the earth sciences. There was a great deal of exchange of ideas in the community of marine geology. Now we all knew each other, and I would never make another telephone call to a European colleague who had been dead for four years. There were a lot of new drill cores, and some sensational scientific achievements.

Handwriting on the Wall

The initial assignment of cruises to different panels for planning was based on a broadly outlined proposal to the U.S. National Science Foundation, which made possible the extension of the various phases of the deep-sea drilling. Drilling ocean crust was to be an emphasis. The success of *Glomar Challenger* had revived the hopes of earlier proponents of the Mohole, who wanted to drill a hole as deep as possible into the ocean crust. The sentiment was best expressed by Frank Press, who was to be the Science Advisor to President Carter and later president of the U.S. National Academy of Sciences. Press was a member of an Academy advisory committee that helped launch IPOD. When I met him at the Massachusetts Institute of Technology in the spring of 1974, he told me excitedly, "They finally are moving away from the 'soft stuff,' and are going to get some hard rocks."

He emphasized the word hard, as if soft oozes, like soft currencies, were less valuable. I was too polite to say that most of the successes of the *Challenger* had been made possible by the "soft stuff."

Press was a geophysicist and one of the foremost experts on determining the structure of the earth's crust through investigations of the transmission of earthquake waves. The crust of the ocean floor has a layered structure. Layer 1, commonly less than one kilometer thick, is the "soft stuff." Layer 2 is the ocean-floor basalt. Below that and above the Moho is Layer 3, consisting of a rock that must be denser than basalt.

From studying ophiolites on land, geologists believe that Layer 3 consists mainly of gabbros, like the gabbro of the Allalinhorn, or the gabbro we drilled into during Leg 13 (see chapter 6). We found the gabbro under a thin sedimentary cover on the Gorringe Bank, because a fracture zone is present there. Under ordinary circumstances, Layer 3 is covered under two or three kilometers of the basalts of Layer 2 (Figure 17.1). Frank Press's hopes were high, as they were later at Orangeburg, that the drill string of *Glomar Challenger* would drill through two or three kilometers, into Layer 3 at least, even if the Moho is too far down to be reached.

Two previous drilling legs had attempted to effect deep penetration into the ocean crust. The pilot studies for IPOD were to gather experience for the IPOD planning. Leg 34 to the Pacific was led by co–chief scientists Bob Yeats of Ohio University and Stan Hart of the Carnegie Institution of Washington, D.C. *Glomar Challenger* left Papeete, Tahiti, on 20 December 1973. She had to steam for 16 days eastward before reaching the first site, only to find that the drilling was slow and difficult. The total length drilled in three holes was less than 100 m, and the maximum penetration at Site 319 was only 59 m, about one percent of the way down to the Moho. The sampling was equally disappointing; a total of 28.75 m of basalt core was the "harvest" of the seven-week cruise. Drillers told us that the Pacific basalts were particularly hard to drill, as many IPOD scientists were to learn later. Compared with the Pacific, we had an easy time in the Mediterranean; Leg 42A penetrated almost 200 m of basalt in one and a half days at our Tyrrhenian Basin Site 393!

The second DSDP Phase III effort on drilling basalt in the Atlantic was a spectacular success, especially compared with the fiasco of the Pacific. Leg 37 was led by Fabrizio Aumento of Dalhousie University, Canada, and Bill Melson of the Smithsonian Institution, Washington, D.C. The target was the so-called FAMOUS area on the Mid-Atlantic Ridge at 36° N, FAMOUS being an acronym for Franco-American Mid-Ocean Undersea Study. Four sites were drilled on this transit cruise between the Antarctic and the Arctic drilling of 1974. The core recovery was almost 10 times that obtained by the Pacific leg,

17.1. Ocean crust. The layering of ocean crust, which has been detected by studies of ophiolite sequences on land. Deep-sea drilling has not yet penetrated from the top to the bottom of Layer 2.

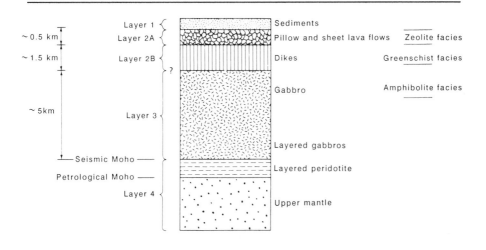

and the maximum penetration reached 582.5 m into basement in a hole at Site 332, which was terminated at 721.5 m depth below the seafloor.

The FAMOUS area was studied by scientific parties on surface ships and by three deep-diving research submarines, *Alvin, Archimede,* and *Cyana.* The area had been thoroughly surveyed before the cruise, and suitable drilling locations were found in the median valley of the Mid-Atlantic Ridge.

The primary goals were to investigate the mineralogy and chemistry of the basalts from Layer 2, and to determine if it is the magnetization of those basalts that has given rise to the lineated magnetic anomalies on the seafloor.

To a layperson, the results seemed simple and straightforward: Layer 2 consists of basalt, as everyone had predicted. Specialists, however, could find subtle differences and see complexities. At Site 332, the scientists could find little in common between the basalt flows encountered in one hole and those in the other, one hundred meters away. Aumento and Melson had the impression that each basalt flow came from a small eruption of a local volcano. Also, there seemed to have been much fracturing and faulting of the seafloor. At Site 335, coarse-grained gabbros and serpentines, like those that constitute the ophiolites of the Alps, were found at the shallow depth of 50 meters below the seafloor. Those supposedly Layer 3 rocks may have been brought up by faulting, which moved the sides of the median valley up while the valley floor sank. Broken fragments from rock falls coming down steep valley walls were mixed with ocean oozes to make up angular breccias, which are sandwiched between successive basalt flows.

The Leg 37 scientists were disappointed that they were not able to solve the problem of magnetic lineations of the seafloor. The upper part of Layer 2 had been supposed to be the source of the magnetic anomalies, but the basalts sampled were not sufficiently magnetized to account for the observed intensities of the linear anomalies on the seafloor. The scientists could only suggest that the sources must lie still deeper in the lower part of Layer 2 or in layer 3.

The technical successes of Leg 37 were sufficient to give the necessary encouragement for an extensive IPOD program to drill into the ocean crust. The drilling problems encountered during Leg 37 were relatively minor. The difficulties of changing drill bits were solved by the new reentry technique. While drilling Hole 332B, the worn-out drill bit was taken out nine times; each time the drill string was able to find the hole again to continue drilling. The "search and reentry" into the hole usually took only a couple of hours; the site was abandoned only on the last trial, when the drill string could not find the hole again.

The operations manager was overly optimistic when he reported that the piling up of drill cuttings around the hole "did not seem to be a problem on this leg." With this assurance, it seemed that shiptime was the only remaining critical factor. Given enough time, and with a little luck, there should be no limit to how deep the drill string would be able to go. The happy planners at Orangeburg were talking about two- or three-kilometer-deep holes, while secretly hoping for a long-shot chance to get a crack at the Moho. To put "all the eggs into one basket," it was decided that one deep hole in the western Atlantic would be given the shiptime of three drilling legs, 51, 52, and 53.

The drill-cuttings problem had not, in fact, been solved. Normally, deep-sea drilling holes are bored into muds and oozes of the deep sea, and the fine sediments would remain in suspension for some time and might even be dispersed by weak bottom-currents. However, when a hole was drilled into a hard rock such as an evaporite or a basalt, drill cuttings brought up by the circulation fluid were dumped on the seafloor, and piled up like an anthill (see chapter 1). During each coring operation, the circulation through the hole had to be stopped for 10 or 15 minutes. Then the chips and cuttings on the seafloor around the hole tended to slump back down, and the friction of such loose debris in the hole would prevent the rotation of the drill pipe. Bill Ryan and I had much experience with this during Leg 13, because we were often drilling into the hard formation of evaporite residues left behind by the desiccation of the Mediterranean. Again and again we were told: "The pipe is stuck."

The pipe was stuck in the sense that it could no longer rotate, and one cannot drill deeper without the rotation of the drill pipe. The usual remedy was to dump some mud down the hole, hoping the circulation of mud would clean up the hole and bring up the loose cuttings again. Sometimes the maneuver was successful. Occasionally nothing helped, and one had to send dynamite down the hole to break the drill string near the bottom of the hole where the pipe was stuck, to be able to take back the long string of pipe segments that were not stuck. Contrary to the impressions conveyed by the optimistic report, Leg 37 operations did encounter such problems, and part of a bottom-hole assembly was lost at Site 332 when the drill string [was] stuck permanently while pulling a core. The drill string was severed . . . leaving the bit, a core barrel, three 8 ¼″ drill collars and a piece of bumper-sub in the hole.

"Stuck pipe," caused by the down-hole slumping of cuttings, was the main culprit that hindered the deep-penetration into the crust of the Atlantic during the IPOD drilling. Several deep holes had to be abandoned because of stuck or broken drill pipes. The unsuccessful experience of trying to drill one hole in three drilling legs proved that it was useless to allot more time for deep penetration before better technology became available.

"The Pipe

Is Stuck"

The drilling of the seafloor in the Atlantic during IPOD came close to what had been achieved by Leg 37, but not once did the drill string penetrate deeper into the basement crust than the critical barrier of 600 meters! The depth of the basement penetrated and cored was 554 m in Hole 395A, drilled during Leg 45, and 542 m in Hole 418B, which required the cooperation of three drilling legs (51–53). It was still a long way to the Moho.

The plan to penetrate deeply into the ocean crust of the Pacific Ocean ended in even more abysmal failure. The performance of Leg 34 was already a bad omen. Soon after Leg 54 began on the East Pacific Rise the same problems had to be faced: the young ocean crust there has a very thin sedimentary cover, and the Pacific basalts are hard to drill. Site surveys had searched for areas where the sediment should be thicker than 100 m. The first site was spudded on 8 May 1977, but the scientists encountered their first surprise when they found that the sedimentary cover was much thinner than had been predicted by geophysicists; the basalt crust was already encountered at 35 m below the seafloor. Under such circumstances, no weight could be put onto the drill bit, and deep penetration was impossible. After a second hole in the general vicinity was drilled into basement at 46 m depth, it was concluded that the geophysicists had somehow misinterpreted their survey data. The co–chief scientists, Bruce Rosendahl of Duke University and Roger Hekinian of CNEXO, Brest, decided to abandon this first target and try their luck elsewhere.

The sedimentary sequence was indeed thicker at Site 420, and the basalt basement was encountered at 118.5 m below the seafloor, a depth at which the bottom-hole assembly could be buried in soft sediments. However, the drilling rate was very slow. The drill pipe, after rotating for nine hours, had penetrated only into 32 meters of basalt crust. In view of the later experiences, perhaps the co–chief scientists should have been a little more patient and kept on trying. However, they were experiencing many difficulties and decided to abandon the hole and drill elsewhere. Another attempt in the neighborhood discovered a basalt basement at a very shallow depth, unsuitable for drilling. At the next site, the basalt crust was buried at 85.5 meters below the seafloor. As our two co–chiefs reported: "Drilling conditions were even worse here than at Site 420, and the drill string began to stick with less than 5 meters of basalt penetrated. The hole was abandoned after only 4 hours and 14 minutes of actual rotation."

After several of the preselected sites had been drilled, *Glomar Challenger* now had to try in areas where the sedimentary cover was even thinner. At the next site, the basement was encountered at 46 m below the bottom. It was decided nevertheless to spud in. For a while, the drilling went smoothly. However, the same difficult conditions as at previous sites were encountered at about 60 m depth. Another hole had to be abandoned, and *Glomar Challenger* had to leave for the next part of her odyssey. The next site (423) did not bring much luck either: the basement was encountered at 38 m beneath the seafloor, and the hole had to be abandoned at 54 m.

At this stage, the members of the JOIDES planning committee began to get worried. Obviously the plans to penetrate deeply into the crust of the Pacific Ocean had to be canceled. Orders were sent to the co–chief scientists to direct *Glomar Challenger* to an area near the Galapagos Islands, where hot solutions coming out of the earth's interior are depositing iron and copper minerals. Leg 54 now had to carry out a reconnaissance to evaluate the possibility of drilling in that area. Reluctantly, the co–chief scientists gave in; they did receive a five-day extension of the cruise to come back to the ill-fated target area to drill a few more holes, but did not have any remarkable successes. The deepest penetration they managed was 52.5 m into the basalt crust at Site 428 before they abandoned the hole when the drill bit wore out. A last 50-meter hole was drilled before the agony of the co–chief scientists could be ended. The experiences of Leg 54 necessitated a reorganization of the schedule of *Glomar Challenger*; further attempts to drill deeply into the Pacific ocean crust were now suspended.

Criss-

crossing

the Oceans

One goal of deep penetration was to study the variation of the ocean crust with depth, and this goal could not be satisfactorily achieved. The other goal was to compare the spatial variation of the ocean crust. For this purpose, the chairman of the panel wrote, "The IPOD Atlantic crustal drilling program involves two transects. The first is an east-west transect . . . The second is a series of [north-south] holes between Iceland and 23° N."

The expectation was that such a grid-like sampling would provide information on the variation in the primary geochemistry of ocean-floor lavas in space and time. Leg 49, led by co–chief scientists Bruce Luyendyk and Joe Cann, left Aberdeen, Scotland, in July 1976 and returned to Madeira in September.

The cruise report was dedicated to the memory of Richard Meadows, who was a rotary helper on the *Challenger*. He was remembered by Luyendyk as "a very special kind of man—as are the others who do this type of job. The labor is exhausting, lasting 12 hours per shift in all varieties of weather. There is also the loneliness of separation from family and friends for 6 months of the year and the boredom that can only be experienced at sea. The dangers on the rigs are present and unpredictable."

"The night of 4 August was cold and windy," Luyendyk told us. "The *Challenger* was drilling on Site 409 just south of Iceland . . . At about 4:30 A.M. rotary helper Richard Meadows was struck suddenly by a falling piece of rig machinery and killed."

Meadows's accident was, as far as I know, the only casualty of operations during the fifteen years of JOIDES/DSDP drilling. Another report was dedicated to John Hinds, the second mate of the *Challenger* during Leg 45 drilling, who was killed in a 1978 helicopter accident on another vessel of the Global Marine company.

Three sites were drilled during Leg 49 in the Reykjanes Ridge area southwest of Iceland, and then four sites in young crust of the Mid-Atlantic Ridge farther south. The grid sampling revealed that the earth's mantle is chemically heterogeneous. Heterogeneities are probably generated, in the view of Joe Cann and his colleagues, by the migration within the mantle of small amounts of interstitial fluid, probably rich in water and carbon dioxide.

A wealth of data was published in several volumes of *Initial Reports of the Deep Sea Drilling Project*, and numerous articles were written on the spatial and temporal variation of basalt chemistry in the Atlantic. There were critics of this grid-sampling approach to using a drill vessel. Jerry van Andel of Stanford University, for example, commented in an article published in *Nature* that the variations found by the drilling in widely scattered parts of the different oceans are about the same order of magnitude as those found by the deep-diving submarines in one small FAMOUS area less than 100 km across. It was indeed difficult for a layman like me to see the significance of such "minute" differences. On the other hand, I appreciated that fact that our knowledge of the processes influencing the geochemistry of the crust and mantle was still rudimentary; random observations in new territory are always necessary before a new set of paradigms can be formulated.

Break-throughs in the Pacific

The first phase of the international phase of deep-sea drilling from November 1975 to October 1979, was to end ocean crust studies by drilling underwater hot springs near the Galapagos region of the East Pacific, on a segment of the East Pacific Rise called the Costa Rica Ridge. Later, an extension was granted, so that the *Challenger* could make one final attempt at deep penetration into the Pacific crust.

The extension was justified in part by the fact that a newly developed hydraulic piston-corer (HPC) would permit in-depth studies of ancient ocean environments, and some shiptime had to be expended to try out the potential of the new equipment (see chapter 20). With the extension, the ridge cruise, the original Leg 68, was rescheduled, and the cruise objectives, Sites 501 and 504, were drilled during Leg 69 and parts of Legs 68 and 70. Joe Cann and S. M. White were co–chief scientists of Leg 68, Cann and Mark Langseth led Leg 69, and Jose Honnorez and Dick von Herzen led Leg 70.

Cann was for many years the British representative on the JOIDES/IPOD planning committee, and he was still an incumbent when I joined the P-Comm in the early 1980s to plan for ODP. Veterans made committee work effective, and I was impressed by Cann's ability to summarize discussions in the P-Comm and in other meetings. He had the talent of breaking through a maze of general discussion and coming up with concrete proposals. A distinguished petrologist from the University of Newcastle-upon-Tyne, he had already had the experience of leading the successful Leg 49 when *Glomar Challenger* departed Puntarenas, Costa Rica, on 5 July 1979.

The major scientific objective of those Pacific legs was to assess the physical and chemical state of young oceanic crust. Hole 501 was drilled during Leg 68 as the pilot hole for reentry. After two months of dry-docking and experiments with HPC, the Costa Rica Rift drilling was resumed, and three holes were drilled during Leg 69 at Site 504, 1°14′ N, 83°44′ W, where the seabottom depth is 3,460 m.

Excellent site surveys were carried out by R/V *Conrad* and

by RRS *Discovery* prior to the drilling. There were no difficulties spudding Hole 504B, even though the crust there is only 5.9 million years of age; the drill string was buried in 275 m sediment when the basement was encountered. Numerous reentries were made to change drill bits, and the hole was drilled to 489 m subbottom depth, or into more than 100 m basement, after some 10 days of operations.

The drilling of Hole 504B was resumed during Leg 70. "Drilling conditions remained excellent throughout," the co–chief scientists reported. Time constraints were such that the hole had to be terminated at 836 m subbottom depth after another 10 days of operations.

The successes of Legs 68, 69, and 70 demonstrated that the difficulties of deep penetration were not insurmountable; the nightmares of Leg 54 could now be forgotten. All we needed now was shiptime. The second phase of the IPOD drilling was to end after Leg 82 at Norfolk, Virginia, in November 1981. At the International Geological Congress in Paris, 1980, a special session was devoted to the achievements of the Deep Sea Drilling Project. My JOIDES friends told me that another extension was in the works. The second extension, in fact, did come through, and the *Challenger* was rescheduled in early 1982 by the P-Comm. We learned from the October 1982 *JOIDES Journal* that Leg 83 would be targeted at deepening Hole 504B.

The *Challenger* left Balboa, Panama, on 14 November 1981 and headed directly toward Site 504. Hole 504B was reentered in the early morning of the twenty-third.

For years after the Leg 37 drilling in 1974, there seemed to be a 600 m barrier to basement penetration. I was among the pessimists who considered the difficulty insurmountable, because of the down-hole slumping of drill debris. The barrier turned out to be merely psychological, and the previous record for penetration was broken by Leg 83 in a few days. Drilling, coring, and reentry operations proceeded smoothly until after the eighth reentry of the hole in a record time of 3 minutes. On the morning of 4 December a drop in the total weight of the drill string "signified that the drill pipe had failed just above the bottom-hole assembly," the co–chief scientists noted. The drill pipe was pulled out of the hole and was found to have been severed 0.5 m below a pipe-thread connection, and the "fish," or the drill-pipe segment to be pulled out, was more than 100 m long.

After two days were lost to "fishing" out the lost pipes, drilling could resume. Drill bit number 8 was lowered down during the sixteenth reentry, and drilling finally penetrated to 1,350 m subbottom on New Year's Day, 1982. One hundred and forty-one cores were obtained during the four drilling legs, while 1,075.5 m of basalt basement was penetrated.

Toward the end of the Deep Sea Drilling Project, relocation of sites and reentry of holes became routine operations. Holes with sonars guiding reentry became deep-sea laboratories. Leg 92, led by Margaret Leinen of Rhode Island and David Rea of Michigan, returned to Site 504 for nine days. They did not have time to deepen the hole, but they sampled interstitial water in basement rocks and made down-hole temperature measurements.

That was not all! Leg 111 of the Ocean Drilling Project was scheduled with the primary purpose of deepening Hole 504. When *JOIDES Resolution* occupied the site from 29 August to 5 October 1986, the drill string made 21 reentries to reach 1,562 m subbottom, 1,288 m into the basement, or about one fourth of the way down to the Moho. In addition to coring, numerous geophysical measurements were carried out.

What did the scientists learn from this tremendous effort of deep penetration? They were able to verify the geologists' model of the ocean crust, which had been established on the basis of study of ophiolite sequences on land (Figure 17.1). Layer 2 at Site 504 consists of two kind of rocks. The upper (2A and 2B) are pillow basalts, some 600 m thick, formed by the extrusion of lavas on the seafloor. The lower (2C) is the so-called sheeted-dike complex: dikes of diabase, a rock of basalt composition but made up of coarser mineral grains, were intruded into previously extruded pillow lavas. There should be Layer 3 gabbro and mantle perioditite farther down the hole at this ocean site, but geologists decided that it would be more economical to drill the ophiolite complex at Cyprus, where they would be able to answer many of the same questions on the origin of ocean crust.

"Smoking Chimneys" and Mineral Treasures in the Deep Sea

While I was describing cores during Leg 3, I noted that the oozes just above the basalt basement are dark maroon in color at several sites; they are red because finely disseminated hematite, a red iron oxide, is present. Checking the literature, I found that in 1966 Kurt Bostrom and Mel Peterson had described similar iron-manganese sediments close to the spreading axis of the East Pacific Rise. Jack Corliss of Oregon State University speculated in 1971 that the iron and manganese could have been derived from hot solutions coming out of newly formed ocean crust.

In 1972 K. Klitgord and John Mudie of Scripps took their

deeply towed geophysical instruments to survey the Galapagos Spreading Center, and they discovered conically shaped features, 5 to 25 meters high and 20 to 50 meters in diameter, which protrude above the seafloor in the area where the ocean crust was formed half a million years ago (Figure 17.2). Reddish-brown iron-rich sediments are found on the sides of the mounds. Scripps scientists also found anomalous seawater temperatures and chemistry in the region, suggesting that hydrothermal solutions were coming out.

About the same time, Jack Corliss and others from Oregon State went down, in the deep-diving submersible *Alvin*, to make observations in the Galapagos Spreading Center, and they found active fields of hot springs in the area. Bob Ballard of Woods Hole then installed giant cameras in an instrument package towed directly beneath the surface ship. With the cameras, the Scripps group was able to find definitive evidence of venting hydrothermal solutions; they even photographed some giant clam shells near the seafloor vents. Deposits of the hot springs are massive sulfides and iron-manganese-rich sediments, forming "chimneys" sticking out of mounds. By 1979, the temperature of the vents was found to be as high as 350 °C.

17.2. Galapagos mounds, conically shaped features, 5 to 25 m high, which protrude above newly formed seafloor in the region near the Galapagos Islands. Hot springs are vented out of chimneys on those mounds, depositing sulfide minerals.

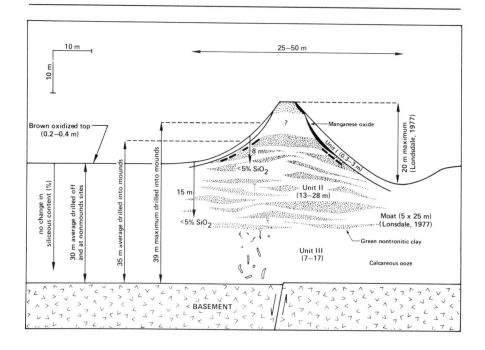

The exiting fluid was clear, but black plumes were induced by entrainment of ambient water. Extremely-fine-grained clouds of sulfide minerals billowed to heights of tens of meters above the vents.

Geologists were excited. Sulfide minerals, especially copper, are common in the sedimentary rocks directly overlying ophiolitic rocks; the copper deposits on Cyprus had been known since the Bronze Age and had given the island its name. Now, for the first time, people were seeing metal sulfides forming directly on the seafloor. The importance of this discovery did not escape the notice of the JOIDES planning committee. The very first plans scheduled a leg to investigate the Galapagos mound area for late 1976.

Mounds were finally drilled during Leg 70; for the first time, the stratigraphy, the physical properties, the chemistry, and the mineralogy of rocks on a mound could be compared with those not on a mound. Hydrothermal sediments accumulated at the mound sites, thinning rapidly away from the summit. Hydrothermal activity was intermittent. Water samples from sediment pore space indicated that hydrothermal solutions are upwelling through the sediments of the mounds' hydrothermal field at Sites 506, 507, and 509, and downwelling in the low-heat flow zone to the south at Site 508.

Site 504 is located within the zone where the basement rocks are altered by hydrothermal solutions, coming up or upwelling from the depth. The normal temperature at the 1,300 m of 504B was 160 °C. The chemical reactions altering the composition of the sheeted-dike complex took place, however, with hot brines at 250 to 350 °C. Higher up in the zone of pillow lavas, the chemical reaction with the hydrothermal solution took place at temperatures of up to 250 °C.

With the down-hole experiments at the Galapagos center, the Deep Sea Drilling Project had gone beyond the verification of revolutionary predictions. The *Challenger* was cruising in new territories, unknown when Hess, Vine, and Wilson started their revolution. Incidentally, the deep penetration of the ocean crust also put one pet idea of Hess's forever in cold storage. The ocean crust is not made of serpentinite, and is not formed by the hydration of the earth's mantle. The ocean crust is manifested by the ophiolites on land, and the magmas are derived from the partial melting of mantle rocks.

Oceanic Hydro-dynamics

Groundwater moves under gravitational potential through rocks in continental crust. It was thought, however, that the water in ocean sediment pore space had to be hydrostatic; there were no pressure differences that would induce the movement of fluids through marine sediments. The presence of sedimentary features indicative of hydrodynamic movement in some rocks of the Alps was thus a great puzzle. One famous sedimentologist went so far as to claim that the Alpine radiolarites were shallow marine sediments, because they all show evidence of rock alterations by moving fluids. Hydrodynamic circulation in shallow marine sediments is possible if they should chance to be exposed shortly after their deposition, but he thought that fluids could never move through deep-sea sediments. Other scientists described neptunic dikes, in which detrital muds were deposited by seawater percolating through fractures.

The discovery of oceanic hot springs and the drilling of the Galapagos mounds have laid to rest another favorite prejudice of geologists on land. We now know that seawater can move through oceanic rocks and oceanic sediments, and the energy of the driving mechanism is thermal. Those hot solutions could leach out iron and copper from ocean crust and precipitate those metals as sulfides on the seafloor. The discovery gave new impetus to geologists searching for metalliferous deposits. It is not random chance that the large iron deposits of Panzhihua in southwestern China and many copper deposits of northern and eastern China have been found in exotic ophiolite blocks of mélanges—in rocks that were first formed as seafloor at the spreading center!

18 Eating Peanuts on Ocean Margins

"Getting shiptime is like eating peanuts. The more you get, the more you want." The planners of the Passive Ocean Margin Panel seemed to be insatiable peanut-eaters, and each JOIDES member nation had its "favorite brand." After a certain area was drilled for the third time, one learns from the report that "the results [of drilling] emphasize the necessity for further drilling and further geophysical studies."

Drilling continental margins seems to have served the purpose of verifying tectonic models that had been developed on the basis of land geology. Nevertheless, we must not forget that Glomar Challenger *has given us a unique opportunity to study processes now in action.*

The Making of a Revolution

A modern approach to the history of science was taken by Thomas Kuhn in 1962, with the publication of his book on the structure of scientific revolutions. Kuhn believed that science does not progress continuously and cumulatively, but by repeated revolutions, in which old ideas, doctrines, and methodologies are replaced by new ones. Of course, in the very beginning, only random observations are made. Gradually, one set of ideas offering an explanation of the randomly collected data would be developed and eventually accepted by the profession as dogma. Kuhn used the word *paradigm* to designate such a constellation of beliefs, values, techniques, and so on which is shared by the overwhelming majority of a scientific community. Paradigm rules science like an absolute monarch rules his realm, but paradigms—like monarchs—are rarely perfect rulers. Eventually, facts or experimental results are obtained that contradict a ruling paradigm. This is the stage of crisis, calling for a revolution. During a scientific revolution an old paradigm, like an old dynasty, is overthrown, and a new one is installed on the throne. After that, the scientific endeavors consist merely of "mopping-up activities," gathering up new trivialities to support the new "king."

The old paradigm of the earth sciences was the historical geology of Roderick Murchison's. Geologists collected observations like philatelists collect stamps, expecting that the "facts" would automatically provide the answers. The process-oriented approaches of Hutton and Lyell were more and more neglected. When I was a graduate student in the late 1940s and early 1950s, one of the biggest projects in North American geology was the publication of a series of correlation charts, with the intention of correlating the hundreds, if not thousands, of formations in the United States. I do not know if the project was ever completed.

Young geologists were discouraged from studying processes. Phenomena are not as solid as rocks. Senior scientists may give a presidential address on the origin of geosynclines or processes of mountain-building. Brave attempts by young upstarts, by Griggs in 1937 or by me in 1955, met with either vocal hostility or dead silence.

The end of World War II was to bring changes, but process-oriented studies in experimental geochemistry were restricted to the Geophysical Lab and a few elite campuses. Many of the best students in the earth sciences became petrologists. We students also observed an unrest among the faculty. The conflicts between diehard Murchisonians and the young Turks were genuine, and not all the races were won by the swift.

Murchison's soldiers went to the colonies to expand the Silurian, which was thought to be the land of gold; they looked for the Carboniferous, where the coal was to keep British gunboats from running out of steam. This philosophy was still prevailing when I worked for Shell. Our methodology was essentially a picture-book approach. Oil is found in structures having a certain three-dimensional geometry. Process-oriented research was tolerated, but seismic definition of traps or "closures" was where the money went. In Murchison's "kingdom of Siluria," discussions on questions such as the permanence of continents and ocean basins were idle speculation. It was thus no coincidence that Heezen was invited to Houston not to talk about his exciting discovery of the mid-ocean rifts, but rather about some trivial features of deep-sea sediments.

The old paradigm was supported by the facts and prejudices of geology on land. With the development of new techniques for studying the ocean floor, new facts kept on popping up, and they rose in opposition to the old paradigm. I have enumerated many of them in this book: the sunken flat-topped mountains of the Pacific, the lineated magnetic anomalies, the youthful geological age of the ocean floor, the seismicity on mid-ocean ridges, the geophysical anomalies in regions of transform faults, the hot-spots, and so on. A geometric interpretation was no longer satisfactory. Yes, the offset of magnetic lineations indicated large faults or fracture zones. These facts can be shown, and were shown, on maps, but mapping is an illustration, not an answer. Questions were asked about the how and the why.

Combined with old enemies of the "regime," the new observations were to start a revolution and force a breakthrough. The old paradigm was utterly defeated and abdicated, and a new paradigm was enthroned. After that, there was much mopping up to be done.

The early deep-sea drilling cruises were exciting because their findings were new and provided the ammunition for the revolution. The earth science community might have had a civil war for years, if the seafloor lineation could not have been dated by geological methods; the first phase of the Deep Sea Drilling Project performed the indispensable experiments. Hot-spots and transform faults might have remained geophysical theories, ignored by land geologists, had they not been investigated "in the field" by drilling.

The second and third phases of the Deep Sea Drilling Project were exciting because the "revolutionary army" entered new territory after the initial breakthrough—for example, there was the discovery of the desiccation of the Mediterranean during Leg 13, the identification of early Cenozoic glaciation in the Antarctic during Leg 28, and the recognition of possible ocean anoxia during Leg 40.

After breathtaking advances by the glamorous "armored divisions," the "infantry" has to move in to exploit the advantages. The International Phase of Ocean Drilling cruises were indispensable follow-ups that were not sensational. But mopping-up was not all—new random observations would be collected for the next revolution.

The drilling into the ocean crust, discussed in chapter 17, betrayed our rudimentary understanding of the genesis of the ocean crust. Lacking a paradigm to guide our planning, we sent *Glomar Challenger* out on a number of hunting expeditions to collect samples at places defined by geographical coordinates. The drilling into the ocean margins, which I shall now discuss, was afflicted with a problem at the other extreme of the spectrum: a new paradigm seemed to have been so well established that only "mopping-up activities" were left for the drilling campaigns.

**Origin
of "Geo-
syncline"**

The term "geosyncline" was an invention of James Hall's in 1840, after he observed the thick Paleozoic sequence in the Appalachian mountains (see Figure 1.2). The older Paleozoic sediments are mainly carbonate deposits, the type found in the Florida-Bahama region of today. Young Paleozoic strata consist, however, mainly of sands, muds, and other erosional debris in the Appalachian mountains. The debris did not come from the interior of North America, as one might have expected. On the contrary, the tremendous amounts of gravels, sands, silts, and clays were brought to the "Appalachian geosyncline" by rivers from an eastern land mass, which seems to have foundered and now lies under the margins of the Atlantic Ocean. During those times the interior of the North American continent was covered by shelf seas and shallow lagoons where carbonate sediments accumulated; the continental interior was not a source of the detritus deposited in the Appalachian region.

When I was an undergraduate in China, I first came across this problem in a monograph by Amadeus Wolfgang Grabau for the Geological Society of China. Grabau had been a well-known professor of geology at Columbia University in New York. Persecuted as an American of German descent by his neighbors during the First World War, Grabau left the United States for China, where he made lasting contributions to the geology of his adopted land before he passed away shortly after the end of the Second World War. In 1924, Grabau made a global review of the geological history of many mountain belts, and he found that the pattern first observed by Hall in the Appalachians is repeated elsewhere: continental margins now submerged under sea were once old land that contributed erosional detritus to sedimentary basins, where the debris was deposited as the sedimentary strata now found in many coastal mountains of today. Apparently continents were being changed into oceans when old land disappeared under the sea to form continental margins.

As I mentioned previously, the continents are different from the oceans not only in their elevation relative to sea level but also in their crustal structure. The continents are underlain by a largely granitic crust some 30 to 50 kilometers thick, whereas the oceans are underlain by a basaltic crust some 5 or 10 kilometers thick. The continental margin is a zone of transition, and is underlain by a granitic crust of intermediate thickness (10–25 km). To change an old land into a submerged continental margin requires a significant change in crustal thickness.

The subsidence is only a surface manifestation of a deeper cause.

Many different hypotheses on the origin of continental margins have been given, and my interest in this matter during the two decades after I first read Grabau prompted me to make a speculative suggestion in the early 1960s. Geophysicists were then finding evidence that the earth's mantle just below the Moho does not have the same density everywhere. In areas where much heat is coming out from the mantle to the surface of the earth, the mantle rocks have been heated up and have expanded, and are therefore less dense than normal. The expansion of the mantle material, I reasoned, should cause an uplift, making mountains and plateaus. Erosion would remove the bulk of materials from the elevated regions, so their underlying crust would become thinner. Eventually, the mantle heat source would be gone, and the mantle materials would cool down again. The former high land, now underlain by a crust thinned by erosion, should then sink below sea level, forming continental margins (see chapter 4).

My idea accounted for the development of margins of continents or of oceans, but not for the origin of the oceans themselves. Vine and Matthews, in their paper published in 1963, the same year mine came out, offered the theory of seafloor-spreading to explain the origin of the ocean basins. Combining the two, one explaining the up and down movements, the other the horizontal displacement, the modern paradigm for the origin of passive ocean margins was established. The word passive is used to distinguish such margins from the active ocean margins, where subduction of oceanic lithosphere along a Benioff zone is taking place.

The word "geosyncline" began to disappear in scientific journals after a "death sentence" was promulgated by the First Penrose Conference on Plate-Tectonics in 1970. Sediments in mio- or lesser-geosynclines are deposits of passive continental margins. We no longer worry about the origin of "geosynclines"; instead, the problem became one of depicting the evolution of passive continental margins. The planning of such endeavors by deep-sea drilling was the mandate of the JOIDES Passive Margin Panel.

Evolution of Passive Continental Margin

The current paradigm depicts several stages in the development of passive continental margins. First, a continent is raised by mantle expansion, and stressed under extension. The continental crust is pulled apart, forming rift valleys. The geothermal gradient in the earth's mantle is steep, causing partial melting of the mantle materials deep down, and the eruption of rock melts as lavas on the surface. The continental crust becomes thinner because of the removal by erosion, and/or because the crust has been stretched by the tension (Figure 18.1). Some regions, such as the Rhein-Graben or the East African rift valleys, arrived at this stage 30 or 20 million years ago, but their further development has been arrested. The Atlantic Ocean underwent the rifting stage some 200 million years ago. The coarse detritus deposited in the rift valleys has been lithified as the New Red Sandstone on both sides of the Atlantic. The lavas pouring out into the rift valleys then can now be seen on cliff exposures, such as those at the Palisades on the Hudson, north of New York City.

In a more advanced stage of the rifting, the continental crust is pulled apart completely. The heat source down in the mantle is still there, and the basalt lavas coming up from the mantle pave the new seafloor and form new ocean crust in the crack

18.1. Evolution of passive continental margins. Passive margins are underlain by continental crust that was thinned during an early rifting phase. This evolution of a passive continental margin had been established by land geology and offshore drilling before the model was verified by the ocean drilling of Atlantic margins.

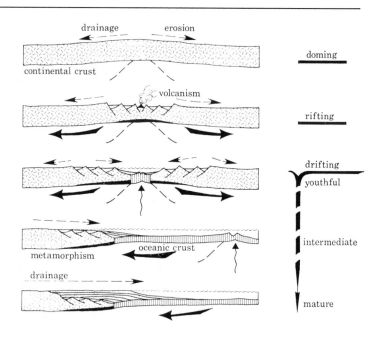

Opposite page:

18.2. West African margin, whose geological history was clarified by ocean drilling.

NOVA SCOTIA SCOTIAN SHELF SW MOROCCO MOROCCAN MESETA ATLAS SEAWAYS (Tethys)

I. LATE TRIASSIC–EARLIEST JURASSIC

CENTRAL ATLANTIC RIFT

NOVA SCOTIA SCOTIAN SHELF SW MOROCCO MOROCCAN MESETA ATLAS SEAWAYS (Tethys)

II. EARLY LIASSIC

SW MOROCCO (ESSAOUIRA W. HIGH ATLAS) MOROCCAN MESETA ATLAS SEAWAYS (Tethys)

Mean Sea Level

III. EARLY DOGGER (AALENIAN)

ESSAOUIRA BASIN MOROCCAN MESETA

IV. LATE DOGGER (BATHONIAN)

ESSAOUIRA BASIN MOROCCAN MESETA

Carbonate platform deposits

Evaporites and associated deposits (salt, anhydrite, dolomite)

Turbidites

Pelagic deepwater deposits

Continental red beds

Basaltic extrusives and intrusives

Coarse Terrigenous clastics (fluvio-marine)

Continental basement (folded Paleozoic)

V. EARLY MALM (CALLOVIAN–OXFORDIAN)

Approximate setting of SITE 416 ESSAOUIRA BASIN HIGH ATLAS

UPLIFT

VI. LATE MALM–EARLY CRETACEOUS

between the "drifting" continents. The Red Sea and the Gulf of California have progressed that far. An axial valley, underlain by ocean basalts, is present in both of these narrow gulfs, which started to spread apart only four or five million years ago. The Atlantic went beyond that stage some 150 million years ago, when the first ophiolites were extruded to form the oldest ocean crust of the Atlantic (Figure 18.2). With continued spreading of the seafloor, the continental margins are moved away from the heat source in the mantle. The ocean widens, with an elevated mid-ocean ridge where the submarine volcanism is active. Meanwhile, continental margins and older ocean crust away from the axis of seafloor-spreading have been conveyed to cooler regions, increasing the density of the mantle materials and causing the subsidence to begin. The subsequent history of the continental margins is then one of sedimentation and of subsidence.

This scenario is a corollary of the seafloor-spreading theory and had been formulated on the basis of geological investigations on land and geophysical investigations of the ocean margins. The paradigm predicts certain geological structures, or certain configurations of geological formations, that could be best investigated by continuous seismic profiling or other geophysical methods. The purpose of the deep-sea drilling was to furnish information on the timing of the various stages of development.

Considerable effort was expended during the first three phases of the Deep Sea Drilling Project to investigate the passive ocean margins, including Legs 1, 10, 11, 12, 14, 15, 36, 38, 39, 40, 41, 43, and 44 in the Atlantic, and Legs 23, 24, 25, and 27 in the Indian Ocean. Later, the International Phase of Ocean Drilling assigned Legs 47, 48, 50, 76, 77, 79, 80, 81, 82, 93, and 95 for further drilling of the Atlantic margin. John Ewing, Maurice's brother and a former chairman of the JOIDES planning committee, once characterized his colleagues' addiction for more shiptime with a metaphor appreciated all too well by oceanographers: "Getting shiptime is like eating peanuts. The more you get, the more you want."

The planners of the Passive Ocean Margin Panel seemed to be good peanut-eaters. Each JOIDES member nation had its "favorite brand." The Americans liked ENA (eastern North America); the Germans, West Africa; the French, the Bay of Biscay; and the British, the Rockall Plateau, a submarine bank in the North Atlantic not far from the British Isles. In 1980, the same areas had been drilled two or three times, and were to be drilled once or twice again during future cruises.

Although I have characterized their efforts as mopping-up activities, our colleagues deeply involved in their projects obviously did not share this view. I was amused to read the following conclusion in one of the 1,000-page official reports after an area was drilled for the third time: "The results [of our drilling] emphasize the necessity for further drilling and further geophysical studies." That was obviously true, because the report did not contain any other conclusions that were exciting enough for a lay reader; a gambler could always hope that the next round of drilling might turn up something! (P.S. In 1990, ten years after the previous paragraph was written, I noted that the *Glomar Challenger* did return and drill one more hole in the area. By then, all excuses of passive-margin drilling had run out; the hole was drilled to investigate ocean history.)

West African Margin

I do not want to give the impression that the geology of passive ocean margins is not important. It is important, because many of our major oil fields are located on present or past passive margins. In the early 1980s, eight U.S. petroleum companies joined the U.S. National Science Foundation to plan an ambitious program to drill the continental margins off North America; it was estimated that the cost would exceed a billion dollars. As with the super-collider in physics, earth scientists had divided opinions about the ocean-margin drilling. I personally was very much against the idea, but my opposition counted little. My friend Jerry Winterer was more effective; he told me that his greatest contribution to scientific ocean drilling was to help kill the proposal for ocean-margin drilling.

Winterer was, of course, being modest. He was not only a science statesman but also a brilliant scientist. He is my candidate for "Mr. JOIDES." Having been involved in the planning even before *Glomar Challenger* was built, Winterer was a member, alternate member, or chairman of the Scripps steering committee for JOIDES/DSDP, 1968–69, JOIDES Pacific Advisory Panel, 1969–74, JOIDES Advisory Panel for Pollution Prevention and Safety, 1975–76, and JOIDES planning committee, 1976–1990. He was chief scientist on DSDP Legs 7, 17, 50, and 79, and served as sedimentologist on Leg 33, after he graciously resigned his chief scientist appointment in favor of Dale Jackson.

There has been some confusion as to the meaning of the acronym JOIDES. S originally stood for sampling, but some persons or institutions thought that S stood for studies. Winterer certainly stood behind sampling and field work. I suspect that he was a main architect of the JOIDES/IPOD policy of continuous coring and of logging every hole. Winterer's Leg 7 record of 1,173 meters cored, with 80% recovery, was to stand for many years.

Winterer and I were fellow graduate students at UCLA, but he was also my instructor in sedimentology. Working part-time for the U.S. Geological Survey, he was the first to recognize turbidity-current deposition in the Ventura Basin. He did not worry about priority, and his USGS Professional Paper was published more than a decade after he completed his work. He was a protégé of Bill Bramlette's, professor of micropaleontology and sedimentology at Scripps, and Winterer was called to Scripps after Bramlette retired.

When Winterer joined Scripps, he seemed to be the only geologist, or at least the only structural geologist, at Scripps. Bill Menard, of course, had a geology degree, but his geology can be mapped on a globe six inches in diameter, as Winterer used to tell us. Menard shared Sam Carey's sentiment that "structural geologists were working on second-order features," and that the theory of plate-tectonics was developed by "outsiders with no credentials in structural problems," by those who "were having the effrontery to say that [structural geology] had little bearing on the gross deformation of the earth." Winterer, of course, disagreed. To understand the processes of rock deformation, gazing at a six-inch globe is useless; one has to climb the great mountain ranges of the world to discover fantastic overthrusting. Winterer could do both; he was as adept at quadrangle mapping as he was at surveying for oceanic fracture zones.

Winterer was a disciplinarian, and his students were scared of him. He could be a "Captain Bligh," and mutiny would have been a distinct possibility on the *Challenger* if she had had to sail for years instead of only 55 days without port call. I could not understand that, because Winterer and I were always the best of friends, and he was one of the kindest and most thoughtful people I knew. Our good friend Sy Schlanger used to say: "Oh, Jerry is all right, if he only takes off his gorilla suit." I saw Winterer with his gorilla suit on only during my short tenure on the JOIDES planning committee. He was uncompromising in his principles and steadfast in his beliefs, and he was right most of the time. But, then, he had his gorilla suit on!

Winterer's interests were centered in the South Pacific during the early phases of the Deep Sea Drilling Project. After his marriage to Jacqueline (a Belgian), they made frequent trips to Europe. Winterer became fascinated by Alpine geology, and he was particularly interested in the early phases of seafloor-spreading that gave birth to the Tethys and the Atlantic (see chapter 6). When the West African margin was to be drilled during Legs 50 and 79, Winterer was a logical choice as co-chief.

Glomar Challenger left Funchal, on the Island of Madeira, on 11 September 1976 and returned on 9 November 1976; the other co–chief scientist of Leg 50 was Yves Lancelot. If the "hit-and-run" Leg 6 personified the impatience of Bruce Heezen, the operation of Leg 50 was a manifestation of Winterer's steadfastness. Only two sites were drilled, and at one site, an 1,624-m-thick sequence was penetrated.

A major goal of passive-margin drilling was to sample very old strata. The margin off Morocco has many features that make it especially suitable for such a study. It is a "starve" margin—starved of sediment supply. The sedimentary cover is sufficiently thin that Jurassic sediments could be reached at about 1,500 m depth, which is within the capabilities of the *Challenger*. Winterer had hoped to drill to 2,500 m depth to sample an old salt formation. The penetration was, however, far short of the goal, because of technical limitations. Rocks were not only hard, slowing down the rate of penetration, but also brittle and prone to cave in.

After the IPOD extension, Leg 79 was sent to drill the Morocco margin again; Winterer and Karl Hinz were co–chief scientists. The *Challenger* steamed out of Las Palmas on 15 April 1981 and returned to Lisbon on 23 May. The four sites drilled during Leg 79 constitute a profile across the continental slope. The most seaward site, Site 546, is located on a salt dome a few kilometers beyond the foot of the slope, at 3,958 m depth. Accepting the *Challenger*'s limitations of penetration, Winterer and Hinz tried an "around-the-end run"; they were able to sample the salt at Site 546.

The samples collected from the two drilling cruises yielded much information for interpreting the evolution of an Atlantic passive margin (Figure 18.2). The overall pattern is one of progressive foundering of the continental margin, from the west toward the east, i.e., from the area of nascent oceanic basin toward the continent. This process began well back in the Triassic. Toward the end of the Triassic period, salt was deposited in this embryonic Atlantic basin. Seafloor-spreading, forming the first ocean crust, started during the middle Jurassic, and the subsidence of the continental margin was accelerated during the late Jurassic.

The early history of the Atlantic is thus very similar to the

Back-Arc-

Basin

Collapse

early history of the Tethys. We were right: the Tethys and the Atlantic were twin sisters that had different destinies. The Tethys died young, and her relics have been heaved up to form the lofty peaks of the Swiss Alps. The Atlantic has been getting older and broader with time.

When Bob Dietz proposed in 1963 that the sediments of the Appalachian geosyncline were passive-margin deposits, I protested because the detritus of the "geosyncline" was not derived from continental interior, but from high land far out on the periphery of the North American continent. The sediments on the present passive margin, be they American, European, or African, are, in contrast, all supplied by rivers draining the continental interior. My observations were correct, and my arguments sound, yet we now all accept Dietz's postulate of equating "mio-geosyncline" with passive margin. Why did I back away from my previous conclusion?

Studies of ocean basins revealed that there are two kinds of passive margins. An ocean such as the Atlantic is bounded on both sides by continents, and both sides are passive continental margins. A back-arc basin is, however, bounded by a continent on one side and by an island arc on the other. In the South China Sea Basin, a back-arc basin, the continental shelf south of the Chinese mainland is a passive margin, but the Philippine Trench on the east side of the South China Sea is an active margin. The ocean floor of the South China Basin is being subducted to the east under the Philippine Arc.

Using this model for comparison, we could now postulate that the Paleozoic Appalachian carbonate sediments were passive-margin deposits of a back-arc basin. The Appalachian clastics were, however, deposited on an active margin behind an island arc. The "mio-geosyncline" of the central Appalachian was thus a back-arc basin comparable to the South China Sea. The late Paleozoic deformation was to effect the elimination of this basin: the ocean lithosphere was subducted under the arc, while the basinal sediments were scraped off to form thrusts and decollements such as those mapped by geologists in eastern New York State and Vermont. When this basin was completely eliminated, the frontal arc (which is called Appalachia) collided with North America, causing the thin-skinned deformation of the Paleozoic carbonate strata of the "Appalachian geosyncline."

This model of elimination of a back-arc basin has been called by Ian Dalziel, chairman of the Tectonics Panel of ODP, "back-arc collapse." The South China Sea is in the process of collapsing: the ocean floor is being subducted under the Philippine

Arc, and China and the Philippines will eventually meet in a collision some 50 or 100 million years from now.

Both Denny Hayes and I sent in proposals for ODP drilling to investigate the tectonic evolution of the South China Sea (see chapter 13). Hayes proposed to test the rate of subsidence on a passive margin, as predicted by the various mathematical models, and I was hoping to collect information to formulate a case history of back-arc-basin collapse on an active margin. Unfortunately, the proposals fell into the no man's land between the Active Margin and Passive Margin panels. They were rejected (see chapter 13).

Accretion-
ary Wedges
on Active
Margins

The Atlantic margins are largely passive, whereas those of the Pacific are largely active. Cruises for a third theme to be investigated by the IPOD were planned by the Active Margin Panel. One would have thought that we knew all about subduction after the Leg 18 studies of the North American margin (chapters 7 and 8). But the fun was just getting started for the experts on the JOIDES Active Margin Panel; they were convinced that they were only beginning to sort out their findings and that there was much mopping up to be done. Solutions to old questions had to be refined, and new questions had to be formulated.

After Leg 18, the *Challenger* circumnavigated the Pacific and drilled the Aleutians on Leg 19, the Japan Trench and the Mariana Trench on Leg 20, and the Tonga Trench on Leg 21, all in 1971. After a detour to the Indian Ocean, the *Challenger* returned to the Pacific for Phase III of DSDP and drilled the Timor Trench on Leg 27 in 1972, and the New Hebrides Trench on Leg 30 and the Nankai Trough on Leg 31, both in 1973. After the start of the international phase, the *Challenger* came to the Pacific for the fourth time to investigate active margins, and drilled the Japan Trench (Legs 56 and 57) in 1977, the Mariana Trench (Legs 59 and 60) in 1978, and the Middle American Trench (Legs 66 and 67) in 1979. She was to come again for a fifth time during the last phases of IPOD to drill during Legs 85 (Mid-American) and 87 (Japan), after a glorious interlude drilling the Barbados margin in the Atlantic (Leg 78A).

The investment of time was considered necessary. Geology is history, and there could be no simple rules. We needed to explore every active margin before we could even start to generalize about subduction tectonics. We had to make new random observations before a new paradigm could be formulated.

Back in the early 1970s, the idea that dominated the profession was that there was an *accretionary wedge* at every active margin. During the underthrusting process, so the rationale went, the ocean lithosphere should be pulled down into the mantle along the Benioff zone, because the ocean rocks have a higher density than those in continental crust. Only the ocean sediments in trenches at the active plate-margin are lighter, and they would stay up. Scraped off their basement, those flysch-like sediments would be wedged between the continent and the downgoing ocean slab. Broken into thousands of pieces in the tight squeeze, slabs of broken sedimentary formations and scattered exotic blocks of ocean crust would form a mélange. This accretionary wedge of mélange tucked under a continent is so called because the wedge of ocean rocks is accreted to the bottom of, and thus becomes a part of, the continental crust (Figure 18.3).

My work on the Franciscan mélange had contributed to the idea of accretionary wedging on active margins. The Coast Range terrain underlain by the Franciscan is about 200 km wide and more than 1,000 km long. We had interpreted the mélange as an accretionary wedge underplating the North American continent while the Pacific Plate was consumed by subduction (see chapter 8). But how did those "accretionary wedges of mélanges" see daylight and come to be exposed on the Coast Ranges?

The simple explanation is that the wedges were jacked up when more wedges were placed under them. According to this "jack-up" model, one should find older mélanges higher up on inner walls of a trench, and the most recent mélanges near the foot of the wall. The drilling of the Middle American Trench during Leg 66 in 1979 did seem to fulfill the prediction. A transect of eight boreholes was drilled. Of those, three were drilled into possible accretionary wedges under the inner wall of the trench. The accretionary wedge at Site 492 at less than 2,000 m seafloor depth is Miocene, some 10 million years of age, that at Site 491 at about 3,000 m depth is Pliocene, 4 million years old, and that at Site 488, near the foot of the trench wall down at 5,000 m depth, is Pleistocene, about a million years old (Figure 18.4). As the theory predicted, older wedges stand higher, and the wedges were jacked up at an average rate of about several hundred meters per million years. In another 10 million years, the Miocene wedge, now still submerged at 2,000 m depth, should be lifted up, eroded, and exposed in a Central American coast range, like the Franciscan mélange of the California Coast Range.

Subduction
Erosion

In 1970, when we drilled into the inner wall of the Hellenic Trench, Bill Ryan and I had a bet. He knew the plate-tectonic model, and he knew of my Franciscan work. He was sure that we would penetrate an accretionary wedge of young Tertiary

18.3. Two types of active margin: accretionary wedge (above); subduction erosion (below).

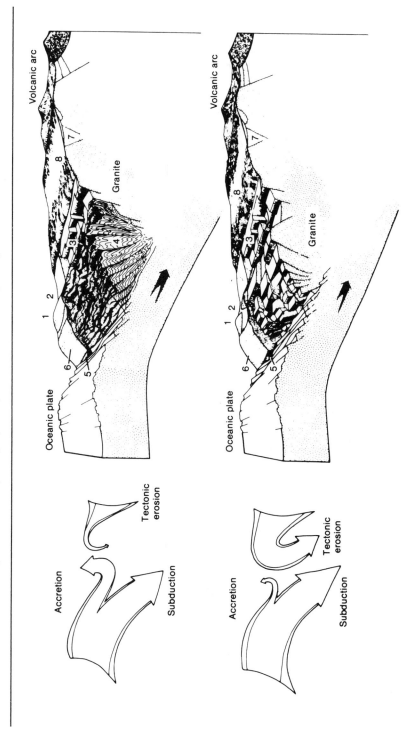

sediments under the trench wall. I was not familiar with the theoretical model then, and I found no reason to assume a wedge down there. Ryan lost his bet.

An accretionary wedge is not present at every active margin; it is not even present in all Pacific coast ranges. Rocks similar to the Franciscan, but of somewhat different age, are indeed exposed on the Oregon and Washington coasts. Farther north, on the British Columbia coast, however, fjords are cut into massive granites. Or going south toward Baja California, the peninsula is also largely underlain by igneous rocks; mélanges similar to the Franciscan are present only on a few offshore islands. In South America, mélanges of former ocean sediments, mixed with ophiolites, are present in the Colombian Andes and in the southern Andes. The central Andes are underlain largely by granitic and volcanic rocks. Those rocks are not accretionary wedges plastered onto the bottom of continents.

In fact, the genesis of mélanges is only one of two corollaries of the theory of plate-tectonics; the other corollary predicts the existence of magmatic arcs at active plate-margins. The magmas were derived from the sediments or the crust of an oceanic lithosphere that had been brought down by underthrusting, heated up, and remelted; they found their way through fissures in the continental crust, and erupted out of volcanoes as andesitic or other lavas. The granitic rocks were rock-melts, or magmas, that were solidified 10 or 20 kilometers beneath the volcanoes; they have been uplifted, eroded, and exposed. The geology of the Central Andes tells us that there are no Cenozoic accretional mélanges on land there. In fact, the edge of the continent has not been jacked up, but dragged down by the downplunging ocean floor.

The situation is similar on the opposite side of the Pacific. Mélanges are present in some coast ranges or on some islands, such as the Kuriles, New Caledonia, and so forth, but absent in others.

My knowledge of geology gave me an unfair advantage

18.4. **Accretionary wedge under the Middle American Trench.**

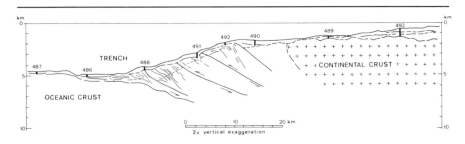

when I made the bet with Ryan. In fact, I was already thinking in those days about the two distinct types of underthrusting margins, the Franciscan type with jacked-up accretionary wedges and the Andean type with down-dragged continental margins. Ryan lost his bet because the margin bordering the Hellenic Trench belongs to the latter type. The island of Santorini was the tipoff. The volcanism there is comparable to that on the Andean type of active margin.

In Hole 127 we drilled into the Cretaceous limestones that had been deposited on an ancient continental shelf near the sea level. Those rocks are now found at 5,000 m below sea level because the southern margin of the European continent has been dragged down, as predicted by the Andean model.

Our conclusion concerning the genesis of the Hellenic Trench was, however, overshadowed by the discovery that the Mediterranean was a desert; few paid much attention to our experiment on *Glomar Challenger* that tested theoretical models of active plate-margins. The members of the JOIDES Active Margin Panel in 1975, like Bill Ryan in 1970, were still entranced by the accretionary-wedge model. Given a boost by Leg 18 drilling, and reinforced by a seismic-profiling study of the Middle American Trench, the model became the ruling theory. But von Huene and his associates were to be surprised, as Bill Ryan was when he lost his bet.

Von Huene was again co–chief scientist in 1977, leading Leg 57 to make a transect of the Japan Trench. Seven holes were drilled, of which six were drilled into the inner slope of the trench. According to the "jack-up" model, the holes should have penetrated into chaotically deformed mélanges of accretionary wedges. Instead, the boreholes penetrated flat-lying undeformed sediments in their normal order of superposition. Only at Site 434, at the foot of the trench wall, was there an indication that the drilling may have penetrated into the deformed sediments of a very small accretionary wedge (Figure 18.5).

A greater surprise was encountered at Sites 438 and 439 on the upper reaches of the inner trench-slope. There the holes ended in gravels and volcanic rocks that were laid down on land some 25 or 30 million years ago. The overlying sedimentary sequence indicates a history of subsidence. An ancient land mass, lying more than 100 km seaward of the present coastline of Japan's Honshu Island, sank slowly down. It sank down to about 1,000 m below sea level 20 million years ago. The seafloor there finally reached its present 2,000-m depth 5 million years ago. The record indicates clearly that the inner wall of the trench has been dragged down by the force pulling down the subduction of the Pacific Plate. Von Huene and his colleague David Scholl were to formulate a "drag-down" model to interpret the drill results of Leg 57. An accretionary wedge, if it is present at all, must be restricted to a very small area at the very base of the inner slope of the trench.

Other drilling cruises also came up with results that favored the "drag-down" model. Leg 56 in the Japan Trench, Legs 59 and 60 in the Mariana Trench, and Leg 67 in the Middle American Trench all failed to find large accretionary wedges. On the other hand, an accretionary wedge was found on the inner wall of a segment of the Middle American Trench, the Aleutian

18.5. Subduction erosion under the Japan Trench.

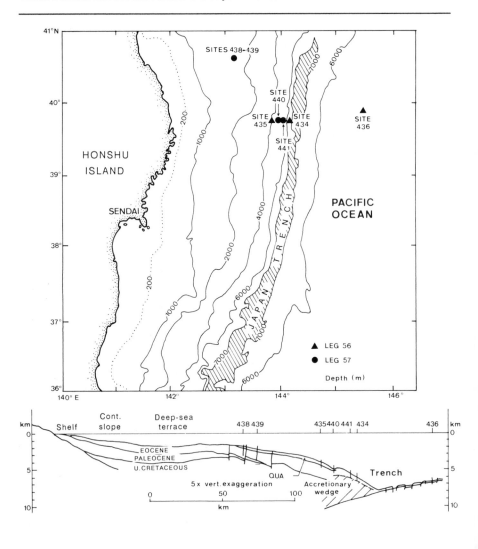

Trench, and the Nankai Trough. It seemed that the score in the Pacific was about even during the first decade of ocean drilling.

One might ask why the continental margin should be lifted up in one place and dragged down in another. There are several factors that must be considered. First of all, an accretionary wedge requires material, namely a lot of sediments in the trench. The Franciscan mélanges, for example, include sandstones and shales thousands of meters thick, made of coarse detritus that had been dumped very rapidly into the trench; those sediments could be tucked under the edge of the North American continent and jack it up. The Pacific Coast of Washington was apparently also being jacked up by the deformed sediments that had been laid down on submarine fans like the Astoria Fan, but older in age (see chapter 8).

The other extreme is illustrated by the Chile-Peru Trench, fringing the Andes. Rivers draining the South American desert carried little debris to the coast, and the little that did get there was trapped by shallower basins above. The trench is thus almost devoid of sediments. Where an ancient crust, covered by little or no sediments, is thrust under the edge of a continent, there would be little accretion. On the contrary, pieces of the continental crust might be broken off and carried down into the mantle by the sinking slab of oceanic lithosphere. Von Huene and Scholl now called this process "subduction erosion" (Figure 18.3). A continental crust that had had its bottom scraped off by such an erosion should sink down, as Ryan and I discovered by drilling the inner wall of the Hellenic Trench, and as Von Huene discovered by drilling the inner slope of the Japan Trench.

This discovery of two types of active margin may explain a well-known geological observation: intrusive granites are common in some mountain ranges, but they are almost completely absent in the Alps. We could now postulate that the southern margin of the Tethys Ocean was a Franciscan type of margin; the accretionary wedge is now exposed in the form of flysches and mélanges that were once jacking up the southern continent. The Alps are more the exception than the rule; the Andean type of margin was more common in the geological record, and the granitic rocks of the mountains were the root of volcanism on an active margin.

Barbados

Venture

I traveled to the island of Barbados in late 1979 to attend a JOIDES panel meeting; we were to finalize the planning of the South Atlantic drilling cruises that were to take place early in 1980. As usual, we took the opportunity to see some local geology. The rocks in the field seemed familiar. There were Oligocene oozes, Miocene marls, and red clays devoid of fossils; I had seen similar sediments during my Leg 3 cruise, and I was to encounter them again on Leg 73. The oceanic sediments had been cut into slices on an ancient active margin, and were jacked up by an accretionary wedge.

Barbados is an island situated between the Lesser Antilles Volcanic Arc and the Barbados Ridge outer arc. The arcs separate the Atlantic from the Caribbean basins. Seismic profiling has shown that the sedimentary cover on the ridge has been cut into a series of thrust slices (Figure 18.6). This type of peeling off is commonly seen, for example in the foreland thrust belt of the Alps or of the Appalachians, and is called thin-skinned deformation, because only a very thin sedimentary cover, detached from its foundation, is deformed. The surface of detachment is called a *decollement* horizon; this French expression has been adopted in geology because it was first used by French-speaking geologists working in the Jura Mountains of Switzerland.

The Barbados Ridge is an active margin in the Atlantic, an anomaly like Denny Hayes's passive margin in the Pacific. Not much attention was paid to it by the Active Margin Panel until the IPOD drilling was extended in 1980. Half of a leg was then added to the schedule to drill the Barbados Ridge, and the rationale was given in the *JOIDES Journal*, February 1981:

The basins of the Caribbean Sea remain an enigma . . . We ask whether the basement of the Caribbean plate consists of old segments of Pacific or of Atlantic crust. Did the Caribbean originate by interarc spreading? Is it best thought of as a plate or as a boundary zone between plates? Drilling in the Caribbean will not only address these tectonic questions, but recovery of the sedimentary record will help define the history of communication of Atlantic and Pacific water masses. Finally, we hope that by drilling the Lesser Antilles fore-arc we can completely penetrate the toe of an accretionary prism, a goal yet to be achieved at any active margin.

Although I was not present during the deliberations of the planning committee, I could envision my friend Jerry Winterer being persuasive. He had always thought that the Caribbean was a fragment of the Pacific crust that had been trapped between Central America and the Antilles Arc, and his intuition seemed to have been verified by the drilling of Nauru Basin, the twin sister on the other side of the Pacific (see chapter 10). As it turned out, the primary goals stated on the eve of departure were hardly addressed. What was added almost as an afterthought turned out to be the greatest achievement of the leg, and was to inspire more drilling during the future.

Leg 78A sailed from San Juan, Puerto Rico, on 11 February 1981, and returned to the same port a month later. The co–chief scientists were Bernard Biju-Duval of the Institut Francais du Pétrole, Paris, and Casey Moore of the University of California at Santa Cruz. At Sites 541 and 542 on the Barbados Ridge, they penetrated what they called offscraped sequence. Although the strata were still coherent, the sequence had been sliced by thrust faults, and even overturned layers were encountered. They came close to penetrating the decollement surface, and noted that disruption of sedimentary layering was not uncommon near the more intensely deformed surface.

Moore and Biju-Duval hoped to drill through the decollement surface, but they failed. There was, as usual, the time constraint, and they had only a half-cruise. The penetration through the detachment surface was done six years later, in July 1986 at Site 671 during ODP Leg 110, on which Casey Moore was again co–chief scientist, the other being Alain Mascle, Biju-Duval's colleague at IFP.

A very significant discovery was the anomalously high fluid pressure in the offscraped sequence. Theoreticians had predicted that high pore-pressure facilitates faulting such as that taking place on the decollement surface. Leg 78A drilling was to provide an actualistic analogue. The abnormal pressure indicates furthermore a movement of interstitial fluids through such an accretionary prism. New ground was being opened, and the question of fluid-migration path was to be a major theme for ODP investigators on *JOIDES Resolution*.

18.6. Deformation of accretionary wedge on the Barbados margin.

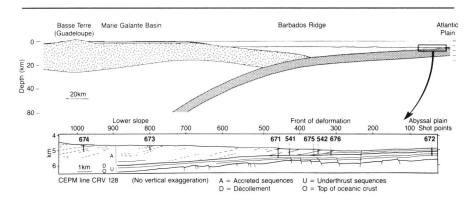

19 What Makes the Ocean Run

The deep-sea drilling gave us paleoceanography. Ironically, it was not what we found, but what we found missing, that told much of the story.

Missing Pages in the History Book

Drilling passive continental margins may not have inspired new and exciting ideas on the origin of the margins, but it did turn up many very interesting facts concerning the movements of water masses in the oceans during past geologic times.

In the spring of 1976, Leg 47 was scheduled to explore the continental rise west of the Spanish Sahara, and my friend Bill Ryan went on *Glomar Challenger* for the third time to serve as co–chief scientist, together with Ulrich von Rad of the Bundesanstalt für Geowissenschaften und Rohstoffe at Hannover. Only two holes were drilled at one location (Site 397) in a little over two weeks' time. One penetrated 1,000 m, the other 1,453 m beneath the 3,000-m-deep seafloor.

The first hole drilled into an unusually thick layer of young sediments. When the hole had to be abandoned because of mechanical difficulties, the drill string had penetrated to the early Miocene, in sediments about 20 million years of age. The scientific staff had a meeting and decided to stay at the site to drill another hole. It was a gamble, because they were not certain if they could drill deeper beyond the Miocene. Besides, the mudstones were giving off hydrocarbon gases. The quantities were minute, but one could never be sure if there would be a sudden increase; there was always the possibility that the second hole would also have to be abandoned, because of the constraints imposed by the JOIDES Panel on Pollution Prevention and Safety. The staff evaluated the prospects and decided to take a gamble.

Toward the end of March, Hole 397A was drilled down to 1,000 m depth, and was to enter into an unknown territory with the beginning of April, but the co–chief scientists were left in limbo for two more days. Again and again the core barrel was raised on deck, and the same Miocene mudstone was found.

However, if I know Ryan, he persevered and von Rad went along. Finally, at 1,300-meter subbottom depth, the 137th core at this site brought them the "pay dirt"! To the great surprise of everyone, they did not find Oligocene, the next oldest formation expected, nor did they find Eocene, Paleocene, or Upper Cretaceous; the core brought up a black shale of early Cretaceous age. Between the Miocene mudstone and the Cretaceous shale was a huge gap in the sedimentary record that spanned 100 million years!

Who had torn away the many missing pages of the history book? Mike Arthur, a sedimentologist on the cruise, made a rough estimate and found that about 10,000 cubic kilometers of sediments had been removed by erosion within the 6,000 km² area. This is phenomenal, especially in view of the fact that much of the materials removed must have been well-lithified rocks.

Large gaps in ancient deep-sea sedimentary records were known to geologists studying the Alpine formations that had been deposited on the northern margin of the Tethys Ocean. Alpine geologists were able to trace such hiatuses in sedimentation from one Alpine valley to another, from one thrust-up rock-mass (called _nappe_) to another. My colleague Hans Bolli worked on such a problem during the Second World War for his doctoral dissertation. A gap is present between the Cretaceous and Tertiary sediments that were laid down on the European continental slope, and the gap was more pronounced in deeper waters, farther and farther away from the European coast. Having no knowledge of any ocean currents that could effectively erode sea-bottom, such a hiatus (or unconformity, a word used by geologists to denote an ancient erosional surface) had to be explained away by a series of very cumbersome assumptions: somehow the deep sea must have become dry land, so that subaerial erosion could take place, and after the removal of a considerable amount of sedimentary record, the same area suddenly sank back into the abyss. The scientists of Leg 47 knew more about ocean circulation; they did not have to appeal to such a farfetched postulate. The movement of bottom waters could reach high enough velocities to erode deep seafloor, so they concluded.

Hiatus in the Sedimentary Sequence

During the eight years of deep-sea drilling prior to Leg 47, considerable evidence had been uncovered to suggest the enormous erosive power of marine bottom-currents, especially the boundary currents flowing along the margins of the Atlantic.

In 1970, when Ryan and I drilled through a similar sequence on the Gorringe Bank west of Lisbon during Leg 13, we encountered lower Cretaceous sediments after we had cored Miocene. We noted this in our cruise report but did not think twice about its significance. Large gaps in the sedimentary record between the Cretaceous and Tertiary sediments were then found in many boreholes on both margins of the Atlantic, and Leg 41 scientists had discussed the meaning of such an interruption of sedimentation. The discovery by Leg 47 was noteworthy only because it was the first complete record of such an enormous erosion.

When did this erosion take place during the 100 million years of missing record?

In the early 1970s, Eugene Seibold of Kiel, Karl Hinz of Hannover, and their associates had carried out extensive investigations of the continental margins west of Africa. Using the continuous-seismic-profiling technique, they discovered two prominent reflecting layers, D-1 and D-2, within the sedimentary sequence underlying the continental slope west of the Spanish Sahara (Figure 19.1). The reflecting layers reflect acoustic signals because the tops of the layers are hard. In the Mediterranean, the acoustic reflectors are the top of the lithified formations of rock salts or evaporites (see chapter 15). In many parts of the North Atlantic and in the Pacific, the acoustic reflectors are the top of hard chert formations (see chapter 9). There should be neither salt nor chert under the ocean margin west of Africa. What are those reflectors, then? Seibold and Hinz had to puzzle over the question.

Leg 41 was scheduled in the spring of 1975, when the Federal Republic of Germany was about to join IPOD, and the cruise has thus been regarded by some as a "dowry" for Germany in anticipation of the forthcoming "marriage." The *Glomar Challenger* was sent to drill offshore west of Cape Bajodor to provide solutions to the riddles encountered by Seibold and Hinz. Leg 41 has also been described as a Franco-German joint

19.1. Gap in the sedimentary record of the West African margin. The seismic profiling record shows a hiatus in sedimentation on the passive margin west of the Spanish Sahara. The gaps in the record, verified by drilling, are now taken as evidence of erosion by bottom-currents.

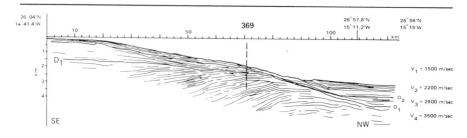

venture, as it was led by Seibold and Yves Lancelot. Lancelot, who was the chief scientist for the Deep Sea Drilling Project during the later years of IPOD, was then a young research assistant from the University of Paris. It was natural that he was a bit apprehensive about having to work with a more senior person like Seibold, who was one of the most prominent geologists of our time. Having been the director of the Geological Institute at Kiel for many years, Seibold was the president of the Deutsche Forschungsgemeinschaft and the president of the International Union of Geological Sciences. However, Lancelot did not have to worry, for Seibold was a most jovial person, as illustrated by one of the many jokes he told me in 1978 during our five-day excursion together across the Dead Sea Desert.

Seibold likes to poke fun at himself, and at his American friends who have difficulty distinguishing _ei_ from _ie_. Being used to p_ie_s and rec_ei_vers, his hosts often introduced Professor S_ei_bold from K_ie_l as Professor S_ie_bold from K_ei_l during his lecture tours across the United States. When that happened once too often during our Israeli trip, Seibold had to tell me a story of his wartime experiences.

After having been wounded at the Russian front, Major Seibold was sent back home for recuperation, and upon his release from the hospital was engaged by an army division trying out new weapons. During those desperate years, the German High Command seemed to be ready to grasp at any straw that might bring victory, and Major Seibold had to try out all kinds of new inventions. More often than not, the new weapons did not work, to the chagrin of those who had been ordered to perform the tests. It was Major Seibold's duty to file reports that served only to disappoint his superiors. Once some bright inventor came up with a crazy idea that artillery shells might be more destructive if they exploded after rebound. So Major Seibold and his men had to spend weeks sweating and wasting their time on something they knew wouldn't work. The idea was a bust. Finally the report on the _Schiessübung_ (shooting exercise) had to be sent in. Major Seibold's report was not too complimentary, of course, but it was properly written, or so he thought. He did not understand the wrath of his superiors or the tongue-lashing he received when he was called to the headquarters of the High Command for a reprimand. Only much later did he realize that a typist had inadvertently substituted "ei" for "ie" in the key word on the front cover of the report. (This conveys the same insult as when an English-speaking person substitutes "i" for "o" in the word shot.) I greatly appreciated his sense of humor, which imparted a more realistic image to the halo of his fame. I could have predicted that Seibold would quickly manage to remove any alarm or diffidence on the part of his young co–chief scientist.

Leg 41 stands as one of the more successful cruises of the Deep Sea Drilling Project. At Site 369, on the continental slope west of Cape Bajodor, Seibold and Lancelot supervised the drilling into the layers that reflect acoustic signals. This site, like Site 397, was also located offshore from Cape Bajodor, but at the shallower depth of 1,770 m below sea level. Drilling indicated that the reflectors D-1 and D-2 are ancient erosional surfaces, or unconformities. These surfaces are hard because erosion has removed soft sediments, while the chemical reaction between the seawater and eroded seafloor makes a hard pavement. The two major time intervals of erosion were late Cretaceous and late Eocene/early Oligocene. The first round of ocean circulation terminated the anoxic event of black-shale deposition, when oxygenated water was brought in by bottom-currents. The second, as I shall now discuss, coincided in timing with the beginning of circulating Antarctic Bottom Water Masses (AABW).

Making use of the information from Site 369 as well as the geophysical data gathered by Seibold and Hinz, Leg 47 scientists concluded that the bottom-current causing widespread erosion on the Atlantic margins was most active during early Oligocene. A sedimentary sequence more than one kilometer thick was removed during an interval of 10 to 20 million years. The erosion stopped abruptly in early Miocene, when the movement of bottom waters became too weak to transport away the cascades of debris slumped down from farther upslope off Cape Bajodor.

Antarctic

Bottom

Water Mass

A submarine bank, the Rio Grande Rise, is separated from the South American margin by a deep-sea channel called Vema. After Leg 41 and Leg 47 drilling, the channel was interpreted as evidence of impressive submarine erosion. The present bottom water masses from the Antarctic, the AABW, flow northward through the channel from the Argentine Basin to the Brazil Basin (Figure 19.2). Drilling during Leg 72 in 1980 indicated that the AABW has been flowing through there for about 40 million years, and the movement deepening the channel was especially active during the early Oligocene, when the continental rise off Cape Bajodor on the other side of the Atlantic was also being eroded.

Tracks left behind by vigorous bottom-circulations of the early Oligocene oceans had been uncovered elsewhere by unconformities, or "missing pages in history books." Leg 21, led by co–chief scientists Robert Burns of Seattle and my shipmate

19.2. Vema Channel off the South American Atlantic margin. The Antarctic Bottom Water Mass (AABW) has been flowing from the Argentine Basin through a gap between the Rio Grande Rise and South America to the Brazil Basin since late Eocene or early Oligocene some 40 million years ago. The gap is called Vema Channel. Note that Antarctic diatoms have been transported north by the AABW.

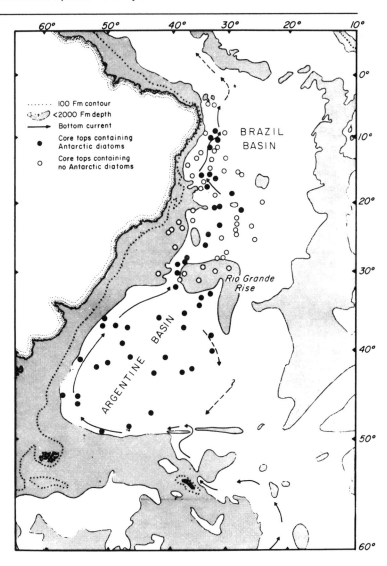

on Leg 3, Jim Andrews of Hawaii, drilled five sites into the floor of the Coral and Tasman Seas. A shipboard paleontologist, Tony Edwards, noted the curious fact that the latest Eocene and early Oligocene sediments are missing at every one of the five sites, corresponding almost exactly in timing to the youngest gap in the sedimentary record west of Cape Bajodor. Scientists on deep-sea drilling cruises to the Indian Ocean had also commented on their strange discovery that the Oligocene sediments are mostly missing on the margins of that vast body of water. When the Southern Ocean south of Tasmania was drilled during Leg 29, the shipboard scientists had enough data to speculate on the occurrence of vigorous bottom-circulations during the late Oligocene.

The history of the oceans could be read back (by ocean drilling) to the Cretaceous period, when the polar regions were ice free. The temperature difference between the equator and high latitudes was small, and there was insufficient difference in water density to cause strong thermo-haline (temperature-salt-driven) circulations, such as those observed today. Denser waters were produced locally, where evaporation increased their salinity. When such salty waters were cooled and sank, they induced a halokinetic (salt-driven) circulation that may have been the dominant pattern during much of the time before the Antarctic glaciers reached the sea. This halokinetic mechanism was not always very efficient, as shown by the repeated deposition of the Cretaceous black shales in poorly ventilated environments of the Atlantic Ocean (chapter 14).

The beginning of the Oligocene was an important date in the history of the oceans. A cooling caused the growth of Antarctic glaciers. The growth of the glaciers and their advance down to sea level cooled the seawater and caused the formation of sea-ice. The surface waters that had become cold and saline sank to form the Oligocene Antarctic bottom waters, and they started their long circulatory journeys as western boundary currents, invading the Southwest Pacific, the Atlantic, and the Indian Oceans (Figure 19.3). The flow of cold Antarctic waters to tropical regions led to a further deterioration of the global climate, which in turn promoted the further expansion of Antarctic glaciers. Thus the circle was complete, and the circulatory responses resulted in a spiraling effect like runaway inflation; the steady state was broken and a drastic cooling set in.

The climatic change at the beginning of the Oligocene also changed the motor for the deep-sea circulation. The efficient production of the Antarctic Bottom Waters promoted vigorous movement, and widespread erosion of the seafloor, especially in regions along the paths of the bottom-currents. The erosional surfaces discovered by the deep-sea drilling west of Africa, within the Vema Channel, in the region of the Coral and Tasman Seas, and on the margins of the Indian Ocean all have a common origin in the cooling of the Antarctic.

Why did the "runaway cooling" stop? Why did the vigorous bottom-circulation, which started in the early Oligocene, slow down during the late Oligocene or early Miocene? Kennett and his associates believed that they had found the answers in their drill cores from Leg 29.

One of the main purposes of that cruise was to determine the history of the Circum-Antarctic Current. This circumpolar circulation is of great oceanographic and climatic importance because it transports more than 200 million cubic meters of water per second, probably the largest volume-transport of any ocean current. Such a circulation was not possible when South America, New Zealand, and Australia were still linked to Antarctica. Kennett thought that the region south of Tasmania

19.3. History of circum-Antarctic circulation. Drill site locations are shown.

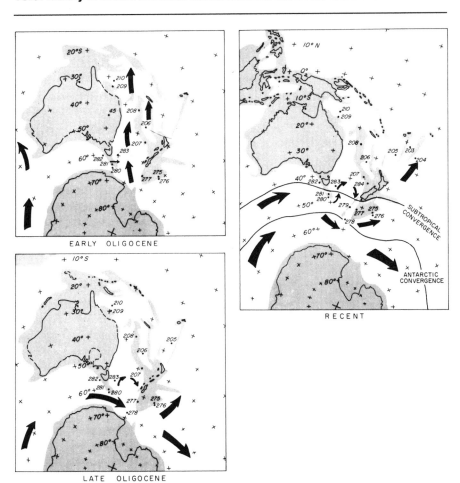

was the last obstacle remaining during the Oligocene after the earlier separations of New Zealand and South America from Antarctica. Australia had also started to march away from Antarctica during the Eocene, 55 million years ago, but Tasmania and the submarine elevation Tasman Rise remained attached to Antarctica for another 15 million years, while the northward motion of Australia was pivoted about the Tasmania axis. The spreading of the seafloor continued relentlessly, and the Tasman Rise was finally ripped away from Antarctica during the late Oligocene, 25 or 30 million years ago, creating a narrow but deep passage there that is floored by an oceanic crust (Figure 19.3). Prior to that, a deep channel through the Drake Passage had also been established (see chapter 13). After those last obstructions to the deep circulation around Antarctica were removed, the Circum-Antarctic Current was born, and no major changes in the circulation patterns have taken place since then.

The theory of the origin of circumpolar circulation explains the changing vigorousness of the boundary-current circulation in middle latitudes. In the early Oligocene, when the Antarctic Bottom Waters were first formed, they were directed northward to the Pacific, to the Atlantic, and to the Indian Oceans (Figure 19.3). After the late Oligocene, the cold bottom waters were largely trapped in the system of circumpolar circulation. Deprived of much of the supply, the bottom-currents in other oceans became too weak to prevent sedimentation on ocean margins. It seems fantastic to assume that the resumption of sedimentation west of Cape Bajodor was related to a happening in the distant Antarctic, but the very essence of science is to find a common cause for apparently unrelated effects.

I have a suspicion that the initiation of the circumpolar circulation may have done more. It may have delayed for millions of years the continued deterioration of the global climate in temperate regions. The records of changing temperatures indicate an early Miocene reversal. We might again seek an answer from Kennett's hypothesis. With all that cold bottom water trapped in the polar regions, while the surface waters of the oceans continued to be warmed in equatorial currents circulating around the globe, the heat exchanges were minimized, resulting in a mild climate for the temperate regions during much of the early Miocene. We are not certain why the cooling trend started anew in the middle Miocene 15 million years ago, and again in the late Miocene 5 million years ago. Wolf Berger and Edith Vincent have hypothesized that the chill was related to the less efficient greenhouse effect, when carbon dioxide in the atmosphere was drastically reduced because of very fertile plankton production in those times.

"Snow Line" in Oceans   I was skeptical of the seafloor-spreading theory in the late 1960s, because conventional wisdom argued against the seemingly fantastic theory. Doc Ewing and his associates had made extensive surveys in the 1960s of the thickness of the ocean sediments in the Atlantic, and the sedimentary blanket is, on the whole, a few hundred meters thick everywhere. We used to believe that the sequence should be thicker on older crust, where there was more time available for sediment accumulation. Ewing's data thus seemed to argue against a theory that predicts a linearly increasing age away from the axis of the mid-ocean ridge.

Well, I was wrong, because sediment thickness depends not only upon the time available, but also on the rate of accumulation. Two kinds of sediments were encountered during Leg 3—calcareous oozes and red clays. Young crust on the ridge crest had as thick a blanket of oozes as the older crust on the flank, where the red clay slowly accumulates. My naive assumption of a constant sedimentation rate was wrong, and I was buoyed by false hopes when I played the "numbers game" on the Leg 3 cruise (see chapter 5).

The red clays are mainly middle Miocene, between 20 and 10 million years of age. Why should those sediments be red clays? Why should any ocean sediments be red clays? During the cruise of HMS *Challenger* more than a century ago, some empirical observations were made: red clays are commonly found on deeper ocean bottoms, and red clays are commonly present away from the tropical and polar regions of high biological fertility. Rational explanations were not slow in coming: the red clays are the residues after all the fossil skeletons have been dissolved away.

Red clays are commonly found on ocean bottoms at about 4 km depth, where plankton fertility is low, and at 5 km depth, where plankton are abundant. Conversely, we can say that white oozes are commonly found on sea bottoms above 4–5 km depth. The depth of the line separating the areas of calcareous oozes from the red clays is called the calcite compensation depth (CCD). Compensation refers to the balance between arriving and dissolving calcareous skeletons. Below the CCD, only the red clays accumulate, because the supply of plankton skeletons is not sufficient to compensate for their dissolution. The calcite compensation depth varies from place to place, both because of variations in plankton fertility and because of the differing corrosiveness of seawater (Figure 19.4).

Wolf Berger of Scripps had used the concept of "snow line"

as an analogy to explain the calcite compensation depth. He asked us to imagine an ocean without water. The underwater elevations would be peaks and ridges with a white cap like snow. The lower slopes, the abyssal plains, and the deep-sea trenches would be covered with dark sediments of red clays or brown muds. The calcite compensation depth separates the white from the dark, as the snow line on land demarcates regions of permanent snow.

The snow line, like the CCD, does not have the same height everywhere; it is a compensation elevation, the elevation at which the annual melting of snow exactly compensates for the annual supply of snow. The snow line is low in polar regions, where there is plenty of precipitation or where the temperature is too cold for much melting; the snow line is high in deserts,

19.4. Carbonate deposition on spreading and subsiding ridges: stipples, calcareous ooze; black area clay; *t*, distance from ridge crest corresponding to time *t* during which carbonate accumulates; *A*, final thickness of carbonate corresponding to time *t* ; CCD zone, range of fluctuation of local CCD level. Upper figure indicates deposition of clay after time *t*, when seafloor subsided below CCD. Lower figure indicates alternate deposition of clay and oozes because of a fluctuating range of CCD.

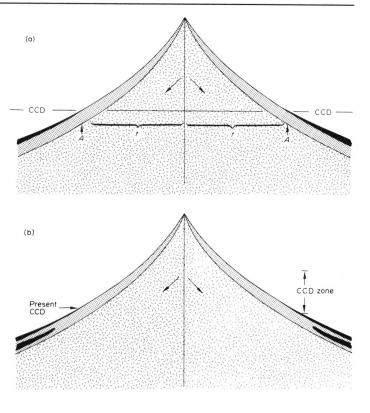

where there is scant precipitation, or in the tropics, where the temperature is high. The calcite compensation depth behaves likewise: the CCD is low in the equatorial zone, where there is abundant precipitation of plankton skeletons, or in regions where the bottom water is not too corrosive; the CCD is high in the middle-latitude ocean "deserts," where plankton production is scanty, or in regions where the bottom water is corrosive.

All this had, however, not been well established when I joined the Leg 3 cruise in 1968. Therefore, I was startled to find a red clay where it did not seem to belong. When I wrote my cruise report, I was still an adherent of substantive uniformitarianism, believing that conditions have remained more or less uniform since the beginning of time. In my ignorance, I chose the Occam's razor and assumed that the CCD had remained all the time at more or less the same depth of 4,500 m. Oozes were not deposited on the flanks of the Mid-Atlantic Ridge as they are now; middle Miocene deposits were red clays because the Mid-Atlantic Ridge was then not 2.6 km subsea, but more than 4 km subsea.

Bill Hay, writing on that problem for the Leg 4 cruise report, rejected my hypothesis. Instead, he postulated that the changing chemistry of the ocean waters causes a rise and fall of the CCD, which in turn determines if a calcareous ooze or a red clay is to be deposited.

Meanwhile, John Sclater and his associates at Scripps came up with a geophysical theory that the Mid-Atlantic Ridge must have stood at about the same site since the present phase of seafloor-spreading started in the Jurassic. Jacked up by the expanded mantle material under the ridge crest, the axial region stands high. As the lithospheric plate is moved away from the hot center, the mantle cools down and the mantle rocks become denser, so the seafloor should sink gradually as the mantle contracts. Using geophysical data, Sclater formulated a mathematical relation between the depth of the seafloor and the age of the underlying ocean crust: the ocean is deeper where the crust is older. Later, when the deep-sea drilling data for the first two phases of JOIDES became available, Sclater and his student Bob Detrick were able to confirm the theoretical prediction. The crest of a mid-ocean ridge should, as a rule, be everywhere 2,500 to 2,700 m deep. During the first 15 million years after an ocean crust is formed, the seafloor above the crust sinks rapidly to 4,000 m depth. The subsidence then slows down, and is halted altogether when the ocean crust has sunk to a depth of about 6,000 m beneath the sea level, after 150 million years (Figure 19.5).

Sclater's model enabled us to determine the bottom bathymetry of the ancient seafloor. The paleobathymetry can be "backtracked" to a former time. Armed with this new theory, Wolf Berger analyzed my Leg 3 data in the early 1970s, and he came up with a verification of Hay's postulate: the ridge topography has not changed, but the ocean chemistry has.

Clinging tenaciously to a bad idea, like many of us tend to do, and hoping that I might not be wrong after all, I continued to collect data that seemed to support my prejudice. I proposed that Sclater's model is applicable only if seafloor-spreading is a steady-state phenomenon. Dating of the magnetic lineation had confirmed the general validity of the assumption, but the available data from the first three phases of the deep-sea drilling were not sufficiently accurate to rule out the interpretation of a temporary slowdown, or even a complete halt, of the seafloor-spreading during the middle Miocene.

Inactive mid-ocean ridges should subside, and the red clays could have been deposited on the ridge if the depth of the ridge crest had sunk below the CCD. After this model was first applied by me to interpret the South Atlantic drill results,

19.5. Path of spreading and subsiding seafloor. The path of a piece of seafloor generated at a ridge, moving away and subsiding as well as moving northward at a uniform rate, is shown in this diagram, constructed by Wolf Berger on the basis of the Sclater Principle. Positions *A*, *B*, *C*, *D*, and *E* correspond to ages of 0, 10, 20, 30, and 40 million years, respectively.

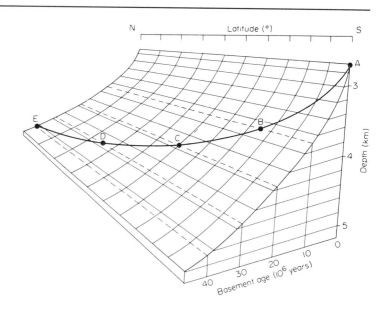

Lucien Leclaire of the Paris Museum of Natural History found a similar increase of $CaCO_3$ dissolution in the middle Miocene sediments of Indian Ocean sites drilled during Leg 25. Leclaire also related a slowdown of seafloor-spreading to red-clay deposition.

When the International Phase of Ocean Drilling started, I responded to invitations and submitted in 1975 a proposal to the JOIDES planning committee to investigate interrelations between paleobathymetry, CCD, and the seafloor-spreading rate in the South Atlantic. The proposal was accepted, and first three and then five South Atlantic cruises were scheduled to study ocean history. After my appointment to the chairmanship of the South Atlantic Working Group, I managed, with the help of Bill Ryan and other colleagues, to work out a drilling program. The question of the changing CCD was one of the themes investigated by the drilling cruises, Legs 71–75 of 1980.

The Leg 73 expedition returned to the region where I first started. A transect of six holes was drilled on the east flank of the Mid-Atlantic Ridge parallel to the 30° S latitude. It was my fifth tour of duty on *Glomar Challenger*, and I served for the third time as a co–chief scientist, sharing my duties with John LaBrecque of Lamont. I was assisted on this cruise by two former students of mine, Judy McKenzie, a veteran of Leg 55, and Helmut Weissert.

Like the accelerator is to nuclear physicists, the drill vessel was becoming indispensable equipment for marine geologists. Experiments were performed to obtain data that could discriminate between alternative hypotheses, the right from the wrong. The two alternative hypotheses on calcite dissolution had been formulated, and now the day of judgment had arrived.

The geological experiment in the South Atlantic was completed in a few weeks during May and June of 1980, and the results were decisive. I was wrong. There was little evidence of a slowdown, and none at all for a halt, of the seafloor-spreading during the middle Miocene. The more intensive dissolution of calcite in the middle Miocene ocean had to be attributed to a rise of CCD.

South Atlantic Ocean History

To acquire a knowledge of past changes in oceans we need samples. The samples of middle Miocene fossils were hard to get, because they are mostly or completely dissolved away in the red clays. Where could we find calcareous oozes of that age in the South Atlantic?

The dissolution of carbonate is everywhere less severe at shallower depths, and the ocean floor is shallower on the crest of a mid-ocean ridge. Sclater's analyses gave us hope; the ridge has stood at the same height during the last 80 million years at least. We could thus expect to sample calcareous Miocene sediments on the crest of the Miocene Mid-Atlantic Ridge. Our efforts were rewarded. Isotope analyses of the South Atlantic samples were to give us a glimpse of its ocean history.

I have told the story, at the beginning of this chapter, that the descent of Antarctic Bottom Waters early in the Oligocene started an epoch of strong current-activity in the Atlantic. The boundary currents seem to have been weaker during the Miocene; some of the cold water-mass was trapped by the Circum-Antarctic Current. With less bottom-water flux from the Antarctic, an interlude of warmer oceans and sluggish bottom-circulations occurred during the early Miocene. Miocene sediments were laid down on an eroded surface upon Eocene or older sediments on Atlantic margins.

A peculiar bottom fauna, dominated by a foraminifera species, *Nuttallides umbonifera*, was the characteristic population living in regions invaded by the Antarctic Bottom Waters. From the record of our cores containing this fauna, Ramil Wright, a shipboard paleontologist on Leg 73, concluded that there was a late Miocene or Pliocene rejuvenation of the bottom-circulation of the South Atlantic. Much Antarctic cold water-mass was produced, and the AABW again spilled beyond the circumpolar circulation. Flowing along the South American margin, the water mass was the strong western boundary current of the Atlantic.

It has been postulated that there should have been a rise of CCD because of the increase in calcite dissolution when the more corrosive bottom water mass was actively circulating. Our South Atlantic data indicate, however, a lowering of CCD at the beginning of the Oligocene, and again at the beginning of the Pliocene, when AABW was active. "Supply-side economics" gives the explanation: Strong bottom-currents brought more nutrients to produce more plankton. With more "snowfall" of calcareous skeletons, the "snow line" (CCD) is depressed to lower elevations.

Paleocean-ography

In the early days of the earth-science revolution, marine geoscientists were either geophysicists or micropaleontologists. They had all the fun, interpreting magnetic anomalies and dating the sediments above the ocean crust, and they were the heroes who upheld the seafloor-spreading theory. Sedimentologists then had the thankless task of describing thousands of meters of pelagic oozes, which all looked alike—a routine that could be done by any high-school graduate. Thinking perhaps that geologists on an oceanographic vessel were as useless as cavalry officers on a battleship, the shipboard staff of Leg 3 included only two sedimentologists. When my colleague went on a hunger strike, I was left alone to handle cores that were piling up at a rate of more than 100 meters per day. This is the reason why many of the Leg 3 cores are still not opened to this day.

Not much was expected, and not much was accomplished. Early in 1970, as the first phase of the Deep Sea Drilling Project was coming to an end, I was asked by the Royal Society to give a talk in London on paleoceanography. I did not know then that such a field existed. Thumbing through the pages of the drilling reports, I could find little of substance. What I finally said could be written down in two sentences: (1) Ancient oceans had a fertile zone of siliceous-plankton production in equatorial waters; (2) Ancient oceans, back to about 100 million years ago at least, had a chemistry similar to that of today. There was not much more to talk about.

The science of paleoceanography got moving in the mid-1970s when a new generation of mass spectrometers, invented by Nick Shackleton, began to analyze samples of very small size. The importance of the technique was not generally appreciated. When I proposed bringing Shackleton to Zürich in 1979 to be a guest professor, my colleagues questioned the merits of his invention. Professors of material sciences preferred to see bigger and bigger samples tested; they saw no need to work with micron-sized specimens. I had to explain that small is beautiful when one needs precision. Isotopic analysis of foraminifers picked out from a 2-m-long interval could give us the average temperature for a quarter or half million years; with glacial and interglacial periods alternating every 0.1 million years, such a sampling interval would give an average of the extremes, but would not detect the changing climate. Now that we could pick out enough foraminifers from a centimeter-sized sample, we could obtain an average ocean temperature over a mere few hundred or thousand years. Shackleton's technique gave us the possibility of seeing very-short-term changes in ancient oceans.

The other indispensable element in paleoceanography is to achieve precision in time resolution. Oozes are stirred by the rotary drill like porridge by a spoon. The ingenuity of Stan Serocki and other DSDP engineers gave us the HPC and undisturbed cores (see chapter 20). Samples could now be separated lamina by lamina, each representing thousands or hundreds of years. Feeding into the new generation of mass spectrometers, we now obtain signals describing ancient oceans that remain in a steady state for at least hundreds of years.

The First Conference of Paleoceanography, held in Zürich in 1973, could be considered the birthday of paleoceanography. As I wrote in the preface to the proceedings volume:

Paleoceanography is a new science of the history of oceans. This book describes what we have found out during the last decade about the past 100 million years of history of the South Atlantic Ocean, thanks largely to drilling by *Glomar Challenger* during five expeditions in 1980. Paleotemperature studies have provided us with a history of climatic variations. Geochemistry of carbon isotopes has provided information on the fertility of planktonic organisms and on the intensity of oceanic over-turns. Correlation of sediment character with changes in oceanic chemistry and fertility has permitted interpretation of the variation of the [CCD] level at which fossil skeletons became dissolved.

Paleoceanographers now had a tool with which to study the bathymetry, the temperature, the current circulation, the chemistry, and the organic production of ancient oceans. At the Zürich conference, I was asked by the plenary body to submit a proposal to the American Geophysical Union to initiate the publication of a specialized journal. This was done, and I was pleased to see volume 5 of *Paleoceanography* as of this writing.

When a final extension of the IPOD drilling was approved in 1981, Legs 81, 85, 86, 89, 90, 93, 94, 95, and 96 were scheduled primarily to study ocean history. The paleoceanographers now had their turn to eat peanuts.

20 The Great Dying

The history of life was thought to evolve by natural selection and the preservation of favored races in the struggle for life; decrepit old species were exterminated by their newly evolved nearest relatives. This Darwinian ideology has been disproved by studies of deep-sea sediments, including those obtained by deep-sea drilling. The last great dying, 65 million years ago, was caused by an environmental catastrophe that resulted from a meteorite hit.

Cuvier's Catastrophism and Lyell's Uniformitarianism

When I was a young man working for Shell, I usually "brown bagged" and joined my friend Alfred Traverse for lunch in his office. Being a paleontologist, he subscribed to the *Journal of Paleontology*. One day he handed me a copy of the latest issue of the journal as I walked in, and said, "Take a look at that article on dinosaurs. Do you think the guy is serious? Or do you think de Laubenfels is a pseudonym of someone who is trying to be funny?"

I had never heard of de Laubenfels either, and I scanned the pages quickly. It was fascinating, but not the usual serious science found in a professional journal. De Laubenfels thought that the dinosaurs were killed by hot air, following the fall of a large meteor. I was particularly amused when he claimed that the crocodiles survived the holocaust because they were mired in the mud, and the turtles did because they held their breath and dived under water to cool off. Now, "hot air" is an American slang expression for "idle talk." Traverse and I suspected that the article was a satire lampooning the many geologists who had made so much "hot air to kill off the dinosaurs." We wondered if a person called de Laubenfels really existed, and if he was indeed serious. After all, the theory of terrestrial catastrophes had long been discredited. Who in his right mind would risk his professional reputation to exhume and dress up the skeletons of Georges Cuvier?

Cuvier, who began his career shortly after the French Revolution, discovered that the mammoths, the elephants of the Ice Age, were anatomically quite different from the elephants of today. Mammoths are, of course, extinct, and as revolution was in the air at the time, Cuvier suspected that those woolly animals were killed off by an "environmental revolution," i.e., by a catastrophe.

Cuvier received the patronage of Napoleon Bonaparte and collected fossils of vertebrate animals from older sedimentary formations all over Europe. He and his co-workers discovered skeletons of primitive mammals in the oldest formations of the Tertiary period, but large reptiles known as dinosaurs were found only in still older formations of the Cretaceous age. The sudden change, from a world inhabited by large dinosaurs to one populated by small mammals, was impressive indeed. Cuvier became more convinced than ever that catastrophes happened again and again in the earth's history.

Cuvier's observations are largely correct, and his conclusions concerning the catastrophic extinction of dinosaurs has also stood the test of time. Unfortunately, the most successful years of Cuvier's career occurred during the Age of Restoration, and his theory was too deeply imbued with the biblical traditions. His idea of catastrophic extinction (catastrophism) was too closely interwoven with the belief in special creation (creationism), so that his theory was easily demolished by Charles Lyell and Charles Darwin, two of his younger contemporaries. Cuvier was, in fact, not a creationist; he postulated migrations to account for the faunal successions of the Paris Basin. Each time the faunas of a region were wiped out, immigrants came in to establish a new faunal dynasty.

Charles Lyell was an Englishman. He has been honored by some historians as the founder of geology, because his principle of uniformitarianism has been the basic tenet, the ruling paradigm, since the middle of the last century. Lyell's uniformitarianism denies arbitrary divine interventions; Lyell believed in the immutability of physical laws. This concept, also known as actualism, had been first applied by a Scotsman of the eighteenth century, James Hutton, to study the earth. Prior to Hutton's time, there was no geology, only "geognosy," and the Bible was the most important source of reference for the naturalists studying the earth.

"Geognosists" believed that all rock formations were precipitated from floods such as the one described in the Bible, even such rocks as granite and basalts. The controversy over the origin of basalt is one of the most famous episodes in the history of geology. The school of "Neptunists," led by the venerated professor at Freiburg an der Oder, Abraham Gottlob Werner, argued for the sedimentary origin of basalts. Nicholas Desmarest, a French amateur, knew better, because he had been to the Auvergne region in central France, where he could trace the basalts to their origin as lava flows from a now extinct volcano.

Hutton assumed the leadership of the school of "Plutonists" when he published in 1775 his book "Theory of the Earth, or an Investigation of the Laws Observable in the Composition, Dissolution and Restoration of Land upon the Globe." The essence of his theory was that one should study the rocks, which have been formed by geological processes in the past, by studying observable processes in the present, because physical laws are immutable. To understand the origin of basalts, one should go to Vesuvius in Italy and observe how basalts are being formed from lavas of that active volcano. Following this suggestion, Leopold von Buch, a favorite pupil of Werner's, went to Vesuvius, and then to the Auvergne. Von Buch observed, and he had to admit that his master was wrong; the basalts have indeed come out of volcanoes. With the victory of the Vulcanists, the Huttonian actualism was more or less established when Lyell started his geological investigations after the turn of the century. Lyell contributed the idea of uniform rates to Hutton's actualism, and called his philosophy uniformitarianism to combat the catastrophism of Cuvier.

Charles Lyell had studied jurisprudence at the university, and he was very persuasive. The idea of uniform rates was a only a belief, a faith, a _Leitbild_, because Lyell had no scientific evidence to verify this assumption. As a brilliant lawyer, Lyell knew he could win his case in the court of public opinion if he attacked the true weakness of his opponent. The weakness of Cuvier's catastrophe theory was its adoption by creationists. Through the work of Charles Darwin, the theory of biological evolution was developed to demolish the postulate of divine creation. Setting up creationism as a straw man, Lyell easily demolished Cuvier and his catastrophism. He was able to persuade Darwin that evolution was a slow, gradual process, and he was able to persuade several generations of geologists that the processes on earth had been proceeding at uniform rates. Catastrophism fell into disrepute and became taboo in the teaching of geology.

In fact, Hutton's actualism did not rule out catastrophes, nor do observations of actual phenomena indicate uniform rates. A few contemporary students of biological evolution, such as Stephen Jay Gould of Harvard University, advocated _punctu-_

ated evolution: nothing much happened for millions of years, and then came a spurt of very rapid changes. A number of geologists are also waking up to the fact that unusual events of catastrophic proportions may have happened more than once during the long history of the earth. One such unusual event that may have greatly influenced the course of biological evolution might indeed be the "hot air" from a fallen meteor of a comet, as suggested by de Laubenfels two decades ago. The cores of *Glomar Challenger* are now being called as witness if a meteor has murdered the dinosaurs.

Cometary Impact?

While writing the first edition of the *Origin of Species*, Charles Darwin was troubled by the sudden extinction at the end of the Cretaceous period. He did not mention dinosaurs, but he did discuss the ammonites. These were floating shellfish, like the nautiloids of the oceans today; the last of the ammonite species became extinct at about the same time as the great reptiles. The change in the biological world was so drastic that the mass extinction of the old and the sudden appearance of the new marked the end of an era, the Mesozoic, and the beginning of a new era, the Cenozoic.

Indoctrinated by Lyell's uniformitarianism, Darwin and many paleontologists after him could not bring themselves to face the evidence of the "great dying." Some specialists emphasized that they found little or no evidence of rapid evolutionary changes during that time. For example, the floras on land, as well as small aquatic animals living in freshwater, did not seem to have undergone many changes from one era to the next. In the marine realm, the animals living on the deep-sea bottom suffered the least, while varying degrees of damage were inflicted upon the faunas of coastal waters. However, when one started to study the fossils of swimming organisms in deep-sea sediments, the changes across the Cretaceous-Tertiary boundary were truly impressive.

The deep-sea sediments, as I mentioned previously, may consist entirely of skeletons of one-celled animals (foraminifera) or plants (nannoplankton). One cubic centimeter may contain many thousands of foraminifera or millions of nannoplankton, belonging to many different species. Therefore, deep-sea sediments provide a statistically valid sampling for determining if the change in ocean life has been evolutionary or catastrophic. As a rule, the changes seem evolutionary. When one studies a sequence of sedimentary strata, or a deep-sea drill core, there is very little difference between samples one meter apart; perhaps one species, constituting less than 2 percent of the total, is present in one sample but not in the other, but the bulk of both samples is made up of the same microfossils, and the samples are indistinguishable except to the most discerning specialist.

The Cretaceous/Tertiary boundary has always been cited as an impressive exception to the rule of slow and gradual changes. In 1960 Hanspeter Luterbacher of Basel and Isabella Premoli-Silva of Milan investigated the foraminifers in rocks exposed near the village of Gubbio in the Northern Apennines. They found a remarkable change in the microfaunas across the Cretaceous-Tertiary boundary. The sedimentary sequence consists of calcareous oozes deposited in an open sea, except for the presence of a centimeter-thick clay at the boundary. Immediately below the boundary clay is a limestone, composed of millions of skeletons of robust foraminifera; many of the disk-shaped shells are larger than half a millimeter in diameter (Figure 20.1). Those fossils belong to the species that lived in the oceans toward the end of the Cretaceous period. Immediately above the same clay is a limestone composed of tiny, thin-shelled foraminiferal skeletons, which are less than a tenth the size of their predecessors. These belong to the species *Globigerina eugubina*, the first Tertiary species, and the ancestor of practically all the foraminifers swimming or floating in the oceans today! The Cretaceous/Tertiary (K/T) boundary was thus placed at the base of the clay, marking the first appearance of Tertiary life forms.

Some twenty years later, Jan Smit of Amsterdam investigated the microfaunas at Caravacas and found evidence for the same catastrophic event. He counted 54 species of planktonic foraminifera in the Cretaceous limestones immediately below a boundary clay. Only one of those species, apparently, survived the catastrophe, and is present in the Tertiary sediments above the clay. The first Tertiary faunas, like those at Gubbio, consist of the immigrant species *Globigerina eugubina* from somewhere outside the disaster area.

When one studies nannofossils, the skeletons of one-celled floating plants, the results are equally impressive. Bill Bramlette of Scripps was one of the first to discover, in 1960, the vast difference between the nannoflora below the K/T boundary clay and that above. Bramlette's finding was to be confirmed later by many nannofossil specialists, studying the Cretaceous-Tertiary boundary on land or analyzing deep-sea drilling cores. Steve Percival, my shipmate on Legs 3 and 42B, studied the nannofloras in such a rock sequence near Caravaca, Spain, and he confirmed Bramlette's observations. He also noted a similar drastic change of nannofloras across the Cretaceous/Tertiary boundary in his studies of the Leg 3 cores.

20.1. Mass extinction 65 million years ago, the terminal Cretaceous extinction, dramatically shown in this study by Luterbacher and Premoli-Silva of the evolutionary history of the foraminifers in sediments near Gubbio. Note the mass extinction of the robust Cretaceous forms, to be replaced, after a barren interval, by a sparse microfauna consisting only of small early Tertiary forms.

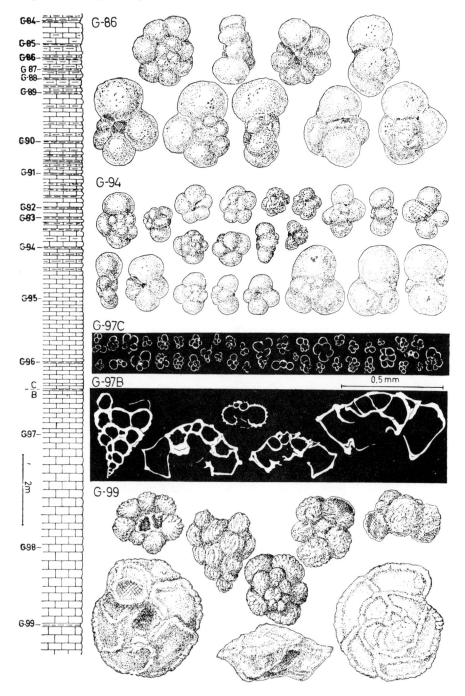

If there had been an unusual event that brought about the great dying, was it the consequence of competition among nearest relatives, or was there an environmental catastrophe? What kind of evidence did we have?

One piece of evidence has already been mentioned. At many locations, as at Gubbio, the sediment at the boundary is a clay—a red, green, or black clay. This clay is almost totally devoid of fossils, but it lies between calcareous oozes or limestones, which contain an abundance of minute skeletons. The clay does not have fossils, because few organisms were living

20.2. Calcite dissolution in boundary clay. Worsley postulated in 1971 that the K/T boundary contains no calcareous fossils because the calcite compensation level in the ocean basin was raised to near the surface at the end of the Cretaceous (model a). Deep-sea drilling has disproved this interpretation. We now believe that the oxygen-minimum zone in the ocean basin expanded at that time, and the coastal waters enriched in carbonic acid caused the dissolution of calcite in the boundary clay (model b). Horizontal scale (distance) is arbitrary.

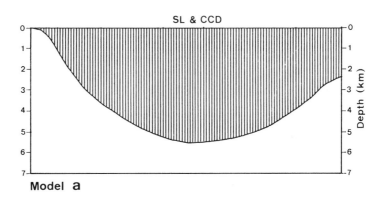

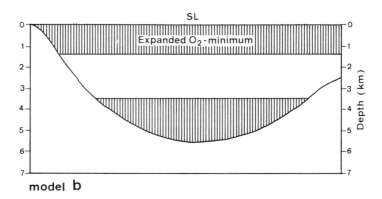

in the oceans then to produce skeletons, and/or because dead skeletons have all been dissolved away.

Since the Cretaceous seas could not have been very deep in regions underlain by continental crust, like at Gubbio or Caravaca, the presence of a clay implies that the calcite compensation depth had risen from the 4,000- or 5,000-meter level to a very shallow depth. This was exactly what Thomas Worsley, a graduate student at the University of Illinois, postulated in 1971; he went so far as to envision an ocean full of corrosive waters right up to the level where photosynthesis of nannoplankton was going on (Figure 20.2). With such a high CCD, calcareous skeletons were produced as fast as they were being dissolved, so that only clays were deposited at the end of the Cretaceous. Worsley's hypothesis was not accepted by all, although most geologists agreed that there must have been reduced production of plankton as well as more corrosive waters in the terminal Cretaceous ocean.

Cyanide Poisoning?

I did not pay much attention to the question of terminal Cretaceous extinction until Nick Shackleton came to Zürich in 1978 to give a talk on the changing temperatures of ancient oceans. Paleobotanists studying plant remains had been arguing for years over whether there were drastic climatic changes at the end of the Cretaceous. They also disagreed as to whether the climate had turned colder or warmer—if there was such a change. Shackleton worked together with Ann Boersma of Lamont, who served as a shipboard paleontologist on Leg 39. Boersma picked out foraminiferal skeletons in samples from a South Atlantic drill site, and Shackleton analyzed the ratio of oxygen isotopes. The results from planktonic foraminifers gave past temperatures (paleo-temperatures) of surface waters, and results from bottom-dwellers gave information on the bottom waters. In his talk, Shackleton presented their newest results in support of the well-known cooling trend during the Cenozoic. However, almost as an afterthought, he mentioned a surprising difference in the oxygen isotopes in the fossils across the Cretaceous-Tertiary boundary. His data could be interpreted as signifying an increase in ocean temperatures of as much as 5 °C in less than a million years' time. In discussions after his talk, he stood by his data, and we agreed that such a change in the heat budget of the oceans could hardly be attributed to terrestrial causes. At that moment, I recalled the incident two decades earlier when Traverse called my attention to the article by de Laubenfels on "hot air." By golly, he was not kidding, and perhaps he was right, too.

I left for a sabbatical term in China shortly after Shackleton's visit, but I was much bothered by that 5 °C rise in ocean temperature, and thought about the puzzle in my idle moments. It occurred to me that the idea of cometary impact was not all that outlandish. Harold Urey, a Nobel Prize winner in chemistry, had also proposed the hot air from the fall of a comet as the cause of the great dying. He first tested the reactions of the scientific community by hiding his unconventional hypothesis in an essay written for the _Saturday Review of Literature_. Only in 1973 did he feel confident enough to publish an article in _Nature_. In that short paper, he made some estimates of the physical consequences of a trillion-ton comet (the size of Halley's comet) hitting the earth. Assuming a collision velocity of 45 km per second, the global air-temperature would rise 190 °C if all the kinetic energy of the impact was absorbed by the atmosphere. Such an air temperature would kill off the dinosaurs, as de Laubenfels had speculated, but it would also kill off everything else on land. Since this was not the case, we can safely conclude that the kinetic energy of the impact was not completely absorbed at once by the atmosphere; much of the energy should have been expended after impact, causing cratering and melting of rocks in the crater. Ocean waters cooling off the crater would have been heated up, and their circulation should have led to increases in ocean temperatures. However, Urey's computation was discouraging; he found that the average ocean temperature would be raised about 0.2 °C, even if all the impact energy was used to heat up the oceans, far less than the 5 °C indicated by Shackleton's data.

Probably it was not hot air after all, but cyanide poisoning. I remembered an article that I had read a few years back on comets in the scientific journal _Nature_. Fred Whipple of Harvard University gave an erudite exposition of his hypothesis that the comet is a heavenly broom, going around the solar system sweeping up cosmic dust. The nucleus of a comet, according to Whipple, is a dirty snowball, made up of cosmic dust and frozen gases. It had been generally assumed that the frozen gases should be ice (H_2O) and dry ice (CO_2), but when the comet Kohoutek came near the earth in 1973, analyses of its optical spectra revealed a great surprise. B. L. Ulrich and E. K. Conklin found cyanides in Kohoutek "at a reasonable abundance." The conclusion was subsequently confirmed by others, and for other comets. I had asked myself when I was reading those reports, What kind of damage would occur to life on earth if a large cyanide-bearing comet should fall into the ocean? Now it seemed that a meteor, perhaps a comet, may indeed have fallen into the ocean at the end of the Cretaceous. Would not the cyanides in such a comet cause mass extinction?

Cyanide poisoning could not be a scientific theory; I could not think of an experiment to prove or disprove this postulate. Besides, the tantalizing fact was that the oxygen-isotope signals suggested temperature change across the boundary. If the impact energy did not impart enough heat, could there be other sources of energy? David Hughes, an astronomer, spoke of possible thermonuclear reactions upon a comet's entry into the atmosphere. My physicist friend, Peter Signer, thought the speculation was outrageous; frictional heating by the atmosphere would never raise the temperature high enough to trigger nuclear fusion. He suggested instead the possibility that a comet falling into the ocean would produce such a pillar of water in the air that the water molecules escaping into the stratosphere would act as a greenhouse. A greenhouse, as every gardener knows, traps the solar radiation bounced off the ground, warming up the inside. The greenhouse effect of the atmosphere, produced by the carbon dioxide in the atmosphere reinforced by additional water molecules, may have caused the increase in temperature observed in the deep-sea record.

After I came back from China, I went to Barbados for a JOIDES meeting. Wolf Berger and Hans Thierstein of Scripps also came, as they were on the same panel. They had just published an article suggesting that the last Cretaceous planktons were killed off by a sudden flooding of freshwater from a giant Arctic lake that had been dammed before the catastrophe. Over cocktails, I told them that their fantasy had no more validity than my cyanide poisoning scenario.

I hesitated to write up my speculations until November 1979, when I came across an article by Victor Clube and Bill Napier of Edinburgh. They presented astronomical arguments for periodic cometary collisions as the cause of biological extinction on earth. I finally gathered up enough courage to sum up my evidence in an article for *Nature*. The essence of my hypothesis was suggesting mass mortality of marine plankton by cyanide poisoning from a fallen comet, and the demise of dinosaurs because of heart failure when the climate became too warm for those giants to maintain viable body temperatures.

Was such a catastrophe likely? Small comets have fallen. On 30 June 1908, the few people living in the thinly populated regions of Central Siberia saw a fireball coming from the east; the heavenly object eventually fell near Tungeska. The heat was so intense that all animal life within a 15-kilometer radius of the site of impact was killed off. Forests were burned. Later, scientists searched for the fallen object and for its impact crater. They did not find any large meteorite, but they found numerous very small rounded objects, commonly less than 1 millimeter in diameter, called cosmic spherules. Some of them were rich in iron; others, potassium. They also did not find one great crater, but many small ones less than 200 meters in diameter. Investigations by Soviet scientists have turned up indications that the fallen object was a comet, and a Czech astronomer, L. Kresak, recently concluded that the comet was a part of the comet Enke, which is now orbiting around the sun as one of the "Apollo objects," a group of small heavenly bodies lurking near the earth.

The Tungeska event was probably a falling comet, but that comet was a small one, only 50 or 100 million tons. A strict follower of Lyellian uniformitarianism would invoke the dogma of uniform rates and deny the possibility of collision with larger comets. Such an argument is obviously false, because we have found many large impact craters on earth, and they must have been produced by collision with extraterrestrial objects much larger than the one that fell at Tungeska. Specialists on impact-cratering processes have made estimates; as one would expect, small craters are far more common than bigger ones. A small crater of about 200 meters in diameter, such as the one at Tungeska, may be created every 350 years, but a large crater of 100 kilometers in diameter is made only once every 14 million years. From such empirical relations, I could see that a Halley-size comet, a trillion-ton heavenly object, should indeed have hit the earth during the last few hundred million years. The fall of such a giant on earth was not only probable, but almost a statistical certainty.

I sent my manuscript to the publisher before Christmas, and a number of copies to my colleagues, including Walter Alvarez of Berkeley. Alvarez returned the favor and sent me a more substantial manuscript that a Berkeley team had prepared for the journal _Science_. I had heard that Alvarez was invoking extraterrestrial causes to explain the mass extinction at the end of the Cretaceous, but I thought he was thinking of a nearby supernova, a hypothesis first proposed by O. H. Schindewolf of Tübingen in 1954. Their manuscript revealed that they were on the same track I was—we both envisioned meteor impact. I suggested a trillion-ton comet, and they postulated a hundred-billion-ton asteroid. While I was speculating on the basis of strange circumstantial evidence, they were painstakingly gathering proof of an extraterrestrial cause of the terminal Cretaceous extinction.

Darkness at Noon?

Walter Alvarez had studied at Princeton University and went on a postdoctoral fellowship to Italy, where he was fascinated by the geology of the Apennines. Later, he went to Lamont and worked with his colleague Bill Lowrie, Isabella Premoli-Silva of Milan, and others to define the magnetostratigraphy (i.e., the chronology of polarity reversal) of an upper Cretaceous and lower Tertiary sequence near Gubbio, which had been studied before by Premoli-Silva and Luterbacher. Now the new team found that the event causing the mass mortality happened at a time when the earth's magnetic field was reversed, during a magnetic-polarity epoch now designated C 29-R.

Alvarez was the geologist on the team. He was now entrusted with the problem of making some estimate of the duration of the magnetostratigraphic epochs on the basis of sedimentation rate of Gubbio rocks. The time intervals represented by the epochs, as Vine and Wilson suggested in 1965, can be correlated with the width of the strips of magnetic lineation on the seafloor. Using this technique, Dennis Kent of Lamont gave an estimate of 0.47 million years for the duration of epoch C 29-R. The event triggering the terminal Cretaceous extinction, a revolution in evolution, must have taken place within a span of time considerably less than half a million years. Alvarez accepted Kent's estimate, but he wanted an independent check. With his physicist father, Luis Alvarez, a Nobel Laureate at Lawrence Berkeley Laboratory, he came up with the idea that the duration of the Gubbio events could be estimated on the basis of the annual flux rate of cosmic dust on earth.

V. M. Goldschmidt had suggested in the 1930s that the amount of platinum-group metals in sediments might indicate whether they contain a substantial component of extraterrestrial material. Such metals include platinum (Pt), iridium (Ir), and osmium (Os), and they are one to ten thousand times more abundant in some meteorites (*chondritic meteorites*) than they are in the earth's crust and mantle. Using a newly developed neutron-activation method of microchemical analysis, J. Barkers and E. Anders in 1968 estimated the annual flux rate of cosmic dust. Ten years later, R. Ganapathy and others found anomalously high concentrations of such metals in tiny spherules buried in red clays of deep oceans; they could thus recognize the extraterrestrial origin of the spherules. Those successes encouraged the Alvarezes to analyze the concentration of platinum-group metals, also known as siderophile elements, in the Gubbio sediments. They hoped that such "measurement might shed light on the time interval represented by the 1 cm thick clay layer that marks the Cretaceous-Tertiary boundary in the Umbrian Apennines."

Assisted by Frank Asaro and Helen Michel, the Berkeley team first determined the iridium concentration in samples from several localities near Gubbio, where the Cretaceous-Tertiary boundary has been most thoroughly studied. The background level of iridium is very steady at about 0.3 parts per billion (ppb) throughout the Cretaceous. This concentration increases by a factor of more than 20, to 6.35 ppb, in the one-centimeter-thick boundary clay. The iridium concentration then falls off rapidly, from rather high values in the first few beds of Tertiary limestones, to the background level in a limestone one meter above the boundary. Later, the Alvarez team analyzed samples from Denmark, and found that the iridium concentration in the boundary clay there rises by a factor of 160 over the background level!

These are sensational results. Does the extraordinary flux of siderophiles indicate an extraterrestrial event at the end of the Cretaceous? What was that extraterrestrial event? In the spring of 1979, the Alvarezes indicated to an audience at the American Geophysical Union that the flux came from a supernova explosion. After studying the plutonium isotopes, the Alvarezes concluded in a November 1979 report that the extraterrestrial material must have come from within the solar system. Thus the event could not have been a supernova, and the most likely candidate was now an "Apollo object" or an asteroid.

Why should an asteroid impact be so deadly? The Berkeley team came up with a "blackout" scenario. The dust thrown out of the impactcrater entered the stratosphere and "turned off" the sunlight on earth for a few years. Plants died because photosynthesis was inhibited, and the dinosaurs starved to death.

The exciting Berkeley discovery was reported by the _Los Angeles Times_ and later by the European edition of the _New York Herald Tribune_. When I came back from the Christmas holidays in 1980, I found a clipping from the _Tribune_ on my desk. My colleague Judy McKenzie thought that I might be interested; they had found the "smoking gun"! Later in January, I got a phone call from the editor of _Nature_. My manuscript was accepted, he told me, but he requested a delay in publication so that my paper and one by Jan Smit and Jan Hertogen on the same subject could be published back to back in the same issue of _Nature_. I agreed to that, and received the galley proof of my article just before I left in April to join the _Challenger_ at Santos, Brazil.

The primary goal of Leg 73 was to collect samples to study the ocean history of the South Atlantic. Dating of events in the history of the earth, as for any history, is a difficult undertaking. As I explained in chapter 4, Allan Cox, Ian McDougall, and others first established the history of the reversal of the earth's magnetic field for the last few million years. Vine and Wilson suggested that the duration of each magnetic-polarity epoch should be correlated with the width of the lineated magnetic anomalies under the seafloor, so that the chronology could be extended back to 150 million years or so on the basis of seafloor anomalies. To establish a reliable chronology of the reversal of the earth's magnetic field, one needs to determine the magnetic polarity of a continuous sedimentary sequence. The deep-sea drill cores were not suitable for such an investigation. First of all, the coring technique was such that the recovery of the core interval is rarely 100 percent; a continuous record is rarely available. Worse still, the heavy weight of the rotary drill often churned up the oozes and disturbed the arrangement of magnetic particles in the sediments (see chapter 13). Such disturbed samples gave no reliable signals.

For many years, we had to rely on the piston-corers first developed by Kullenberg to provide the samples for studying remnant magnetism in sediments. To obtain older and older records, longer and longer piston-corers had to be developed. However, there is a limit to the depth that a piston-corer driven by weight can penetrate in the ocean bottom; a twenty-meter sedimentary core was about the upper limit. With such long cores, deep-sea research has managed to extend the chronology of the reversal of the earth's magnetic field back to about 15 million years.

The engineers of the Deep Sea Drilling Project, whom I maligned so much in a previous chapter, did come up with a new instrument, the *hydraulic piston-corer*. *Glomar Challenger* would drill a hole, and the hydraulic piston-corer (HPC) would be lowered to the hole and would be driven by water pressure into the sediments beneath the bottom of the hole. Using this technique, continuous coring could be assured, because the recovery of piston-corers was almost always 100 percent. Also, the sediment samples were undisturbed.

The new gadget was first tried successfully in 1979 during the Leg 64 cruise to the Gulf of California. A short cruise, Leg 68, carried out further tests at two sites on opposite sides of the Isthmus of Panama. When John LaBreque and I went to Brazil to join the *Glomar Challenger* in the spring of 1980, we were told that the magic tool was at our disposal. We had a new mission, which had not been foreseen during the planning—we had now been given the opportunity to extend the history of the magnetic field reversals farther back in geological time. For this task, we were assisted by three experts, Nicolas Petersen of Munich, Peter Tucker of Edinburgh, and Lisa Tauxe of Lamont. They worked day and night, on two 12-hour shifts, spinning the magnetometer, an instrument that determines the magnetic polarity of sediments. Meanwhile, four paleontologists, Steve Percival from Mobil Oil in Dallas, with whom I worked on Legs 3 and 42B, Ramil Wright of Florida State University, a veteran with me on Leg 42A, Dick Poore of Menlo Park, California, and Andy Gombos from Exxon Oil, Houston, determined the ages of sediments by studying the microfossils. Before the cruise ended at Cape Town, our team had succeeded in extending the chronology of the reversal of the earth's magnetic field back to 75 million years before present. We had also succeeded in a three-way correlation of the history of polarity reversal, the width of linear magnetic anomalies under the seafloor, and the paleontological zonation of sedimentary sequences.

The cruise had been planned long before I became interested in the problem of the terminal Cretaceous catastrophe. Just the same, I secretly hoped as I was joining the ship that I could obtain some samples across the Cretaceous-Tertiary boundary at one South Atlantic site. At the last planning session in La Jolla in February 1980, I had heard reports from Dieter Futterer of Kiel, who had been on a 1979 cruise to the South Atlantic to make preparatory surveys and to obtain piston cores near the proposed drill sites. One of the cores from the Cape Basin, west of Cape Town, produced a surprise. In a water depth of about 5,000 m, a 10-m core was found to contain sediments as old as the Paleocene, or about 60 million years of age. That would be an excellent place to investigate the terminal Cretaceous event. Unfortunately, the information came too late; the priorities had already been assigned.

Some years ago, Isabella Premoli-Silva had sent us a proposal to study the Cretaceous-Tertiary boundary at a site in the Cape Basin. Preoccupied with our interest in the paleoceanography of the Mid-Atlantic Ridge area, we on the JOIDES Ocean History Panel gave only a third priority to the Cape Basin site. That was a polite form of rejection, because second-priority sites were rarely drilled, and third-priority ones never were. Out of courtesy to Premoli-Silva, we did send the proposal to the JOIDES Safety Panel, and it was cleared. After the *Glomar Challenger* left Santos, Brazil, we had a long cruise across the western half of the South Atlantic. We scheduled daily meetings of the scientific staff to discuss the drilling program and to exchange ideas. During one of those morning sessions I brought up the question of Premoli-Silva's third-priority site again. My colleagues had read a copy of the galley proof of my article on cometary impact. The paleontological record at various boundary localities had indicated catastrophic extinction within a very short period of time. But how short? And did the "great dying" take place in the South Atlantic at exactly the same time it did elsewhere? We had the tools to answer these questions, but we needed samples. My persuasiveness was effective, and we all agreed that we should drill in the Cape Basin, if we could accomplish our assigned tasks on the Mid-Atlantic Ridge ahead of schedule.

Having always been a little on the pessimistic side in my planning, I had made a schedule that left room for maneuvering. Also the weather had cooperated, and we were ahead of schedule throughout the cruise. At 10:30 A.M. on 19 May we were drilling near the bottom of Hole 523. Our goal was to reach the basalt crust under the seafloor, and we estimated that we had drilled to a depth of some 15 or 20 meters above the basalt. At that critical juncture, the drilling superintendent came to tell us that the barrel of the hydraulic piston-corer had broken, down in the hole. We had to take out (to "fish" out) the broken pipe. Much time was wasted, without any success. We then tried to drill ahead and bypass the broken pipe. We were able to do that, and two more cores were taken. We were now practically down to our goal. Again, the barrel got stuck down the hole. That was the end, and we had to pull up the drill pipe.

We had a staff meeting. Less than a week was still available for drilling. We could try to drill another hole here down to basalt, as originally scheduled. We could drill another site, down to basalt, as called for in a contingency plan. Or we could go to the Cape Basin. All of us decided that we should wire the headquarters at La Jolla to drill the Cape Basin, and the permission was granted while we were steaming eastward.

Site 524 was located at 3 ° E and 30° S, at 4,000-m water depth. I was writing the seventeenth chapter of this book when we began drilling. We had bad luck at the start, but managed to achieve our goal after several moments of suspense. I thought it might be amusing reading to reprint a few pages of our operations report, which I filed after the hole was drilled:

20: The Great

Dying

At 2048Z May 21, the vessel was put under automatic control for dynamic positioning over Site 524. At 2054Z, the crew started to run the drill pipes down. Toward the morning of May 22, winds reached 40 miles per hour and were increased to 45 miles per hour 10 minutes later when the vessel started 6° rolling. At 0600Z operation on the rig floor was suspended to wait for the weather to improve. At 1015Z the weather improved, and the drilling crew resumed operations.

At 1200Z, May 22, the pipes were down to the seafloor and the crew was ready to spud in. But wait! An engineering test to monitor the heave of the ship motion had to be conducted, even though a more sensitive monitoring had been registered by "Green Flash" Wright, who suffered seasickness. The engineering package was pumped down. It was apparently tightly embraced by the drill bit and could not free itelf. The "fishing" expeditions to pull out the package were all in vain, as was expected. At 1730Z, May 22, the crew prepared to pull the pipes out of the hole.

A prize, also known as the "Alex" Alexander Tripping Trophy, was awarded to the innocent testing engineer on board by the loving roughnecks, who had the pleasure of making three "trips" of assembling and disassembling the drill strings, and of taking a cold shower on the rig floor every time a pipe-segment was unscrewed. The featherbrained designers had wisely stayed home, having the foresight to avoid cries of "bonzai" by the frustrated co-chiefs.

The morale of the scientific staff reached an all-time low in the afternoon of May 22nd, when numerous scientists of both genders went on a sit-down strike. Poore had the wisdom to organize a "study session" of "art magazines" and "art films" between 1350Z and 1730Z. A solemn occasion, and the epitome of concentration was recorded for posterity by our Scripps representative.

At 1147Z, May 23, almost exactly 24 hours after the round-trip "pleasure trip" by the engineering package, the drill bit again felt bottom and was ready to be spudded in. The co-chief on duty was at a loss for words when the first core came up and was found empty, because no amount of persuasion could dissuade our good-natured operations manager from his generous gesture of awarding us with a barrel of water during the baptismal journey of the core barrel to a newborn hole. He made us such a present at every site!

Meanwhile, the captain of the vessel showed up in the core lab. After we had lightheartedly given away the precious hours of calm sea for "fishing" and "pleasure trips," Neptune was to manifest his waves of anger. We were threatened by an invading front due to arrive at 0600Z, May 24, and the temptation was to bury the bottom-hole assembly before the storm should arrive. The co-chief on duty was glad to be relieved, so as to bury his head under a bed cover when the wrath of Neptune should arrive.

It never came. After having made the threatening gesture, the front was ordered to turn right at a position 60 miles south of the vessel.

With sunshine and calm sea, and good coring recovery, the scientific staff slowly recovered their wits and received a further boost in morale when a telegram came from the "beach" [La Jolla] authorizing the use of rented cars after their arrival at Cape Town on a Calvinistic Sunday; the generosity was touching. Throughout the day of May 24th cores came up regularly, to be dated regularly by N.N.P. [Nothing-New Percival] as NP 3, always three zones above the much-sought-after Tertiary-Cretaceous boundary. The core recovery was good, but we were kept in suspense whether the treasured boundary would be found in the next empty barrel.

At 0105Z, May 25, Core 19 was hauled up. The co-chief off duty hurried to the paleo lab with a lump of red mud. N.N.P was reluctant to disappoint, and pronounced a Cretaceous age for a Tertiary sediment as a substitute for tranquilizer pills to put the off-duty co-chief in bed.

The boundary was eventually found in Core 20. Although no one could find traces of a comet that should have fallen 65 million years ago, still the presence of nearly sterile sediments induced a great excitement and resulted in a telegraphic message to DSDP requesting Baslerwasser [a strong alcoholic beverage] or similar solvent to bypass the stringent doctrines of the Dutch Reformed Church.

When it became obvious to the co-chiefs that the prominent reflector of seismic echoes at this site might not be the much-dreaded chert that could seal off a rich hydrocarbon accumulation, considerable telegraphic exchanges took place between the vessel and the beach concerning the acceptability of drilling beyond the 100-meter limit below the reflector. It became a moot point, however, when the prominent reflector was encountered at about 280 meters below seafloor (as predicted successfully for the first time during the cruise) and was found to be a basalt rock. There was neither a smell of hydrocarbon, nor shiptime for scientists to venture into that "risky realm" below the hard rock. After 24 hours of basement drilling was carried out, the drill bit was "conveniently" worn out. Hole 524 was terminated at 348.5 m—68.5 m below the prominent reflector, or some 5,000 m above the Moho. Petrologist Carman, alarmed by the horrendous prospect of getting his 100 meters of basalt requisitioned by the JOIDES planning committee, and Yeoman Collins, who was to type his monographic treatise, were both more than pleasantly surprised when the last barrel brought up "dirty green sands" to be described by our super-efficient team of sedimentologists.

Having terminated drilling and coring for Hole 524 at 2119A, May 26, the drill string was pulled clear of the mudline at 0000Z, May 27.

We got the Cretaceous-Tertiary contact, and we got good recovery, cores with very little disturbance. Our team of experts could tell us, even before we finished drilling the site, that a mass extinction had occurred here, and that the "great dying" took place during the epoch C 29-R, 65 million years ago, exactly the same time as that at Gubbio, Italy!

Environmental Crisis

What happened at the end of the Cretaceous? Did the global temperature rise? By heat of impact or through greenhouse warming? Was there a mass mortality of the ocean? If so, why should there be a mass extinction? Could not all the plankton bloom again to repopulate the ocean a few months or a few years after the catastrophe? What were the changes triggered by the event? How long a period did the environmental degradations last? A few years, a few thousand years, or a few million years? To answer these questions, we needed deep-sea cores, because sediments on land have been reacting with groundwater to produce changes that modified the fossil record and made the signals of isotope geochemistry difficult to interpret. The ocean sediments are rarely altered, and those we obtained at Site 524 are particularly suitable, because they were accumulated very rapidly, at a rate of 3 cm per thousand years. By taking a sample every centimeter, we could decipher the changes in time intervals of 300 years.

We took our samples home, and set up a team of interdisciplinary investigators. Already on *Glomar Challenger*, the shipboard sedimentologists Ken Picciuto (of Scripps), Anne Marie Karpov (of Strasbourg), Helmut Weissert, and Judy McKenzie had found the boundary clay. Mr. Q. X. He, a guest investigator at our institute from the People's Republic of China, analyzed the calcium-carbonate content of the samples across the boundary, and found that the clay contains less than 5% $CaCO_3$.

A sample of the boundary clay was sent to Urs Krayenbühl at the University of Bern, who has the equipment to make trace-metal studies. We were not surprised when Krayenbühl reported finding an iridium anomaly—the iridium concentration of the clay is 3.6 ppb, or about 10 times the background.

The foraminiferal fossils from our cores have been much dissolved, but the nannofossils are well preserved. Steve Percival and Katharina Perch-Nielsen made independent studies and came up with identical results. For million of years toward the end of the Cretaceous, the nannofloras remained little changed. Suddenly, at the level of the boundary clay, old species were decimated, and four new species first appeared. Those species had actually existed before, but they were so rare that they could hardly be found in a normal-sized sample. Those "new" species stood out after the catastrophe, because the number of individuals of other species had been drastically reduced. The nannoplankton suffered a mass mortality within a very short time, probably less than a few hundred years. However, both paleontologists believed that the extinction of the various old species took place successively in an interval of time not less than 30,000 years; there was a decline in the population of the old species and an increase in the new during this span of time.

He Qixiang, assisted by Hedy Oberhänsli and Judy McKenzie, determined the ratios of the oxygen and carbon isotopes of the nannofossils. The paleo-temperature record shows an initial decline of about 5 degrees, followed by a subsequent increase of about 10 degrees; the maximum was reached 40,000 years after the boundary event. The flash-heating during the entry of the comet could not have left a record in the sediments. The initial cooling probably registered the insulation effect of the ejecta (from the crater) in the stratosphere, which cut down the influx of incoming solar radiation. This impact-winter scenario had been proposed by Cesare Emiliani and Gene Shoemaker, also in a 1980 paper. The climatic warming could be attributed to the greenhouse effect.

Good samples from the Cretaceous/Tertiary boundary were eventually obtained by other DSDP cruises. Studies of samples from Legs 74 and 86 and earlier legs and from sections on land verified, as a rule, earlier findings: the mass extinction took place with the C 29-R interval. The boundary deposit contains less calcium carbonate than overlying and underlying sediments. An iridium anomaly could be identified. Only the oxygen-isotope results could not be reproduced. The erratic results could be interpreted as a manifestation of an unstable climate—global temperatures oscillated so rapidly during the environmental crisis triggered by the terminal Cretaceous catastrophe that no simple trend could be revealed by the random sampling of the boundary interval.

We also found a remarkable carbon-isotope anomaly. Unlike the oxygen signal, the carbon-isotope signal is very consistent (Figure 20.3). What is the significance of that?

Strangelove

Ocean

Carbon has three common isotopes. Carbon-14 is a radioactive isotope, which decays spontaneously. The abundance of carbon-14 in a carbon-bearing substance, such as a piece of wood, can be used to determine the age of that substance, and the method is now a standard technique in archaeology and in the study of very young sediments. The other two isotopes, carbon-13 and carbon-12, will not decay; they are stable isotopes, like oxygen-18 and oxygen-16. Commonly, carbon-13 atoms constitute a little over one percent of the total carbon atoms; the rest are carbon-12. A standard is chosen, and all other analyses are compared with that standard, and expressed in terms of parts per thousand deviations from the standard (δ^{13}C ‰). For example, say a standard has 1,110 carbon-13 atoms per 100,000 carbon atoms; then a sample with a δ^{13}C of plus 1‰ would have 1,111 carbon-13 atoms and a sample with a δ^{13}C of minus 1‰ would have 1,109 carbon-13 atoms in a total of 100,000 carbon atoms.

Different carbon-bearing compounds have different proportions of carbon-13 atoms as compared with carbon-12. Organic substances, such as wood, or protoplasm of living organisms, have much fewer carbon-13 atoms, whereas the carbonate skeletons of foraminifera have more or less the same proportion of carbon-13 atoms as the standard, because the standard itself is the skeleton of a marine fossil.

J. C. Brennecke and T. F. Anderson of the University of Illinois, Urbana, analyzed deep-sea drilling cores across the Cretaceous/Tertiary boundary. The samples consist almost exclusively of nannoplankton, indicative of conditions in surface waters of the oceans. They were the first to note, in 1977, that the carbon-isotope composition changes drastically at the

20.3. Geochemical anomalies across the Cretaceous-Tertiary boundary, which indicate unusual environmental changes when a biotic crisis occurred at the end of the Cretaceous. The carbon-isotope anomaly indicates that the reproduction of ocean plankton was largely suppressed during the time of crisis. (See text for details.)

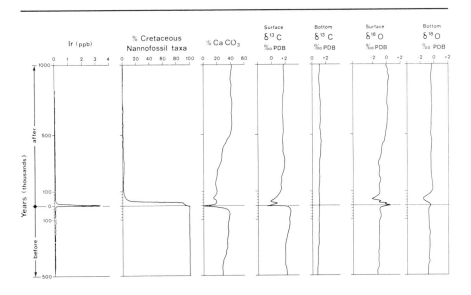

boundary. A negative anomaly (i.e., a decrease) of one to three parts per thousand carbon-13 atoms was reported.

When Judy McKenzie and I carried back our precious samples from the South Atlantic, we gave them to He Qixiang for isotope analyses. He found two anomalies—a sharp perturbation right at the boundary, and a systematic decrease of more than three parts per thousand carbon-13 atoms in the nannofossils during the 40,000 years after the boundary event (Figure 20.3).

A negative anomaly means an excess of carbon-12 atoms in ocean water. Where did this excess come from? When the results were brought to me, my first reaction was to postulate excessive supply of carbon-12. Where could that excess come from? Could it be extraterrestrial? No, it was not possible; even a trillion-ton comet would not bring in enough carbon-12 atoms to significantly alter the carbon-isotope composition of ocean water. In our cruise report and in our article published in _Science_, we had to assume an excessive supply from land, but could suggest no experiment to test the hypothesis.

In the spring of 1983, Werner Stumm, a colleague at ETH, called me. He had heard that Wally Broecker was taking a sabbatical leave in Germany and suggested that we invite him down to give a talk. At that time, there had been a most exciting scientific discovery based on the analysis of carbon dioxide in fossil air trapped in gas bubbles of ancient ice: Hans Oeschger of Bern had found that the atmosphere during the last glacial stage contained only about three fourths as much carbon dioxide as the air does now. Broecker had an explanation, Stumm told me.

Broecker is a brilliant isotope-geochemist, and is known for his pioneering work in chemical oceanography, using carbon-14 as a tracer. He found that ocean water descended from the surface would circulate the globe as bottom-currents and come back in less than 2000 years' time. When he came to Zürich, he was getting involved in working with carbon-13.

Aquatic organisms have soft parts and hard skeletons. The skeletons, composed of calcium carbonate, will have a carbon-12 composition more or less the same as that of the calcite precipitated inorganically from the ocean under equilibrium conditions. Cells of soft parts, however, have a preference for the lighter carbon atoms, C-12. Judy McKenzie studied this phenomenon in a Swiss lake and found a depletion of C-12 atoms in the dissolved carbonate ions during the summer months of plankton blooms. This process is called "isotope fractionation." In winter months, when production is almost nil, there is no fractionation. Carbon-12 atoms are then not tied up in the cells of living organisms, and the dissolved carbonate in the lake thus has a seeming excess of C-12. Marine geochemists use the same principle to explain the changing carbon-isotope composition of seawater. At times of more fertile plankton production, more C-12 atoms are tied up in the living cells. At times of lower fertility, fewer C-12 atoms are tied up, or more lighter carbonate is retained in seawater. Such changes during the past are monitored by the fossils' skeletons.

Living organisms not only prefer C-12 atoms, they also take carbon dioxide from the atmosphere for their photosynthesis. At times of higher production, more carbon dioxide is consumed, and thus the atmospheric carbon dioxide is depleted compared with times of lower production. It is a well-known fact that the glacial stages were times of higher production. Therefore, as Broecker explained to us, the carbon dioxide was depleted during those times.

The hypothesis can be verified by studying the carbon-isotope composition of plankton skeletons, Broecker added. The prediction is that the present ocean, being less fertile, should have an excess of C-12 atoms in surface waters compared with to the ocean of the glacial stages, because of phytoplankton's preferential use of that isotope. On the ocean bottom there are few living organisms to fractionate C-12, and the isotope composition of the water there should have been the same during the glacial stage as it is now; there should always be an excess of C-12 in bottom waters, during both glacial and interglacial stages. We can express this phenomenon another way: the difference in the carbon-isotope composition between the skeletons of plankton and those of bottom dwellers was greater during the glacial stages of greater fertility than during interglacial stages of lower fertility.

Broecker's explanation intrigued me. If a reduced plankton productivity should decrease the difference in carbon-isotope composition between plankton and benthic skeletons, I wondered, what would happen if there had been no productivity at all? So I asked.

"Oh, you are asking me about the Strangelove effect," Broecker answered in high spirits. "An ocean without plankton would have no gradient of carbon isotopes. There would be no isotope fractionation, and the carbon-isotope composition would be the same from sea surface to sea bottom. The ocean would be a Strangelove ocean!"

I had never seen the movie *Dr. Strangelove*, but I understand that the fictional Dr. Strangelove wanted to wipe out all life on earth with a nuclear holocaust. Half in jest, Broecker had picked a very picturesque term.

Judy McKenzie and He Qixiang asked Ramil Wright to send them some samples of benthic foraminifers, sorted out of the oozes of Site 514. Their analysis showed that there was little difference between the isotopic compositions of the planktic and benthic fossils of the boundary clay. There was a Strange-love ocean at the end of the Cretaceous.

We now know, of course, that the earth's history is not all that simple. Not all the signals could be explained by the "Strangelove effect." Under certain circumstances, the phyto-plankton production could be very fertile, but the surface water still has a negative carbon-isotope anomaly, because the tiny picoplankton cells are decomposed by bacteria to release the excess C-12 atoms. Such conditions are, however, abnormal, and are thus also a manifestation of a severe environmental crisis.

Ten years after my first rush of enthusiasm, I have long given up the naive postulate of cyanide poisoning; cyanides in com-ets, if there had been any, would have been dissociated and thus detoxified in the mushroom cloud after the impact. It is useless to argue whether dinosaurs died of hunger or of the cold; we shall never know. There is not much sense in arguing whether the impact was caused by a comet or an asteroid, or whether there was no impact at all but only explosive volcanism. The important fact is that an event triggered an environmental catas-trophe, causing the long-term destruction of the habitats of many animal and plant species. Mass extinction was a conse-quence of environmental devastations.

Geology, Evolution, Ideology, and Darwinism

The two great scientific theories of our civilization are the Cop-ernican and the Darwinian theories; they changed man's view of his place in the universe. Karl Popper warned us, however, that "a scientific theory may become an intellectual fashion, a substitute for religion, and entrenched ideology." The Darwin-ian revolution has indeed become an entrenched ideology. The notion of the struggle for existence and survival of the fittest has bred fear in our psyche, to such an extent that our legislators vote happily for a "defense budget" that accounts for a great share of our gross national product.

Homo sapiens is motivated by fear. What are we afraid of?

We were afraid of the supernatural. Religion was a defense against that, and tax money was spent to build temples and ca-thedrals. Then came the Age of Enlightenment; we no longer worried about the supernatural, and transferred our angst to a fear of other human beings, or other groups of human beings. In collective defense against our enemies, we have had nationalism, imperialism, colonialism, racism, capitalism, Marxism, socialism, and National Socialism, and now we have national liberation movements and terrorism. The nineteenth-century gentleman Charles Darwin invented none of these. He was horrified when Karl Marx wanted to dedicate _Das Kapital_ to him. Nor would Darwin ever have imagined that his theory was to be misappropriated by a German biologist, Ernst Haeckel, as the "scientific" basis for an evil ideology that ended up with an Adolf Hitler. Yet the quick acceptance of Darwin's theory, as historians have pointed out, had much to do with the fact that his idea of "natural selection" was politically palatable to the masses because of its ideological impact. We forget now that Darwin's theory has long lost much of its scientific basis. Neo-Darwinists have brought the theory of evolution to new heights, but what has survived of Darwin's theory in public memory is the ideology of natural selection, or the preservation of favored races in the struggle for life—the subtitle of Darwin's masterpiece.

The origin of Darwinism can be traced to two favorite postulates of Darwin's time: Malthus's theory of population, and Lyell's substantive uniformitarianism. They have been accepted as axioms to interpret the history of life on earth.

To understand the growth of ideas on evolution, we should go back to Carl Linnaeus, the founder of modern taxonomy. Linnaeus was still imbued with the Christian dogma of creationism: God created a fixed number of species, and the species are immutable. Lamarck made the first break from this tradition when he proposed lineage evolution; species are not immutable. He continued, however, to believe in the dogma that "God does not take away what he has given"—the number of species remains fixed, and there was no extinction. Lamarck was thus not only an evolutionist, but also a creationist.

Cuvier investigated the faunas of the Paris Basin. He recognized four distinct fauna assemblages in succession, and he postulated catastrophic extinctions. Cuvier did not speculate on evolution, but he was not a creationist, at least not in his scientific writings; he postulated migrations to explain the sudden appearances of new faunas in France. Darwin completed the biological revolution when both of Linnaeus's basic postulates were overturned. Darwin accepted the paleontological record that species not only evolved, their number also proliferated, and he accepted extinction.

This fear of unlimited proliferation was expressed in the following passage from Darwin's _Origin of Species_:

There is no exception to the rule that every organic being naturally increases at so high a rate, that if not destroyed the earth would soon be covered by the progeny of a single pair. . . . In looking at Nature, it is most necessary to keep the foregoing considerations in mind. . . . Lighten any check, mitigate the destruction ever so little, and the number of the species will almost instantaneously increase to any amount. The face of Nature may be compared to a yielding surface, with ten thousand sharp wedges packed close together and driven inwards by incessant blows, sometimes one wedge being struck, and then another with greater force.

Here and in later passages, Darwin left no doubt that he applied the Malthusian principle: the growth of the number of individuals and of species has been checked by means of intraspecific and interspecific competitions. Individuals have to die from enemies, and species have to become extinct.

Darwin accepted extinction as a historical fact, but he did not accept Cuvier's theory of catastrophe. Darwin dismissed the problem when he wrote that "the extinction of old forms is the almost inevitable consequence of the production of new forms," and that "each new variety or species, during the progress of its formation, will generally press hardest on its nearest kindred, and tend to exterminate them. We see the same process of extermination . . . through the selection of improved forms by man." Extermination was invented by Darwin, not by Hitler! It is, in fact, a favorite expression of Darwin's, and he used it at least 11 times in the first edition of his *Origin*. What is Darwin's evidence for extermination? He gave none, except the Malthusian argument. His conclusion has, however, been taken for granted. In 1984, a hundred years after the publication of Darwin's manifesto, John Maynard-Smith expressed in a *Nature* article his surprise that

the paleontologists read the fossil record differently. The dinosaurs, they believe, became extinct for reasons that had little to do with competition from the mammals. Only subsequently did the mammals, which had been around for as long as the dinosaurs, radiate to fill the empty space. The same general pattern, they think, has held for other major taxonomic replacements.

This well-known geneticist, whose work has been central to the development of the modern synthesis of Darwinian theory with genetic mechanism, had thought otherwise; he had expected a major cause of extinction to be competition from other taxa.

Reviewing the evidence concerning contemporary extinctions, expert ecologists and paleontologists at a Dahlem Conference concluded that competition does not appear to play a major role in contemporary extinctions. I would go a step further and challenge anyone to come up with a single recorded instance in which a species was exterminated by someone other than *Homo sapiens*!

Darwin's theory of natural selection was "grounded on the belief that each new variety, and that ultimately each new species, is produced and maintained by having some advantage over those with which it comes into competition, and the consequent extinction of less-favoured forms almost inevitably follows."

Darwin, charmed by the Malthusian postulate, believed "that the production of new forms has caused the extinction of about the same number of old forms," and that "the improved and modified descendants of a species will in general cause the extermination of the parent-species; and if many new forms have been developed from any one species, the nearest allies of that species, i.e., the species of the same genus, will be liable to extermination."

Darwin gave no scientific facts to support the belief that is the foundation of his theory of natural selection. Is there any more substance to support the dogma of "the preservation of favored races in the struggle for life" other than the racism of nineteenth-century England?

Modern evolutionary biologists recognized the importance of physical environment in natural selection, but that is not Darwinism. Tony Hallam pointed out in 1983 that Darwin overemphasized the role of biotic interaction when he wrote:

As species are produced and exterminated by slowly acting causes and not by miraculous acts of creation, and as the most important of all causes of organic change is one which is almost independent of altered and perhaps suddenly altered physical condition, namely the mutual relation of organism to organism—the improvement of one organism entailing the improvement or extermination of others.

If one still has any doubt on that score, he can read Darwin's statement on page 325 of the first edition of *Origin of Species*:

It is, indeed, quite futile to look to changes of currents, climate, or other physical conditions, as the cause of these great mutations in the forms of life throughout the world.

Why did Darwin make the mistake of relying upon biotic interactions? He was perhaps influenced by an urge to present a theory different from that Lamarck, who emphasized environmental influence. At any rate, he was misled by the philosophy of his geologist friend, Charles Lyell, who claimed that there has been no environmental catastrophes in the history of the earth. If the earth's environment has remained a steady state, it could not exert selection pressure no matter how much geological time has gone by.

Why did Lyell come to his false assumption? Perhaps he was confused by the word "uniform" in Hutton's theory of the earth. Hutton talked of uniformity of physical laws, but Lyell made the unwarranted assumption of uniformity of rates. When we had no physical method for measuring geological time, the Occam's razor was a justifiable approach. Since the discovery of radiometric dating in the first part of this century, we know better. Defined as grossly accelerated rates of fluxes of materials and/or energy across boundaries of atmospheres, hydrospheres, and lithospheres, catastrophes do exist. I have documented several such instances in my Dahlem report on environmental changes in times of biotic crisis.

How often do natural catastrophes occur? I proposed in 1983 the concept of actualistic catastrophism: anything that can happen will happen, given enough time. What is enough time? We have witnessed some of the larger earthquakes in our lifetime, and the biggest possible earthquake has probably already happened in historical time. Lyell was right when he postulated that some geological phenomena could be interpreted on the basis of processes observable by man. On the other hand, the biggest meteorite impact takes place only every 100 or 200 million years. It is thus not surprising when we learn from David Raup that the "waiting time" for successive mass extinctions is on the order of 10^8 years. It epitomized the arrogance of nineteenth-century England when Lyell insisted that the history of the earth has to be explained in terms of phenomena observable by _Homo sapiens_; the age of the species is two orders of magnitude less than the "waiting time" needed to observe a natural event that could trigger a mass extinction.

Nevertheless, we are witnessing a process that is triggering mass extinction. It is not a natural event, but the anthropogenic environmental degradation. We are facing problems of ozone destruction, acid rain, deforestation, ocean pollution, greenhouse warming by increased carbon dioxide—all of which were culprits in the terminal Cretaceous extinction. The current rate of species extinction is comparable to, if not in excess of, the rate of terminal Cretaceous events.

What should we be afraid of? Who is our enemy? Our enemy is the disastrous disregard of the manmade catastrophe that threatens to exterminate many species on earth, including, eventually, _Homo sapiens_. Our neighbors are not our enemy, nor are we likely to be exterminated by our nearest relatives. This evil aspect of Darwin's theory has, fortunately, been disproved by the geological record and should not be permitted to survive.

On this philosophical note, I conclude my personal history of the Deep Sea Drilling Project, 1968–1983.

Epilogue

The deep-sea drilling cruises ended on 8 November 1983 when the *Challenger* returned to Mobile, Alabama, after Leg 96 drilling of the Gulf of Mexico. The drill vessel was dismantled for scrap iron. JOIDES survived.

There was a flurry of activity to explore the possibilities of future scientific drilling. When I met Mel Peterson in Paris during the 1980 International Geological Congress, he told me that the plan was for the United States scientists to work with the American petroleum industry to drill the North American continental margins; JOI Inc., representing the U.S. JOIDES institutions, were actively promoting the Ocean Margin Drilling Program, OMDP. I had dinner then with Yves Lancelot, Lucien Montadert, and other French friends, and we Europeans were very concerned to see this renewal of American isolationism.

The OMD plan was also actively pursued by the NSF, and a Division of Ocean Drilling was created. The NSF announced early in 1981 that the OMD had formally begun on 1 October 1980. The NSF had issued a five-volume report dealing with the preliminary scientific and technical planning that had gone into development of that program. Scientists at large were surprised by the development; few of them knew what was going on.

There was parallel planning by JOIDES. The U.S. National Science Board (NSB) passed a resolution to extend the *Challenger* drilling for two years, until the end of fiscal year 1983. It would cost 26 million dollars, of which 10 million were to be provided by IPOD partner countries. JOIDES had also explored other options for possible continuation of ocean drilling. During its meeting in Atlanta, 18–19 November 1980, the JOIDES executive committee proposed that there should be a Conference on Scientific Ocean Drilling (COSOD) "to review and reassess goals and directions of scientific drilling." The COSOD was not to replace the ongoing operational planning efforts for OMD, but the JOIDES executive committee favored the idea of continuing to drill with the *Challenger* for up to five years after 1983. Jerry Winterer and Mel Peterson were to prepare a draft proposal for post-1983 drilling.

In the early summer of 1981, as I was preparing for a lecture tour in Australia, I got a phone call from Roger Larson of the University of Rhode Island. He had been asked by JOIDES to invite me to chair a COSOD committee on ocean history, and to prepare a White Paper for the COSOD. The necessity of submitting a proposal to the NSF by November 1981 for post-1983 drilling meant that the COSOD had to be convened in October. It was not possible for me to take over the chairmanship; I recommended Wolf Schlager of Miami instead, although I agreed to serve as a committee member. Schlager and I finalized our White Paper when we met in an airport hotel in Boston in late September 1981.

The Ocean Margin Drilling Program continued to develop, and the United States announced that non-U.S. participation in the OMD was not invited. On 22 July 1981 representatives of NSF, JOI, and petroleum companies contributing to OMD met in Houston to discuss funding. The industry was asked to help finance conversion of the _Glomar Explorer_ for OMD, starting in FY 1984. Five regional planning advisory committees (PACs) were constituted by the OMD Science Advisory Committee, and each PAC was to hold a workshop in August and September 1981 to plan drilling in (1) spreading ridges in the Atlantic and Pacific, (2) the Pacific Coast active margin, (3) the Atlantic passive margin, (4) the Gulf of Mexico, and (5) the South Atlantic/Antarctic. It seemed that OMDP had goals very similar to those of COSOD. The people most interested in this program were those from JOI.

The OMD panels met as planned, and JOI submitted an "Initial Science Plan" to the OMD Science Advisory Committee in September 1981. All that planning had been done while JOIDES was convening its COSOD. The JOI optimism was misplaced, however, and the Ocean Margin Drilling Program was killed when the contributing companies informed the National Science Foundation that they would not be able to participate in the OMD. That was that. The remaining co-mingled NSF/industry funds for OMD science were to be used for regional synthesis studies; there would be no OMD drilling.

The NSF's new acronym, AODP, stood for Advanced Ocean Drilling Program. Lockheed was asked in late 1981 to do the preliminary design and cost analysis for converting the _Explorer_ for scientific drilling. Allen Shinn of the NSF took over the planning for AODP. He reported to the JOIDES planning committee in February 1982 the three options considered by the NSF:

1. Scrap scientific ocean drilling at the end of the current program (end of FY 1983);

2. extend *Glomar Challenger* drilling for as long as possible; or

3. replace the *Challenger* with a converted *Explorer* capable of operating for up to 20 years.

Since the *Challenger* was getting old, the NSF favored the last option and was considering the possibility of owning a converted *Explorer*. The cost of conversion was, however, apparently prohibitive.

After industry backed out of the OMD, the NSF was again looking for international support. A delegation of NSF, JOIDES, and DSDP officials visited Europe in late 1981. I am not certain if they came to Switzerland, but I became aware of such activities when I was contacted by Jan Stel of the Netherlands Council of Ocean Research. He asked if Switzerland was interested in becoming affiliated with JOIDES. Mindful of the unhappy experience of the much-ado-about-nothing in 1975, when the NSF backed down from a bilateral agreement on Swiss associate membership in JOIDES, I politely declined to get involved.

I was not to be left alone. Peter Fricker, the general secretary of the Swiss Nationalfond, called me up in March 1982 and asked me to represent him at the NSF's JOIDES Council meeting. I did not want to go, but April in Washington, D.C., was very tempting, especially as I would be able to visit my son Andrew there.

The meeting, as I recall, was not particularly encouraging. In addition to the Dutch and the Swiss, representatives from other European countries, Canada, Australia, China, and a few other countries were present. None of us foreigners were in a position to offer anything. The NSF invited us to visit the *Explorer*, but only the Chinese delegate and I took up the offer; he was making a stop on his way home and I let my curiosity get the best of me.

Glomar Explorer was the notorious spy-ship allegedly built by Lockheed for a CIA project. It was said that the vessel was designed to mine manganese nodules, but newspaper reports claimed that its real purpose was to raise a sunken Soviet submarine from the Pacific bottom. Whether the venture was successful was an official secret; the fact remained that the U.S. government had a "white elephant" on its hands, and was trying to find some use for the expensive toy.

The vessel was not built for drilling, and the cost of conversion was so expensive that industry was scared away. The United States was now looking for help with the financing. The two Chinese, one from Peking and the other from Zürich, were accompanied by Shinn of the NSF, and we took a motorboat ride up the Sacramento River to where the *Explorer* was moored. The vessel was impressive indeed. The *Challenger*'s moon pool was less than a dozen meters across; the *Explorer* had a hole amidships the size of a football field—large enough for a submarine to be raised aboard.

Preparations for the conversion of *Glomar Explorer* continued, and ambitious plans for new laboratories were proposed. Jose Honnorez of Miami, who had succeeded Winterer as chairman of the JOIDES planning committee, emphasized again the need for international cooperation in AODP. The scientific community in Europe was, however, discouraged, because the cost of future ocean drilling was getting beyond our financial capacity. Some French colleagues were agitating for a European venture.

The good news came suddenly, first through the grapevine and later through an official communication by JOIDES, that a "drilling platform D/V SEDCO 472 had become available at a competitive price because of the presently depressed mobile drilling rig market." This unexpected option surfaced just in time to be taken into consideration by an ad hoc NSF committee, chaired by Charles Drake, to decide on the future of ocean drilling. The committee met on 3–4 February 1983 and unanimously recommended that scientific ocean drilling continue, and that additional foreign participants be invited to join. Drake, I might add, was then the president of the International Union of Geophysics and Geodesy, and he was a former student of Doc Ewing's.

We now had a whole new ball game, after the discouraging forays with OMD and *Explorer*. I was quite willing to represent Peter Fricker when the Swiss Nationalfond was again invited to the JOIDES Council meeting in April 1983. The atmosphere was quite different. Mike Kane came from Canada, and Jan Stel was again present. Olaf Eldholm, Kurt Bostrom, and Enrico Bonatti, if I remember correctly, came to represent Norway, Sweden, and Italy, respectively. Meanwhile, the JOIDES executive committee had decided that no associate membership was to be allowed; smaller or financially weak countries had to get together to apply for a consortium membership. Kane told us that Canada was either to go alone, or to form a consortium together with a country from the British Commonwealth, meaning Australia. We five European representatives met over lunch and decided to call a meeting in Zürich of representatives from European funding agencies.

I gave my report to Fricker after my return, and the Swiss

Nationalfond hosted a meeting in the early summer of 1983. The Netherlands, Italy, Norway, Sweden, and Switzerland were definitely committed to working toward a consortium; several other countries such as Finland, Spain, Greece, and Belgium were also interested. But all were very, very reluctant when the hat was passed around, and we saw little chance of coming up with more than half of the funding. France and Germany had no interest in the consortium; they were to continue their national membership in JOIDES. Tony Meyer came to represent the United Kingdom and gave us an outside chance that the United Kingdom might take up the dues of half a consortium membership. The possibility of having either Australia or Canada was not completely ruled out at this first meeting. Bernard Munsch came to represent the European Science Foundation (ESF). It was decided that the best possible route was to organize through the secretariat of the ESF. Two committees were constituted; one consisted of representatives from funding agencies, and the other was the Scientific Committee.

The ESF committees met again in December 1983 and were able to come up with funds for the JOIDES Candidate Membership of 250,000 dollars for the fiscal years 1984 and 1985, when no drilling was scheduled. I was elected to represent the ESF Consortium at the next JOIDES planning committee meeting in Texas. After more than ten years of effort, Switzerland finally became a "stockholder" of JOIDES.

The rest was smooth sailing for JOIDES and ocean drilling, now called ODP. The original nine U.S. institutions, plus the University of Texas at Austin, were the U.S. members. France, Germany, and Japan continued their memberships. The United Kingdom hesitated for a few months before they too came up with the funds. Canada became a new member, thanks largely to the able maneuvering of Bill Hutcheson, who was then assistant deputy minister in the cabinet. (Hutcheson was the general secretary of the International Union of the Geological Sciences when we visited China together in 1977, and he died of cancer at the height of his career when he was president of IUGS.)

The problem of funding by the smaller or financially weaker countries of Europe was not unexpectedly difficult, especially when the membership fee was increased to 2.5 million dollars at the start of the drilling in 1985. We had eleven meetings during the next three years, in Amsterdam, Strasbourg, Göteborg, Milan, Athens, and so on, but we were not able to come up with the funds. We could no longer look to Canada, now a full member. Australia gave us much hope that it would pay for one third of the consortium membership. All was set for the intercontinental marriage, but the budget request was rejected by the Australian parliament. The British were never seriously interested in getting involved with five little dwarfs of the European earth sciences. At the October 1984 Athens meeting, we were at the end of our rope. I went to the January JOIDES P-Comm meeting in Austin in 1986, and had to announce in humiliation that we were out.

The unexpected came later, in the spring of 1986. The Norwegians, with their petroleum industry and offshore oilfields, had been more interested in earth sciences than their GNP had indicated. They came through with a proposal that the five Nordic countries, Norway, Sweden, Denmark, Finland, and Iceland, were to contribute half of the membership dues, and the non-Nordic countries, Italy, the Netherlands, Switzerland, and others, were to come up with the other half. We met in Oslo on 17–18 June 1986. The representatives of funding agencies from non-Nordic countries agreed on the amount to be contributed by each country. I made my last contribution to the JOIDES cause by chairing the ad hoc group that assigned panel memberships to scientists from non-Nordic consortium countries; it was one of the most difficult assignments of my adult life.

With Switzerland now firmly anchored in the European Consortium, I resigned my membership on the ESF Scientific Committee. Hans Thierstein, who had left Scripps to join the ETH, was nominated to take my place. I was to serve two more years on the JOIDES Tectonics Panel before I was finally relieved of all my JOIDES and ODP duties.

The drill vessel was named *JOIDES Resolution*, at the suggestion of Sy Schlanger, who was then chairman of the U.S. Scientific Advisory Committee (USSAC) for JOI. I had proposed the name of another of Captain Cook's vessels, *Endeavor*, which I thought would be more representative of the effort to be expended by ODP, but my proposal was turned down because there was another oceanographic vessel of that name currently in operation.

Now that funding was ensured, it was "peanuts time" again. The emphasis for ODP was "bare-rock" and "high-latitude" drilling, making use of technical capabilities not available to the *Challenger*. The shiptime was divided five ways, to five regions, or regional panels—Atlantic, Southern Oceans, Indian Ocean, Southwest Pacific, and Central and East Pacific. There were three thematic panels—Ocean Crust, Tectonics, and Ocean History.

There were "lollipops" too, as my colleagues on the planning committee described my Mediterranean proposals on behalf of ESF scientists. Two Mediterranean legs were scheduled when the ESF was a candidate member, but were cut to half a leg after ESF failed to come up with funding at the start of ODP drilling. The consortium money in 1986 came too late for a change in the schedule; the Atlantic plans had been finalized, and the drill vessel would soon be heading for the Pacific.

Any denial that politics had nothing to do with scientific drilling could be countered with the example of the Makran proposal, a British "brand of peanuts." It was labeled highest priority when a British scientist chaired the Tectonics Panel, and it was dropped from the schedule when, early in 1986, the British failed to cough up the dough. The proposal was back on its pedestal when the British rejoined JOIDES and the British representative was reinstated. The proposal was dropped again in 1987 when the said person was rotated out of his membership on the panel.

This is not a singular example. I met another person who had been a member of the JOIDES planning committee for years. She told me that in her opinion, Denny Hayes's proposal to drill the South China Sea was the best proposal ever submitted to JOIDES for drilling a passive margin. It was not drilled, however, because Hayes did not have the charm of his teacher, Doc Ewing.

I could understand all the politics, but many of us could never forgive the NSF for denying Maria Cita's nomination to be a co-chief on Leg 107 to the Mediterranean. Cita had served the JOIDES from the very beginning; she was shipboard micropaleontologist on Leg 2. Peter Briggs, a journalist, described Maria Cita as "the Sophia Loren" of Leg 13, referring to not only her feminine charm and personal glamour, but also her star quality as a scientist. She was a coauthor of the Hsü-Ryan-Cita model of the Mediterranean desiccation. Cita served again on Legs 42A and 47A. She was professor of geology and geophysics at the University of Milan; she was the prima donna, the foremost earth scientist in Italy. She was a member of the Linnaean Society, and an Honorary Fellow of the Geological Society of America. Having worked on various JOIDES panels, she was from 1984 to 1986 a member of a JOIDES working group to plan for the Mediterranean drilling. She was thus the logical choice for co-chief of the forthcoming Leg 107, and was nominated by JOIDES; Cita was looking forward to crowning her career with that expedition. Unfortunately, her appointment came at the critical juncture in early 1986 when the ESF was scrambling for funds for a consortium membership. The complete opposite of the precedent set in 1975, when a designated co-chief for Leg 42B was asked to step aside to make room for a Soviet scientist even before the Soviet Union had joined

JOIDES, occurred in 1986, when the authorities ruthlessly denied the post of co-chief to an ESF scientist, because the ESF had not *yet* managed to come up with the membership dues! Cita was not one to complain; she now serves JOIDES faithfully as the ESF representative on the planning committee.

I would have been more chivalrous had I kept my opinion of ODP to myself. However, in 1984 I was bombarding my JOIDES friends with letters and recommendations concerning the future of ocean drilling, now that the project was soon to terminate. I wanted to see a continuation of scientific ocean drilling. But we were back to a crisis, as we had been ten years earlier. There were competing ideas, and a French or a Japanese drill vessel was being discussed.

My colleagues Judy McKenzie and Daniel Müller had just come back from the *JOIDES Resolution* Leg 134, drilling the Great Barrier Reef, and they were bubbling with enthusiasm. I have noticed that the ODP activities never got reported by the news media. Gone are those glorious days when the findings of Leg 13 were announced simultaneously at press conferences in Paris and New York, when the story of the Mediterranean desiccation made headlines in *Figaro*, *Le Monde*, the *New York Times*, and papers all over the world, when I woke up one morning in Kuching, on the island of Borneo, to see a newspaper cartoon of a camel caravan heading from Marseille to Algiers. Tom Wiley, the press officer for the DSDP, used to compose flowery prose for the press releases before and after each drilling cruise. That was one post which the new ODP operator at Texas A & M did not seem to need.

Are the earth sciences so dead that there are no more headlines to be made?

No, I do not think so. The discovery of extraterrestrial catastrophes, which killed dinosaurs, caused mass extinctions, and influenced the course of biological evolution, has been making headlines in the decade of ODP drilling. The problems of global environmental change, climatic history, and natural catastrophes are making headlines every day. The earth sciences are more relevant than ever. But what is happening in ocean drilling?

I could imagine the following newspaper article appearing sometime in June 1987:

Impact Crater of Killer Comet Found

Ocean drilling in the Indian Ocean has discovered an impact crater 250 km across in the Amirante Basin, buried under 2 km of ocean oozes, Leg 115 scientists of the Joint Oceanographical Institutions

Deep Earth Sampling Program announced on the 30th in Colombo, Sri Lanka. A comet, weighing a trillion tons, hit the ocean 65 million years ago, wiped out the dinosaurs, and caused the extinction of three fourths of all species living in that prehistoric world . . .

The nature of the ODP planning structures had ensured that no such headlines were to appear. A proposal to drill the candidate crater was not even reported out of the panel, and my attempt to bring up the matter in the planning committee was considered out of order—it was said that the committee should discuss only panel reports, not ideas or proposals that had not been processed through the proper channels.

The problem facing ODP is that they have a 40-million-dollar blank check to be apportioned to competing proposals, all good and solid like the ones that irritated Walter Munk and Harry Hess into proposing the Mohole. Marine geoscientists seem to think that the revolution is over; the only remaining task is the mopping up by "normal science."

Isn't it about time that we have another Munk, another Hess, another Cesare Emiliani to "think big"? We could sample mélanges in subduction zones, look for magma chambers on spreading ridges, drill impact craters in the Indian Ocean or the Caribbean. We could even reach the Moho, if we really wanted to.

Deep-Sea Drilling Legs

The list was compiled on the basis of information contained in the Initial Reports of the Deep Sea Drilling Project, vols. 1–96, 1969–1986, and in the *JOIDES Journal*, vols. 1–11, 1975–1985. The dates of some cruises are not explicitly given in the reports; estimated dates are marked by asterisks.

Leg	Ports	Dates	Co-Chiefs	*Planning Panels Sites*
Phase 1				
1	Orange, Texas Hoboken, New Jersey	20.07.68 23.09.68	Ewing, M. Worzel, J. L.	Gulf, Atlantic 1–7
2	Hoboken, New Jersey Dakar, Senegal	01.10.68 24.11.68	Peterson, M. Edgar, T.	Atlantic 8–12
3	Dakar, Senegal Rio de Janeiro, Brazil	01.12.68 24.01.69	Maxwell, A. von Herzen, R.	Atlantic 13–22
4	Rio de Janeiro, Brazil Cristobal, Panama	27.01.69 22.03.69	Bader, R. Gerard, R. D.	Atlantic 23–31
	Transit			
5	San Diego, Calif. Honolulu, Hawaii	12.04.69 05.06.69	McManus, D. A. Burns, R. E.	Pacific 32–43
6	Honolulu, Hawaii Apra, Guam	11.06.69 03.08.69	Fisher, A. G. Heezen, B. C.	Pacific 44–60
7	Apra, Guam Honolulu, Hawaii	08.08.69 02.10.69	Winterer, E. L. Riedel, W. R.	Pacific 61–67
8	Honolulu, Hawaii Papeete, Tahiti	08.10.69 02.12.69	Tracy, J. I. Sutton, G. H.	Pacific 68–75

Appendix A

Leg	Ports	Dates	Co-Chiefs	Planning Panels Sites
9	Papeete, Tahiti Balboa, Panama	06.12.69 27.01.70	Hays, J. D. Cook, H. E.	Pacific 76–84

Phase 2

Leg	Ports	Dates	Co-Chiefs	Planning Panels Sites
10	Galveston, Texas Miami, Florida	13.02.70 05.04.70	Worzel, J. L. Bryant, W.	Gulf, Atlantic 85–97
11	Miami, Florida Hoboken, New Jersey	08.04.70 01.06.70	Hollister, C. D. Ewing, J. I.	Atlantic 98–108
12	Boston, Mass. Lisbon, Portugal	19.06.70 11.08.70	Laughton, A. S. Berggren, W. A.	Atlantic 109–119
13	Lisbon, Portugal Lisbon, Portugal	13.08.70 06.10.70	Ryan, W.B.F. Hsü, K. J.	Mediterranean 120–134
14	Lisbon, Portugal San Juan, Puerto Rico	09.10.70 01.12.70	Hayes, D. E. Pimm, A. C.	Atlantic 135–144
15	San Juan, Puerto Rico Cristobal, Panama	05.12.70 02.02.71	Edgar, N. T. Saunders, J. B.	Atlantic 146–154
16	Cristobal, Panama Honolulu, Hawaii	02.02.71 30.03.71	van Andel, G. R. Heath, G. R.	Pacific 155–163
17	Honolulu, Hawaii Honolulu, Hawaii	30.03.71 25.05.71	Winterer, E. L. Ewing, J. I.	Pacific 164–171
18	Honolulu, Hawaii Kodiak, Alaska	29.05.71 20.07.71	Kulm, L. D. von Huene, R.	Pacific 172–182
19	Kodiak, Alaska Yokohama, Japan	24.07.71 11.09.71	Creasy, J. S. Scholl, D. W.	Pacific 183–193
20	Yokohama, Japan Suva, Fiji	17.09.71 10.11.71	Heezen, B. C. MacGregor, I. D.	Pacific 194–201

Deep-Sea Drilling Legs

Leg	Ports	Dates	Co-Chiefs	Planning Panels Sites
21	Suva, Fiji Darwin, Australia	15.11.71 06.01.72	Burns, R. E. Andrews, J. E.	Pacific 203–210
22	Darwin, Australia Colombo, Ceylon	11.01.72 05.03.72	von der Borch, C. C. Sclater, J. G.	Indian Ocean 211–218
23	Colombo, Ceylon Djibouti, F.T.A.I.	08.03.72 01.05.72	Whitmarsh, R. B. Weser, O. E. Ross, D. A.	Indian Ocean 219–230
24	Djibouti, F.T.A.I. Port Louis, Mauritius	04.05.72 26.06.72	Fisher, R. L. Bunce, E. T.	Indian Ocean 231–238
25	Port Louis, Mauritius Durban, South Africa	28.06.72 21.08.72	Simpson, E.S.W. Schlich, R.	Indian Ocean 239–249

Phase 3

Leg	Ports	Dates	Co-Chiefs	Planning Panels Sites
26	Durban, South Africa Fremantle, Australia	06.09.72 30.10.72	Davies, T. A. Luyendyk, B. P.	Indian Ocean 250–258
27	Fremantle, Australia Fremantle, Australia	01.11.72 09.12.72	Veevers, J. J. Heirtzler, J. R.	Indian Ocean 259–263
28	Fremantle, Australia Christchurch, New Zealand	20.12.72 23.02.73	Hayes, D. E. Frakes, L. A.	Antarctic 264–274
29	Lyttelton, New Zealand Wellington, New Zealand	02.03.73 18.04.73	Kennett, J. P. Houtz, R. E.	Antarctic 275–284
30	Wellington, New Zealand Apra, Guam	24.04.73 13.06.73	Packham, G. H. Andrews, J. E.	Pacific 285–289
31	Apra, Guam Hakodate, Japan	16.06.73 05.08.73	Karig, D. E. Ingle, J. C.	Pacific 290–302
32	Hakodate, Japan Honolulu, Hawaii	16.08.73 10.10.73	Larson, R. L. Moberly, R.	Pacific 303–313

Appendix A

Leg	Ports	Dates	Co-Chiefs	Planning Panels Sites
33	Honolulu, Hawaii	02.11.73 (!)	Schlanger, S. O.	Pacific
	Papeete, Tahiti	17.12.73	Jackson, E. D.	314–318
34	Papeete, Tahiti	20.12.73	Yeats, R. S.	Pacific, crust
	Callao, Peru	02.02.74	Hart, S. R.	319–321
35	Callao, Peru	13.02.74	Hollister, C. D.	Antarctic
	Ushuaia, Argentina	30.03.74	Craddock, C.	322–325
36	Ushuaia, Argentina	04.04.74	Barker, P.	Antarctic
	Rio de Janeiro, Brazil	22.05.74	Dalziel, I.W.D.	326–331
37	Rio de Janeiro, Brazil	31.05.74	Aumento, F.	Atlantic, crust
	Dublin, Ireland	27.07.74	Melson, W. G.	332–335
38	Dublin, Ireland	29.07.74	Talwani, M.	Atlantic
	Amsterdam, Netherlands	26.09.74	Udintsev, G.	336–352
39	Amsterdam, Netherlands	09.10.74	Supko, P. R.	Atlantic
	Cape Town, South Africa	16.12.74	Perch-Nielsen, K.	353–359
40	Cape Town, South Africa	17.12.74	Bolli, H.	Atlantic, margin
	Abidjan, Ivory Coast	15.02.75	Ryan, W.B.F.	360–365
41	Abidjan, Ivory Coast	20.02.75	Seibold, E.	Atlantic, margin
	Málaga, Spain	10.04.75	Lancelot, Y.	366–371
42A	Málaga, Spain	14.04.75	Hsü, K. J.	Mediterranean
	Istanbul, Turkey	21.05.75	Montadert, L.	372–378
42B	Istanbul, Turkey	21.05.75	Ross, D. A.	Black Sea
	Istanbul, Turkey	11.06.75	Neprochnov, Y. P.	379–381
	Transit			
43	Delgada, Azores (!)	27.06.75	Tucholke, B. E.	Atlantic, margin
	Norfolk, Virginia	12.08.75	Vogt, P. R.	382–387

Deep-Sea Drilling Legs

Leg	Ports	Dates	Co-Chiefs	*Planning Panels Sites*
44	Norfolk, Virginia	15.08.75	Benson, W. E.	Atlantic, margin
	Norfolk, Virginia	30.09.75	Sheridan, R. E.	388–394

International Phase 1

Leg	Ports	Dates	Co-Chiefs	*Planning Panels Sites*
45	San Juan, Puerto Rico	30.11.75	Melson, W. G.	Crust
	San Juan, Puerto Rico	20.01.76	Rabinowitz, P. D.	395–396
46	San Juan, Puerto Rico	28.01.76	Dmitriev, L.	Crust
	Las Palmas, Canary Is.	12.03.76	Heirtzler, J.	396
47A	Las Palmas, Canary Is.	20.03.76	von Rad, U.	Passive margin
	Vigo, Spain	12.04.76	Ryan, W.B.F.	397
47B	Vigo, Spain	12.04.76	Sibuet, J.-C.	Passive margin
	Brest, France	10.05.76	Ryan, W.B.F.	398
48	Brest, France	12.05.76	Montadert, L.	Passive margin
	Aberdeen, UK	13.07.76	Roberts, D. G.	399–406
49	Aberdeen, UK	16.07.76	Luyendyk, B. P.	Crust
	Funchal, Madeira	07.09.76	Cann, J. R.	407–414
50	Funchal, Madeira	12.09.76	Winterer, E. L.	Passive margin
	Arrecife, Canary Is.	29.10.76	Lancelot, Y.	415–416
51	San Juan, Puerto Rico	20.11.76	Donnelly, T. W.	Crust
	San Juan, Puerto Rico	17.01.77	Francheteau, J.	417
52	San Juan, Puerto Rico	22.01.77	Bryan, W. B.	Crust
	San Juan, Puerto Rico	08.03.77	Robinson, P. T.	417–418
53	San Juan, Puerto Rico	12.03.77	Flower, M.F.J.	Crust
	San Juan, Puerto Rico	21.04.77	Salisbury, M.	418
54	Cristobal, Panama	30.04.77	Rosendahl, B. R.	Crust
	Long Beach, Calif.	18.06.77	Hekinian, R.	419–429

Leg	Ports	Dates	Co-Chiefs	Planning Panels Sites
55	Honolulu, Hawaii	14.07.77	Jackson, D. E.	Crust
	Yokohama, Japan	06.09.77	Koisumi, I.	430–433
56	Yokohama, Japan	10.09.77	Langseth, M.	Active margin
	Yokohama, Japan	20.10.77	Okada, H.	434–438
57	Yokohama, Japan	21.10.77	von Huene, R.	Active margin
	Yokohama, Japan	05.12.77	Noriyuki, N.	438–441
58	Yokohama, Japan	11.12.77	deVries Klein, G.	Active margin
	Okinawa, Japan	29.01.78	Kobayashi, K.	442–446
59	Okinawa, Japan	03.02.78	Kroenke, L.	Active margin
	Apra, Guam	15.03.78	Scott, R.	447–451
60	Apra, Guam	21.03.78	Husson, D.	Active margin
	Apra, Guam	16.05.78	Uyeda, S.	452–461
61	Apra, Guam	22.05.78	Larsen, R.	Crust
	Majuro, Marshall Is.	28.07.78	Schlanger, S. O.	462
62	Majuro, Marshall Is.	29.07.78	Thiede, J.	Paleoenvironment
	Honolulu, Hawaii	07.09.78	Vallier, T.	463–466
	Transit			
63	Long Beach, Calif.	09.10.78	Yeats, R. S.	Paleoenvironment
	Mazatlán, Mexico	26.11.78	Haq, B.	467–473
64	Mazatlán, Mexico	01.12.78	Curray, J.	Passive margin
	Long Beach, Calif.	14.01.79	Moore, D. G.	474–481
65	San Pedro, Calif.	20.01.79	Lewis, B.T.R.	Crust
	Mazatlán, Mexico	13.03.79	Robinson, P.	482–485
66	Mazatlán, Mexico	18.03.79	Watkins, J. S.	Active margin
	Manzanillo, Mexico	04.05.79	Moore, J. C.	486–493

Deep-Sea Drilling Legs

Leg	Ports	Dates	Co-Chiefs	Planning Panels Sites
67	Manzanillo, Mexico Puntarenas, Costa Rica	08.05.79 27.06.79	Aubouin, J. von Huene, R.	Active margin 494–500
68A	Puntarenas, Costa Rica Willemstad, Curaçao	05.07.79 20.07.79	Cann, J. R. White, S. M.	Crust 501
68B	Willemstad, Curaçao Guayaquil, Ecuador	13.08.79 17.08.79	Prell, W. L. Gardner, J. V.	Paleoenvironment 502–503
69	Guayaquil, Ecuador Balboa, Panama	18.09.79 29.10.79	Cann, J. R. Langseth, M. G.	Crust 504–505

International Phase 2

Leg	Ports	Dates	Co-Chiefs	Planning Panels Sites
70	Balboa, Panama Balboa, Panama	03.11.79 18.12.79	Honnorez, J. von Herzen, R. P.	Crust 504, 506–510
	Transit			
71	Punta Arenas, Chile Santos, Brazil	10.01.80 20.02.80	Ludwick, W. J. Krasheninnikov, V. A.	Paleoenvironment 511–514
72	Santos, Brazil Santos, Brazil	26.02.80 08.04.80	Barker, P. F. Carlson, R. L. Johnson, D. A.	Paleoenvironment 515–518
73	Santos, Brazil Cape Town, South Africa	13.04.80 01.06.80	Hsü, K. J. LaBrecque, J.	Paleoenvironment 519–524
74	Cape Town, South Africa Walvis Bay, South Africa	06.06.80 22.07.80	Moore, T. C. Rabinowitz, P. D.	Paleoenvironment 525–529
75	Walvis Bay, South Africa Recife, Brazil	27.07.80 06.09.80*	Hay, W. W. Sibuet, J.-C.	Paleoenvironment 530–532
	Transit			

Appendix A

Leg	Ports	Dates	Co-Chiefs	Planning Panels Sites
76	Norfolk, Virginia Fort Lauderdale, Florida	11.10.80 21.12.80	Sheridan, R. E. Gradstein, F. M.	Passive margin 533–534
77	Fort Lauderdale, Florida San Juan, Puerto Rico	27.12.80 01.02.81	Buffler, R. T. Schlager, W.	Passive margin 535–540
78A	San Juan, Puerto Rico San Juan, Puerto Rico	06.02.81* 12.03.81*	Biju-Duval, B. Moore, J. C.	Active margin 541–543
78B	San Juan, Puerto Rico Las Palmas, Canary Is.	12.03.81* 10.04.81*	Hyndman, R. Salisbury, M. H.	Downhole exp't 395
79	Las Palmas, Canary Is. Brest, France	15.04.81 25.05.81*	Hinz, K. Winterer, E. L.	Passive margin 544–547
80	Brest, France Southampton, UK	30.05.81 22.07.81	de Graciansky, P. C. Poag, C. W.	Passive margin 549–551
81	Southampton, UK Ponta Delgada, Azores	27.07.81 16.09.81*	Roberts, D. G. Schnitker, D.	Passive margin 552–555
82	Ponta Delgada, Azores Balboa, Panama	21.09.81* 10.11.81*	Bougault, H. Cande, S.	Crust 556–564

International Phase 3

Leg	Ports	Dates	Co-Chiefs	Planning Panels Sites
83	Balboa, Panama Balboa, Panama	14.11.81 05.01.82	Anderson, R. N. Honnorez, J.	Crust 504
84	Balboa, Panama Manzanillo, Mexico	11.01.82* 26.02.82*	von Huene, R. Aubouin, J.	Active margin 565–570
85	Manzanillo, Mexico Honolulu, Hawaii	08.03.82* 02.05.82*	Meyer, L. Theyer, F.	Paleoenvironment 571–575
86	Honolulu, Hawaii Yokohama, Japan	05.05.82* 19.05.82*	Heath, G. R. Burckle, L. H.	Paleoenvironment 576–581

Deep-Sea Drilling Legs

Leg	Ports	Dates	Co-Chiefs	Planning Panels Sites
87	Yokohama, Japan Hakodate, Japan	25.06.82 18.08.82	Kagami, H. Karig, D.	Active margin 582–584
88	Hakodate, Japan Yokohama, Japan	19.08.82 20.09.82	Duennebier, F. Stephens, R.	Downhole exp't 581
89	Yokohama, Japan Nouméa, New Caledonia	11.10.82 29.11.82	Moberly, R. Schlanger, S. O.	Paleoenvironment 462, 585–586
90	Nouméa, New Caledonia Wellington, New Zealand	02.12.82 11.01.83	Kennett, J. von der Borch, C.	Paleoenvironment 587–594
91	Wellington, New Zealand Papeete, Tahiti	16.01.83 20.02.83	Menard, H. W. Natland, J.	Downhole exp't 595–596
92	Papeete, Tahiti Balboa, Panama	23.02.83 19.04.83	Leinen, M. Rea, D.	Crust 505, 597–602
93	Norfolk, Virginia Norfolk, Virginia	03.05.83 17.06.83	Van Hinte, J. E. Wise, S. W.	Paleoenvironment 603–605
94	Norfolk, Virginia St. John's, Newfoundland	17.06.83 17.08.83	Kidd, R. B. Ruddiman, W. F.	Paleoenvironment 606–611
95	St. John's, Newfoundland Fort Lauderdale, Florida	17.08.83 26.09.83	Poag, C. Watts, A.	Passive margin 603, 612–613
96	Fort Lauderdale, Florida Mobile, Alabama	29.09.83 08.11.83	Bouma, A. Coleman, J.	Passive margin 614–624

Bibliographical Notes

For readability, I have not followed the usual format of footnoting for scholarly treatises. Instead, I acknowledge here the references I consulted while composing the various chapters, as well as the sources of citations in the text.

PREFACE   TO THE CHINESE EDITION

The statement attributed to d'Aubuisson de Voisins is cited in Geikie's *Founders of Geology*, 2nd ed. (London: Macmillan, 1905), p. 243.

CHAPTER 1.   MOHO AND MOHOLE

The history of the Mohole project is reconstructed mainly on the basis of my personal memory, checked against C. Emiliani's article "A New Global Geology," in *The Sea*, vol. 7, *The Ocean Lithosphere* (New York: Wiley, 1981), 1738 pp.

CHAPTER 2.   ICE AGE AND LOCO

I consulted the excellent book *Ice Ages: Solving the Mystery* (Short Hills, N.J.: Enslow, 1979) by J. Imbrie and K. P. Imbrie, and the proceedings of the Schweiz. Naturforschende Gesellschaft, 1820–1840, in composing my narrative on the theory of the Ice Age. The history of the LOCO project is told in Emiliani's article, cited under chapter 1, above.

CHAPTER 3.   THE *CHALLENGER* GOES TO SEA

The description of the drillship is based upon unpublished press releases by the Deep Sea Drilling Project, distributed to the public in the 1960s and 1970s. I looked up the *Initial Reports of the Deep Sea Drilling Project* (*IRDSDP*), vol. 1, (Washington, D.C.: U. S. Government Printing Office, 1969), 672 pp., for specific details concerning the findings of the inaugural cruise. Ted Bullard's comment about Ewing cited in the text was made during an address given at a dinner held in memory of Maurice Ewing on 29 March 1976 at Columbia University and was the dedication in *Island*

Arcs, Deep Sea Trenches and Back-Arc Basins, ed. M. Talwani and W. C. Pitman, Maurice Ewing Series no. 1 (Washington, D.C.: American Geophysical Union, 1977), 470 pp.

CHAPTER 4. THE EARTH SCIENCE REVOLUTION

I relied upon my memory in making this summary of the history (until 1968) of the earth science revolution, but I had to verify various specific details by consulting the following three excellent books: *Continents in Motion* (New York: McGraw-Hill, 1974) by Walter Sullivan, *The Road to Jaramillo* (Stanford: Stanford University Press, 1982) by William Glen, and *The Ocean of Truth* (Princeton: Princeton University Press, 1986) by H. W. Menard. Major contributions by Runcorn, Elsasser, Hess, Dietz, Mason and Raff, Menard, Heezen, Maurice Ewing, Tuzo Wilson, Bullard, Revelle and Maxwell, Reynolds, Everdeen, Cox, Doell *et al.*, Vine and Matthews, LePichon, Morgan, McKenzie and Parker, Sykes, Pitman *et al.*, Oliver *et al.*, Heirtzler *et al.*, and others discussed in the text are now common knowledge. The relevant articles are cited in the three books above, and most are also cited in the acknowledgments of this book.

The description of the hostility to Griggs's theory of convection current in the earth's mantle is quoted from Griggs's acceptance speech when he received the Arthur Day Medal; *Geological Society of America Bulletin* 85 (1975): 1342–1342.

CHAPTER 5. A GAME OF NUMBERS

Again I relied upon my memory; I verified specific details in *IRDSDP* vol. 3 (1970), 806 pp.

CHAPTER 6. ATLANTIC AND TETHYS

The articles "Seafloor spreading in the North Atlantic" by Walter Pitman and Manik Talwani, *Geological Society of America Bulletin* 83 (1972): 619–643 and "Origin of the Alps and Western Mediterranean" by K. J. Hsü, *Nature* 233 (1971): 44–48 form the basis of the scientific narrative in this chapter. The drilling of the Gorringe Bank was described in *IRDSDP* vol. 13 (1973), 1447 pp.

CHAPTER 7. ARC AND TRENCH IN THE MEDITERRANEAN

An elementary textbook, *Principles of Geology*, by James Gilluly, Aaron Waters, and A. O. Woodford (San Francisco: Freeman, 1951), 691 pp., gives an excellent account of the history of the evolving concept of isostasy. The development of the idea since then has been summarized in two of my articles in the *American Journal of Science*: "Isostasy and a theory for the origin of geosynclines," vol. 256 (1958): 305–327, and "Isostasy, crustal thinning, mantle changes, and the disappearance of ancient land masses," vol. 263 (1964): 97–109. Argand's classic _La tectonique de l'Asie_ was published in _Proceedings of the 13th International Geological Congress_ (Brussels: 1922), pp. 171–372. The circumstances leading to the formulation of the theory of plate-tectonics by Jason Morgan, by Oliver and Isacks, and by McKenzie and Parker are common knowledge, and I was able to refresh my memory by reading the excellent review by Menard in his _The Ocean of Truth_. The drilling of the Hellenic Trench took place during Leg 13 (see _IRDSDP_ vol. 13).

CHAPTER 8.   SWALLOWING UP OF THE OCEAN FLOOR

A more scholarly treatise on the history of the mélange concept is my article for the Centennial Special Volume 1 of the Geological Society of America (1985), entitled "A basement of mélanges: A personal account of the circumstances leading to the breakthrough in Franciscan research," pp. 47–64. I depended very much on Robert A. Stafford's _Scientist of Empire_ (Cambridge, England: Cambridge University Press, 1989) for my portrait of Roderick Murchison. The drilling of the Northeast Pacific was first carried out during Legs 5 (_IRDSDP_ vol. 5 [1970], 827 pp.) and 18 (_IRDSDP_ vol. 18 [1975], 1017 pp.).

CHAPTER 9.   MARGINAL SEAS

I asked Dan Karig, when he visited me in 1988, how he got his idea on the origin of marginal seas. The short biographical profile of Bruce Heezen was sketched on the basis of personal knowledge, supplemented by the account in Menard's _The Ocean of Truth_. The Pacific back-arc basins were drilled by the _Challenger_ first during Leg 6 (_IRDSDP_ vol. 6 [1971], 1329 pp.), again during Leg 31 (_IRDSDP_ vol. 31 [1975], 927 pp.), and finally during Legs 59 (_IRDSDP_ vol. 59 [1981], 820 pp.) and 60 (_IRDSDP_ vol. 60, [1982], 929 pp.).

CHAPTER 10.   HOPE AND FRUSTRATION IN NAURU

The mid-plate volcanism was investigated by the _Challenger_ during Legs 17 (_IRDSDP_ vol. 17 [1973], 930 pp.), 33 (_IRDSDP_ vol. 33 [1976], 973 pp.), 61 (_IRDSDP_ vol. 61 [1981], 885 pp.), and finally 89 (_IRDSDP_ vol. 89 [1986], 678 pp.). The final announcement of finding the Jurassic Pacific was made in the _JOIDES Journal_, vol. 14 (1990). The article by Schlanger and Hsü, "Thermal history of the upper mantle and its relation to crustal history of the Pacific basin," was published in _Proceedings of the 23rd International Geological Congress_ (Prague: 1968), vol. 1, pp. 91–105.

CHAPTER 11.   HAWAIIAN HOT-SPOT

My sources for the description of Jackson's cruise are shipboard operations report for Leg 55; *IRDSDP* vol. 55 (1980), 868 pp.; and personal communication, Judith McKenzie, Zürich.

CHAPTER 12.   INDIA'S LONG MARCH

Suess's concept of Tethys was reviewed by Hugh Jenkyns and K. J. Hsü in their introduction to the book *Pelagic Sediments on Land and under the Sea* (Oxford: International Association of Sedimentologists, special publication no. 1, 1974), pp. 1–10. Menard gave, in his *The Ocean of Truth*, an excellent account of his discovery of the oceanic fracture zones and of Wilson's theory of transform faults. The Ninety-East Ridge was drilled during Leg 22 (*IRDSDP* vol. 22 [1974], 890 pp.).

CHAPTER 13.   ANTARCTIC ADVENTURES

Reference is made to shipboard operations reports and to *Initial Reports of the Deep Sea Drilling Project* (vol. 28 [1975], 1017 pp.; vol. 29 [1975], 1197 pp.; vol. 35 [1976], 923 pp.; vol. 36 [1976], 1079 pp.). The cited dedication to Ewing was published in *IRDSDP* vol. 35.

CHAPTER 14.   MID-CRETACEOUS ANOXIA

I learned of the activities during Leg 40 mainly through conversation with Bill Ryan, and the facts were checked by referring to the shipboard operations report and *IRDSDP* vol. 40 (1978), 1078 pp. The drilling of the Angola Basin was reported in *IRDSDP* vol. 75 (1983), 752 pp.

CHAPTER 15.   WHEN THE MEDITERRANEAN DRIED UP

This chapter is a summary of my book *The Mediterranean Was a Desert* (Princeton: Princeton University Press, 1983), 197 pp. The quotation from Dietz and Woodhouse was published in *Geotimes*, May 1988.

CHAPTER 16.   THE BLACK SEA WAS NOT ALWAYS BLACK

The scientific results of the Black Sea drilling are reported in *IRDSDP* vol. 42B (1978), 1248 pp. My work on relic back-arc basins was published in *New Perspectives in Basin Analysis*, ed. K. L. Kleinspehn and C. Paola (New York: Springer Verlag, 1988), pp. 245–263.

CHAPTER 17.   GETTING STUCK IN OCEAN CRUST

This review of the ocean-crust drilling during the International Phase of Ocean Drilling (IPOD) is based mainly upon the following volumes of *IRDSDP*: vol. 38 (1976), 1256 pp.; vol. 49 (1979), 1020 pp.; vol. 51/52/53 (1980), 1613 pp.; vol. 69 (1983), 864 pp.; vol. 70 (1983), 481pp.; vol. 83 (1985), 539 pp.; vol. 92 (1986), 617 pp.

CHAPTER 18.   EATING PEANUTS ON OCEAN MARGINS

My ideas on the origin of "geosynclines" are contained in the two articles cited under chapter 7 above. I consulted Jerry Winterer and referred to *IRDSDP* vol. 50 (1980), 863 pp., and vol. 79 (1984), 934 pp., in writing up the summary on drilling the West African margin.

   This summary of active-margin drilling during IPOD is based mainly upon *IRDSDP* vol. 13 (1973), 1447 pp.; vol. 56/57 (1980), 1417 pp.; vol. 66 (1982), 864 pp.; vol. 67 (1982), 793 pp.; vol. 78 (1984), pp. 1–630.

CHAPTER 19.   WHAT MAKES THE OCEAN RUN

This summary of the achievements in ocean paleoenvironments is based upon *IRDSDP* vol. 41 (1978), 1259 pp.; vol. 47A (1979), 835 pp.; vol. 72 (1983), 1024 pp.; vol. 73 (1984), 798 pp. The proceedings of the First Conference of Paleoceanography were published as *South Atlantic Paleoceanography*, ed. K. J. Hsü and H. Weissert (Cambridge, England: Cambridge University Press, 1986), 350 pp.

CHAPTER 20.   THE GREAT DYING

This short history of the recognition of a K/T catastrophe and its implications for evolution is based on my full-length book on this theme, *The Great Dying*, (San Diego: Harcourt Brace Jovanovich, 1986), 292 pp. The German translation, *Die letzten Jahre der Dinosaurier* (Basel: Birkhäuser, 1990), 270 pp., gives selected references. The epoch-making article by Luis Alvarez, Walter Alvarez, Frank Asaro, and Helen Michel was entitled "Extraterrestrial Cause for the Cretaceous-Tertiary Extinction," *Science* 208 (1980): 1095–1108. The quotations from Darwin's *Origin of Species* are taken from the first edition (London: John Murray, 1859), 513 pp. John Maynard Smith's article "Paleontologists at High Table" appeared in *Nature* 309 (1984): 401–402, and A. Hallam's article "Plate Tectonics and Evolution" is in *Evolution: From Molecules to Men*, ed. D. S. Bendall (Cambridge, England: Cambridge University Press, 1983), pp. 367–386.

EPILOGUE I relied on my memory in reconstructing the events leading to the initiation of the Ocean Drilling Program, 1984–1994, and the organization of a European Science Foundation consortium membership in JOIDES, with the dates verified by my personal files and by notices in *JOIDES Journal*, vols. 6–10 (1980–84). The book by Peter Briggs is entitled *200,000,000 Years beneath the Sea* (New York: Holt, Rinehart and Winston, 1971), 228 pp.

Index